Lecture Notes in Mathematics

Edited by A. Dold and B. Eckmann
Series: Mathematisches Institut der Universität Bonn
Adviser: F. Hirzebruch

628

Hans J. Baues

Obstruction Theory

on Homotopy Classification of Maps

Springer-Verlag
Berlin Heidelberg New York 1977

Author
Hans J. Baues
Sonderforschungsbereich 40
„Theoretische Mathematik"
Mathematisches Institut der Universität
Wegelerstr. 10
5300 Bonn/BRD

AMS Subject Classification (1970): 55-02, 55A05, 55A20, 55B10, 55B25, 55B45, 55C25, 55C30, 55DXX, 55EXX, 55GXX, 55H05, 55H15, 55H99

ISBN 3-540-08534-3 Springer-Verlag Berlin Heidelberg New York
ISBN 0-387-08534-3 Springer-Verlag New York Heidelberg Berlin

Printed in Germany

Printing and binding: Beltz Offsetdruck, Hemsbach/Bergstr.
2140/3140-543210

Für Charis und Barbara

Contents

CHAPTER 4. OBSTRUCTION THEORY FOR CW-COMPLEXES

CHAPTER 5. POSTNIKOV HOMOTOPY DECOMPOSITIONS

CHAPTER 6. ITERATED PRINCIPAL FIBRATIONS

Introduction

The homotopy classification of maps, and closely related to it the study of extension and lifting problems, is a central topic in algebraic topology. Steenrod writes in [122]:

> "Many of the basic theorems of topology, and some of its most successful applications in other areas of mathematics, are solutions of particular extension problems. The deepest results of this kind have been obtained by the method of algebraic topology. The essence of the method is a conversion of the geometric problem into an algebraic problem which is sufficiently complex to embody the essential features of the geometric problem, yet sufficiently simple to be solvable by standard algebraic methods. Many extension problems remain unsolved, and much of the current development of algebraic topology is inspired by the hope of finding a truly general solution."

Obstruction theory is an attempt at such a general solution. This theory had its origins in the classical works of Hopf, Eilenberg, Steenrod and Postnikov around 1940 and has been developing ever since, albeit in an uncoordinated fashion. Portions of obstruction theory appear in most textbooks on algebraic topology, for instance in Steenrod's book on fiber bundles [120], or in the books by Spanier [116] and G.W. Whitehead [130]. These treatments often differ in approach and in the assumptions under which the theorems are proved, such as simply-connectedness, or that the fundamental group operate trivially, or that a fiber bundle be given instead of a fibration, or that only maps be considered instead of sections and retractions.

In this book we dispense with such restrictions wherever possible and so prove classical theorems in their full generality, for instance theorems on the Postnikov decomposition of a fibration, on primary and higher-order cohomology obstructions, and on the homotopy classification of maps that, as we show, apply to sections and retractions as well. Theorems of this kind are known to algebraic topologists, at least in a rough and ready way, and are commonly used. However, no self-contained exposition of obstruction theory has appeared.

We have here endeavored to give a systematic presentation of the subject, integrating the different approaches found in the literature. The essential tool for this is Eckmann-Hilton duality, which divides the presentation into two parts leading in parallel to the same goals. We have also systematized in another way by generalizing in stages. That is, we develop in 4 parallel stages the homotopy classification first of

1)	maps	using	principal cofibrations,
2)	sections of fibrations	using	principal cofibrations,
3)	maps	using	relative principal cofibrations,
4)	sections of fibrations	using	relative principal cofibrations

and then, as dual to these, the homotopy classification of

1')	maps	using	principal fibrations,
2')	retractions of cofibrations	using	principal fibrations,
3')	maps	using	relative principal fibrations,
4')	retractions of cofibrations	using	relative principal fibrations.

Stage 1) is a special case of 2) and 3), which are themselves special cases of 4). The reader need not shrink from having eight versions sung to him of the same old song, since in fact we develop only 2) and 2') thoroughly, in other stages often omitting details in explicitly formulating dual theorems, generalized ones, or their proofs. In the simpler stages we always point up the basic ideas clearly. We feel that the reader profits more from stagewise generalization of the theory, than if we had begun with the complicated versions 4) and 4') and only later moved on to the special cases, which are important in their own right.

In the literature most attention has been paid to approaches 1), 2) and 1'). We will show that classification theorems of Barcus-Barratt [5] for 1), and dually of James-Thomas [57] for 1'), are special cases of general classification results which we formulate using spectral sequences and which are valid for 2) and 2') also. Well-known in the context of 1) is the Puppe or cofiber sequence, as is in 1') the dual fiber sequence. We generalize these sequences at every stage to long exact classification sequences, and construct from them exact couples yielding spectral sequences for homotopy classification. These we will study in some detail. The classification sequences can also be derived from cofiber and fiber sequences in the category of ex-spaces, see (2.6).

However, we will construe them in terms of properties of primary obstructions and differences, which are already intimated in the classical works and have a natural significance in obstruction theory. Nonetheless, the importance of the ex-space category and relative methods as in 3), 4) and 3'), 4') is made clear by the existence of the principal reduction of CW-complexes and of Postnikov decompositions. These two existence proofs are a main result of this book. Relative methods in obstruction theory have been developed in the last ten years by James, Thomas, McClendon, Larmore, Becker and others. This book can be regarded as a systematic foundation of, and motivation for, these relative methods.

In the course of our presentation it was often necessary to introduce new notations because of the uncoordinated state of the theory in the literature. The text also contains various new, amplifying results. Cross-references to the literature for our results and definitions are contained under 'Remarks'. The bibliography is not intended however to encompass the entire subject.

I would like to acknowledge the support of the Sonderforschungsbereich 40 Theoretische Mathematik towards the completion of this book, in particular I am grateful to Mrs. Motee Spanier for typing the final version. I especially thank Stuart Clayton for translating the German manuscript.

H.J. Baues

CHAPTER 0. CONVENTIONS AND NOTATION

(0.0) Maps and homotopy. Excision theorems

The following conventions and notation will be maintained throughout the book. Let Top be the category of topological spaces and continuous maps, and let Top^{o} be the category of pointed topological spaces. Unless expressly stated to the contrary, from now on all spaces are pointed, that is, they all have distinguished base points * (starting with Chapter 1 we in fact require all spaces to be well-pointed, see (0.1.2)). Furthermore, all maps and all homotopies (denoted by $\simeq$) preserve the base points. The set of homotopy classes of base point-preserving maps $f: X \to Y$ will be denoted by $[X,Y]$. In this set 0 denotes the class of nullhomotopic maps. We denote by 1, 1_X or id the identity map. We will often use the same symbol to refer to a map and the homotopy class it represents. It will be clear from the context whether a map or a homotopy class is meant. The composition of maps or homotopy classes $f: X \to Y$ and $g: X \to Z$ will be denoted by $g \circ f$ or gf . Composition induces mappings of sets

$$g_* : [Y,X] \to [Y,Z] \quad \text{with} \quad g_*(f) = gf ,$$

$$f^* : [X,Z] \to [Y,Z] \quad \text{with} \quad f^*(g) = gf .$$

There are corresponding mappings for homotopy classes of maps between pairs.

Let $A \times B$ be the topological product of A and B , and let $A \vee B \subset A \times B$ be the one-point union or wedge of A and B , that is $A \vee B = A \times \{*\} \cup \{*\} \times B$. We have the canonical bijections

$$[A \vee B, X] = [A,X] \times [B,X]$$

$$[X, A \times B] = [X,A] \times [X,B] .$$

From pairs of maps we obtain maps $(f_1,f_2): A \vee B \to X$ and $(g_1,g_2): X \to A \times B$. The maps $c = (1,1): X \vee X \to X$ and $d = (1,1): X \to X \times X$ are called the folding map and the diagonal, respectively. We have

$$(f_1,f_2) = c \circ (f_1 \vee f_2) \quad \text{and} \quad (g_1,g_2) = (g_1 \times g_2) \circ d \quad .$$

Let X^Y be the space of non-pointed continuous maps $Y \to X$ with the compact-open topology. Then

(0.0.1) Exponential Law : For locally compact K the map

$$\vartheta : X^{K \times Y} \to (X^K)^Y$$

with $(f)(y)(t) = f(t,y)$ for $y \in Y$, $t \in K$ is a bijection. ϑ is a homeomorphism if Y and K are hausdorff, see [24] .

Related to the exponential law are the following facts.

(a) Let K be locally compact. Then the evaluation map $X^K \times K \to X$ with $(f,t) \mapsto f(t)$ is continuous, see (4.14) of [24] .

(b) Let K be locally compact and let $q: A \to B$ be an identification map. Then $q \times 1 : A \times K \to B \times K$ is also an identification map, see (4.13) of [24] .

(c) As an extension of (b), let $A \subset X$ and let A be compact and let $q: X \to X/A$ be the identification map. Then for any space Z the map $q \times 1 : X \times Z \to (X/A) \times Z$ is also an identification map.

(d) Let $i: A \to B$ be an embedding. Then for any space Z the map $i^Z: A^Z \to B^Z$ is an embedding, see 4.6 of [24] .

(e) In connection with (c) we can say the following. Let the map $p: A \to B$ be surjective and such that for every compact subset

L of B there is a compact subset $K \subset A$ with $p(K) = L$. Then for any space Z the map $Z^p: Z^B \longrightarrow Z^A$ with $f \mapsto f \circ p$ is an embedding. An example of such a map p is the map q in (c).

(f) For Z hausdorff and arbitrary X, Y we have

$$(X \times Y)^Z = X^Z \times Y^Z \quad .$$

(g) For arbitrary X, Y, Z we have

$$X^{Y+Z} = X^Y \times X^Z \quad ,$$

where $Y+Z$ is the topological sum, that is, the disjoint union.

In the commutative diagram of unbroken arrows

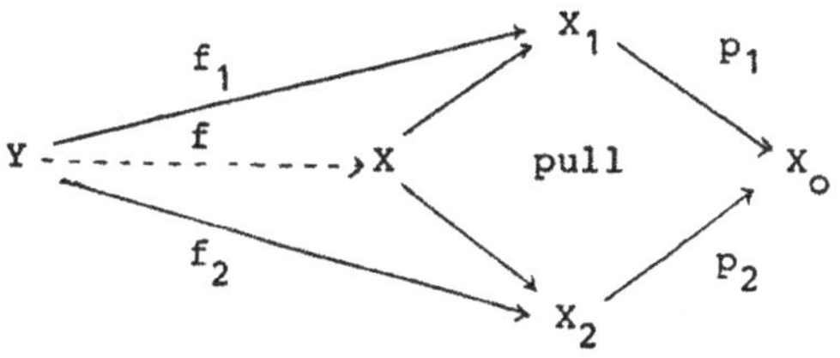

the subdiagram 'push' is called a cocartesian square or <u>pushout</u> when to every pair of maps f_1, f_2 there exists exactly one map $f = f_1 \cup f_2$ extending the diagram. By 'extending the diagram' we always mean 'commutatively'. Such an X is uniquely determined up to homeoporphism. There exists a cocartesian square for i_1 and i_2, since we can take

$$X = X_1 \cup_{X_o} X_2 = (X_1 + X_2)/\sim$$

where the equivalence relation in the disjoint union $X_1 + X_2$ is generated by $i_1(x) \sim i_2(x)$ with $x \in X_o$. X is given the quotient topology. If i_1 is an inclusion, X is called an adjunction space.

There is a dual definition to the preceeding one. In the commutative diagram

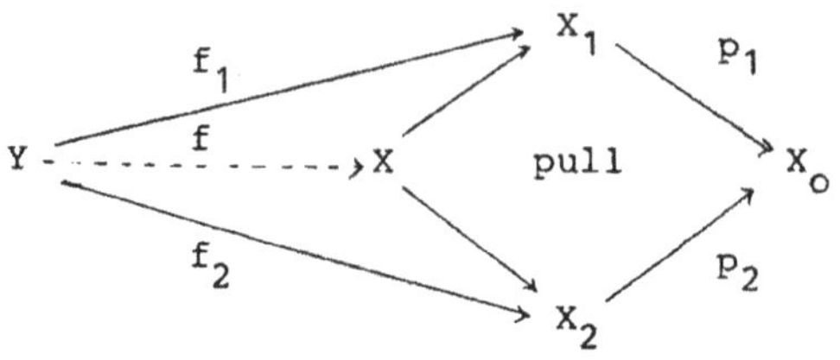

the subdiagram 'pull' is called a cartesian square or <u>pullback</u> when to every pair of maps f_1, f_2 there exists exactly one map $f = f_1 \underline{\times} f_2$ extending the diagram. Such an X is uniquely determined up to homeomorphism. There exists a cartesian square for p_1 and p_2, since we can take

$$X = X_1 \times_{X_o} X_o = \{(x,y) \in X_1 \times X_2 \mid p_1(x) = p_2(y)\}$$

with the subspace topology from $X_1 \times X_2$.

(0.0.2) Cocartesian and cartesian diagrams can be combined in the following way. Consider the commutative diagrams

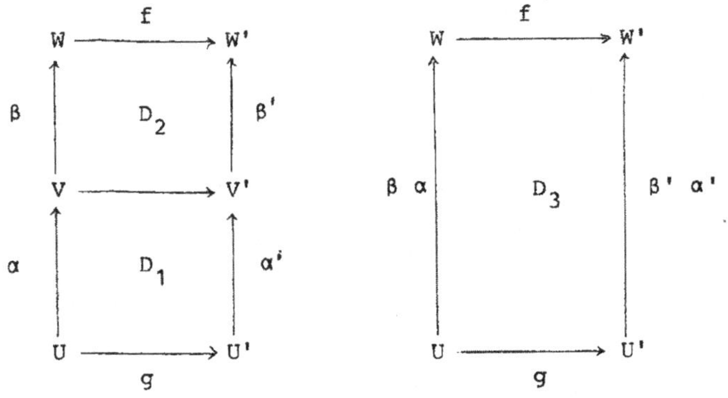

If D_1 is cocartesian, then

$$D_2 \text{ is cocartesian} \Leftrightarrow D_3 \text{ is cocartesian.}$$

If D_2 is cartesian, then

$$D_1 \text{ is cartesian} \Longleftrightarrow D_3 \text{ is cartesian} .$$

(0.0.3) Cocartesian and cartesian squares are compatible with products and mapping spaces in the following ways. Let a commutative diagram

$$\begin{array}{ccc} A & \longrightarrow & Y \\ \downarrow & (D) & \downarrow \\ X & \longrightarrow & B \end{array}$$

be given. Then

(I) If (D) is cocartesian and K is locally compact, the diagram

$$(D) \times K : \quad \begin{array}{ccc} A \times K & \longrightarrow & Y \times K \\ \downarrow & & \downarrow \\ X \times K & \longrightarrow & B \times K \end{array}$$

is also cocartesian, as follows easily from (0.0.1)(b) . An extension of this will be described in (0.5.3) .

(II) If (D) is cocartesian and X, Y, B are compact hausdorff (or alternatively if the quotient map $X+Y \rightarrow B$ satisfies the condition in (0.0.1)(e)), the diagram

$$Z^{(D)} : \quad \begin{array}{ccc} Z^A & \longleftarrow & Z^Y \\ \uparrow & & \uparrow \\ Z^X & \longleftarrow & Z^B \end{array}$$

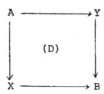

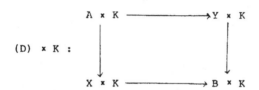

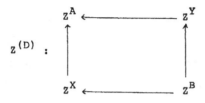

is cartesian, as follows easily from (0.0.1)(e) and (g) .

(III) If (D) is cartesian and Z is hausdorff, the diagram

$$(D)^Z : \quad \begin{array}{ccc} A^Z & \longrightarrow & Y^Z \\ \downarrow & & \downarrow \\ X^Z & \longrightarrow & B^Z \end{array}$$

is also cartesian, as follows easily from (0.0.1)(d) and (f).

We now define some further homotopy concepts. Let $I = [0,1]$ be the unit interval and let $H: f_0 \simeq f_1 : X \longrightarrow Y$ be a homotopy. That is, $H: I \times X \longrightarrow Y$ is a map with $H_0 = f_0$ and $H_1 = f_1$, where for $t \in I$ the pointed map $H_t: X \longrightarrow Y$ is defined by $H_t(x) = H(t,x)$ for $x \in X$. The map H gives us the adjoint map $\bar{H}: X \longrightarrow Y^I$ with $H(x)(t) = H(t,x)$, $q_0\bar{H} = f_0$ and $q_1\bar{H} = f_1$ (see (0.0.1)), where we define $q_t(\sigma) = \sigma(t)$ for $\sigma \in Y^I$. Conversely, every such map H gives us a homotopy $H: f_0 \simeq f_1$. H is a pointed map, with the trivial map $0 \in Y^I$ as base point in Y^I .

Given the maps (in Top^0)

$$\begin{array}{ccc} A & \xrightarrow{\;g\;} & Y \\ {\scriptstyle i}\downarrow & \nearrow {\scriptstyle g_0, g_1} & \\ X & & \end{array}$$

we call $H: g_0 \simeq g_1$ a <u>homotopy under</u> A when for all $t \in I$ we have $H_t \circ i = g$. The set of homotopy classes under A is denoted by $[X,Y]^g$ or $[X,Y]^A$. It will also be referred to as the homotopy set relative g , especially when i is an inclusion. If g is the identity, the homotopy set under A will also be called the <u>retraction homotopy set</u>

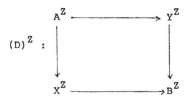

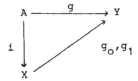

for i , denoted by $\langle X,A\rangle$. Every homotopy set under A can be regarded as a retraction homotopy set in the following way. Let

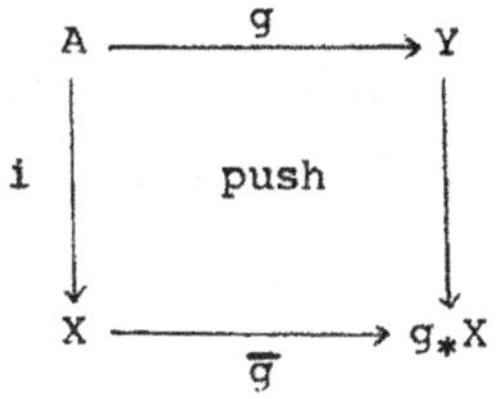

be the cocartesian diagram for (i,g) . Then $\bar{g}$ induces a bijection

(0.0.4) $$[X,Y]^g = \langle g_*X,Y\rangle \quad .$$

Dual to 'homotopy under' is the concept of 'homotopy over', defined as follows. Given the maps (in Top^o)

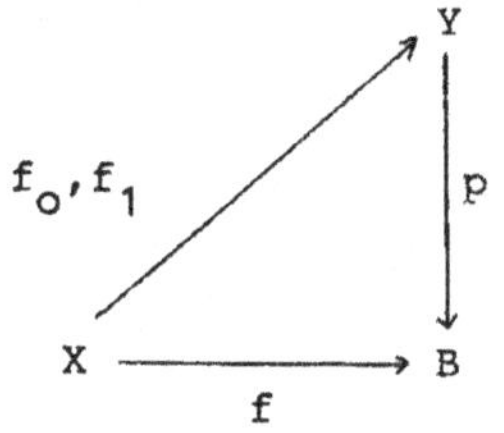

we call $H: f_o \simeq f_1$ a <u>homotopy over</u> B when for all $t \in I$ we have $p \circ H_t = f$. The set of homotopy classes over B is denoted by $[X,Y]_f$ or $[X,Y]_B$. It will also be referred to as the homotopy set of liftings of f , especially when p is a fibration. If f is the identity, the homotopy set over B will be called the <u>section homotopy set</u> for p , denoted by $\langle B,Y\rangle$. It will always be clear from the context whether $\langle\ ,\ \rangle$ denotes a section homotopy set or a retraction homotopy set. Every homotopy set over B can be regarded as a section homotopy set as follows. Let

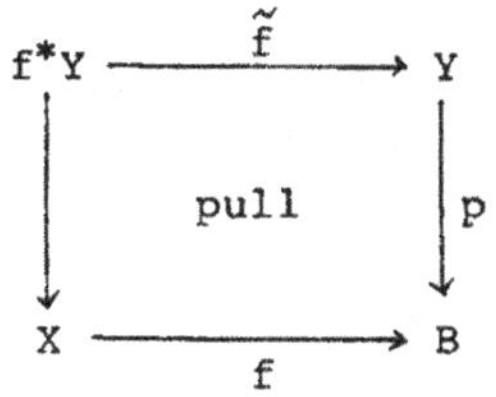

be the cartesian diagram for (f,p) . The map f induces a bijection

$$[X,Y]_f = \langle X, f^*Y \rangle \quad . \tag{0.0.5}$$

We will frequently use the following properties of homotopy sets 'over' and 'under'. In the commutative diagram

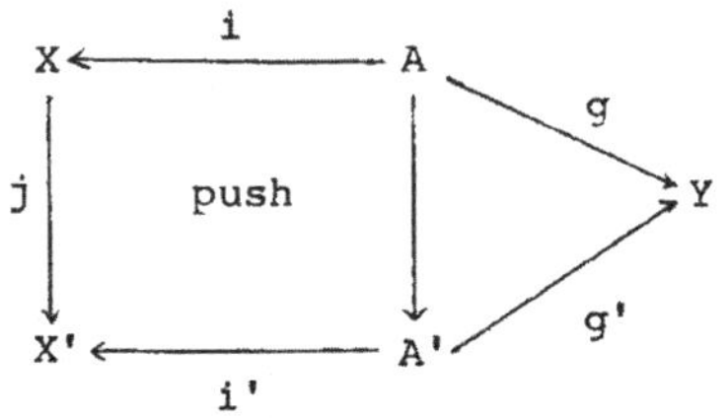

the map j induces a bijection

$$[X',Y]^{g'} = [X,Y]^{g} \quad \text{with } g_0 \rightarrow g_0 \circ j \tag{0.0.6}$$

of homotopy sets under A' and A . This follows as before from (0.0.3), or from (0.0.2) and (0.0.4) .

In the commutative diagram

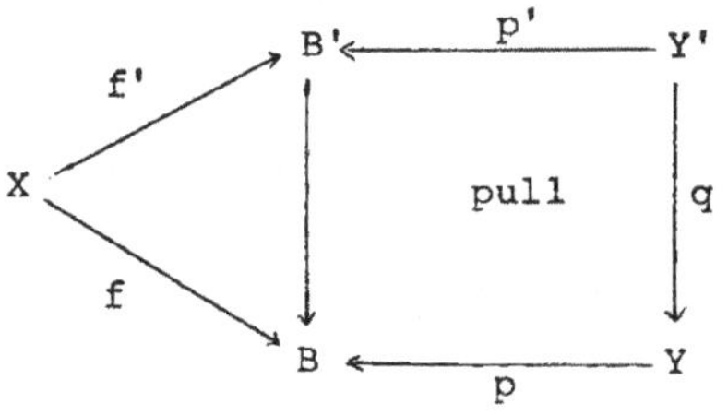

the map q induces a bijection

$$[X,Y']_{f'} = [X,Y]_f \quad \text{with } f_0 \rightarrow q \circ f_0 \tag{0.0.7}$$

of homotopy sets over B' and over B . This follows from (0.0.3), or from (0.0.2) and (0.0.5) .

Derived from the concepts of 'homotopy over' and 'homotopy under' is that of 'homotopy under and over' . Given the maps in the commutative diagram

(0.0.8) 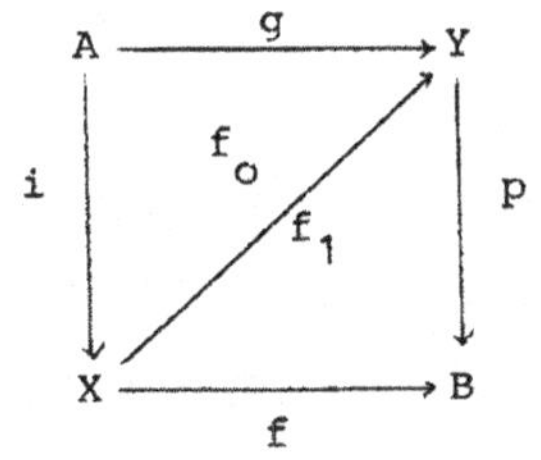

we call $H: f_1 \simeq f_2$ a <u>homotopy under</u> A <u>and over</u> B when for all $t \in I$ we have $p \circ H_t = f$ and $H_t \circ i = g$. The corresponding set of homotopy classes is denoted by $[X,Y]^g_f$ and by $[X,Y]^A_B$. It will be called the homotopy set of relative liftings, especially when i is an inclusion and p a fibration. We will be particularly interested in those cases where g or f is the identity. (If both g and f are identities, i = p is a homeomorphism with inverses $f_o = f_1$).

We now describe the case where f is the identity. In the cartesian square

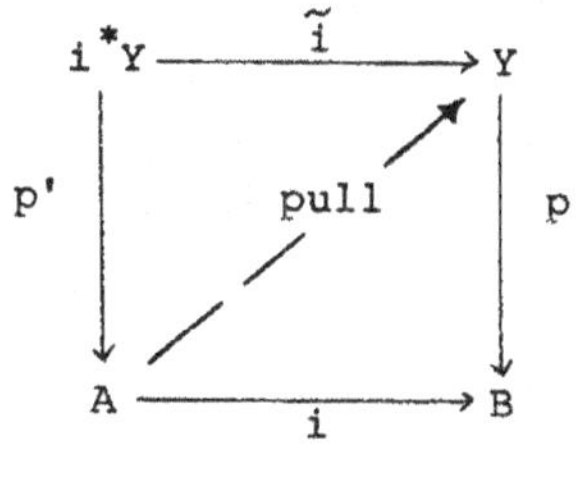

the map g gives rise to a section $g: A \to i^*Y$ of p , namely $g \underline{x} 1_A$. We say that a section $g_o: B \to Y$ of p is a <u>section extension</u> of g , and that $g = i^*g_o$ is the <u>induced section</u> of g_o , when the diagram

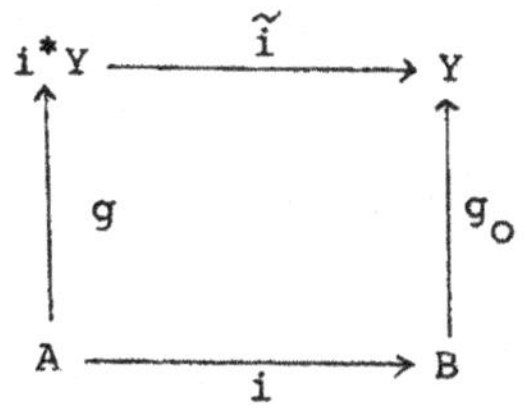

commutes. Let g_1 be another section extension of g. Then g_0 and g_1 are section homotopic relative to g when there is a section homotopy $H: g_0 \simeq g_1$ such that for all $t \in I$ H_t is a section extension of g. The corresponding set of homotopy classes is denoted by $\langle B,Y\rangle^g$ and called the <u>section homotopy set relative</u> g. A section g of the projection $p: Y = B \times U \to B$ corresponds to a map $g: B \to U$, and we have $\langle B,B \times U\rangle^g = [B,U]^g$.

Now we describe the dual case where g is the identity. In the cocartesian square

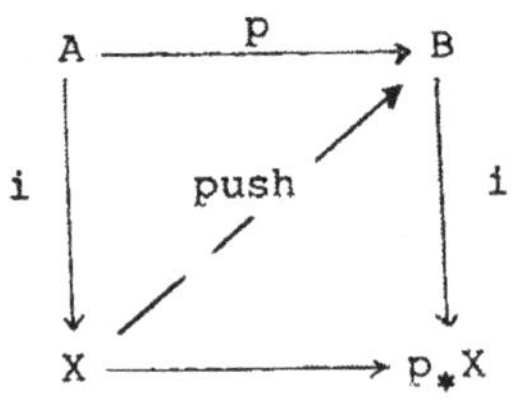

the map f determines a retraction $f: p_*X \to B$ of i, namely $f \cup 1_B$. We call a retraction $f_0: X \to A$ for i a <u>retraction lifting</u> of f, and $f = p_*f_0$ the <u>induced retraction</u> of f_0, when the diagram

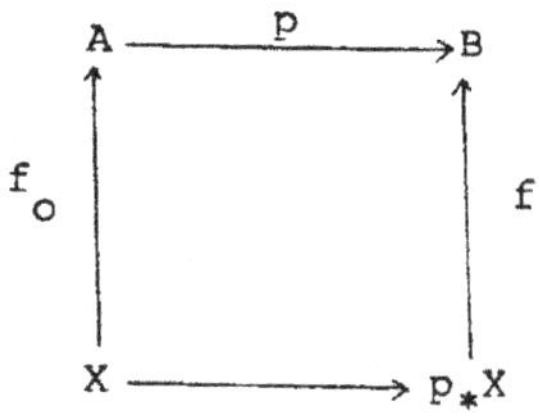

commutes. Let f_1 be another retraction lifting of f. Then f_0 and f_1 are retraction homotopic relative f when there is a retraction homotopy $H: f_0 \simeq f_1$ such that for all $t \in I$ H_t is a retraction lifting of f. The corresponding set of homotopy classes is denoted by $\langle X,A\rangle_f$ and called the <u>retraction homotopy set over</u> f. A retraction of the inclusion $i: A \subset X = A \vee U$ corresponds to a map $f: U \to B$, and we have $\langle A \vee U, A \rangle_f = [U,A]_f$.

The diagram (0.0.8) without the arrows f_o, f_1 is equivalent to each of the diagrams

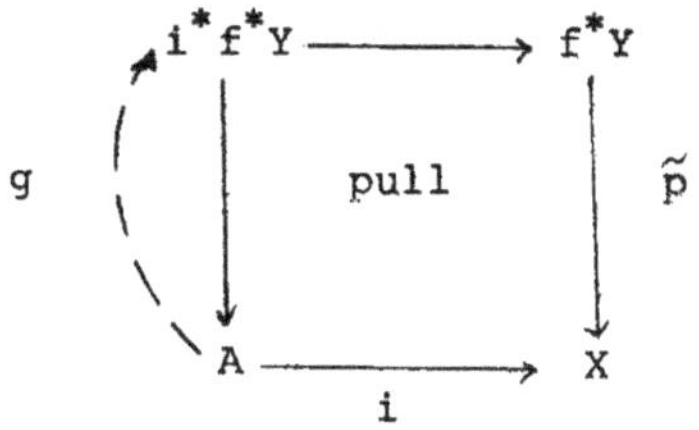

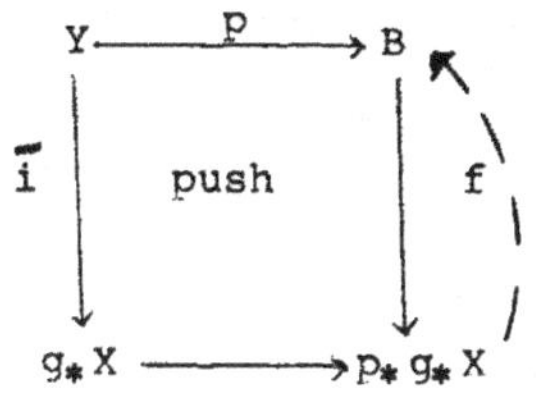

Here g is a section and f is a retraction, for which , in generalization of (0.0.4) and (0.0.5),

$$(0.0.9) \qquad [X,Y]_f^g = \langle X, f^*Y \rangle^g = \langle g_*X, Y \rangle_f$$

holds. We now describe generalizations of (0.0.6) and (0.0.7) for the homotopy sets $\langle\ ,\ \rangle^g$ and $\langle\ ,\ \rangle_f$. It will be seen that many properties of the homotopy sets $[\ ,\ ]^g$ and $[\ ,\ ]_f$ are shared by the homotopy sets $\langle\ ,\ \rangle^g$ and $\langle\ ,\ \rangle_f$ mutatis mutandis. Thus the familiar fibration and cofibration exact sequences for $[\ ,\ ]^g$ and $[\ ,\ ]_f$ can be adapted for $\langle\ ,\ \rangle^g$, $\langle\ ,\ \rangle_f$ (see (2.4) and (2.5), where we use these generalizations of (0.0.6) and (0.0.7)). First we generalize for homotopy sets of the form $[\ ,\ ]_f^g$. In the commutative diagram

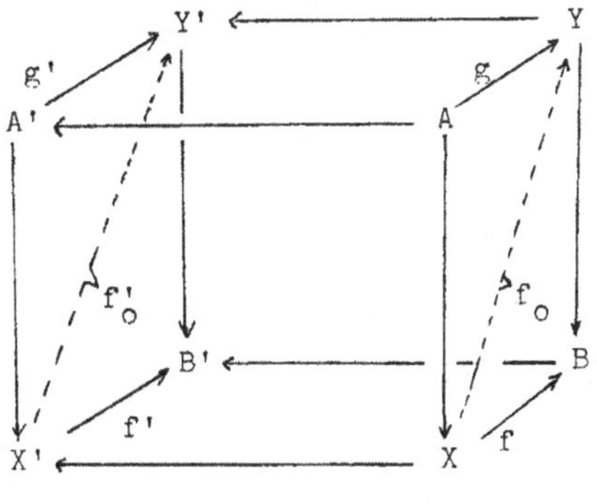

the two lateral faces correspond to the diagram (0.0.8) . The homotopy sets $[X,Y]_B^A$ and $[X',Y']_{B'}^{A'}$ are defined for these squares.

(0.0.10) Theorem: If the back face in the diagram is cartesian and the front face is cocartesian, then

$$[X,Y]_B^A = [X',Y']_{B'}^{A'} \quad .$$

Proof: Corresponding to the map f_o there is a map f_o' uniquely defined by g' and $X \xrightarrow{f_o} Y \longrightarrow Y'$, since the front face is cocartesian. Conversely, given the map f_o', the map f_o is uniquely defined by f and $X \longrightarrow X' \xrightarrow{f} Y'$, since the back face is cartesian. Corresponding statements hold for 'homotopies under and over' because of (0.0.3) . □

From (0.0.10) we derive two corollaries. In the commutative diagram

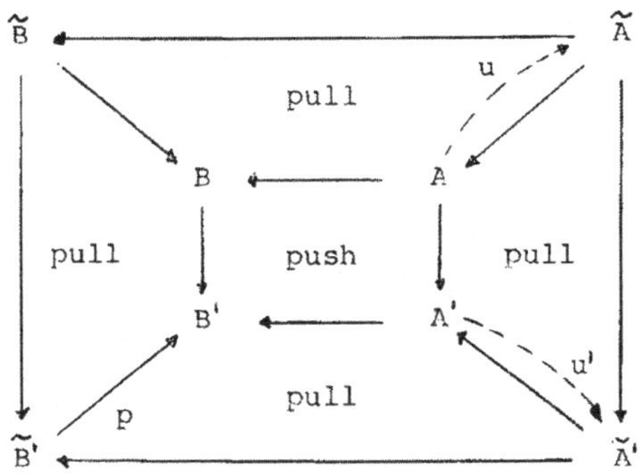

let u' be a section extension of u . Then for the section homotopy set relative u and u' we have the excision theorem

(0.0.11) Corollary: $$\langle B,\tilde{B}\rangle^{u} = \langle B',\tilde{B}'\rangle^{u'} \quad .$$

If p is the projection $B' \times U \rightarrow B'$, (0.0.11) is equivalent to (0.0.6) . The meaning of Corollary (0.0.11) can be illustrated when B'

is the pushout of inclusions $A \subset B$ and $A \subset A'$, so that $B' = B \cup A'$ and $A = B \cap A'$. The schematic picture

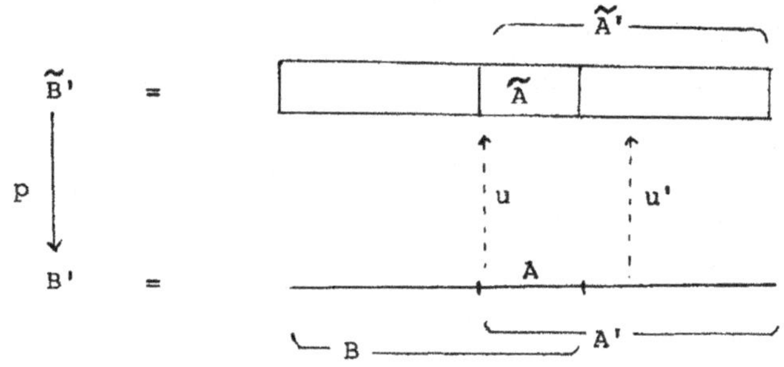

suggests that section extensions of u are equivalent to section extensions of u', since that part of u' over $A' - A$ is irrelevant to the classification of section extensions of u' and so can be 'excised'. We will see that the above excision theorem in the form (0.0.6) corresponds to the excision axiom for the cohomology functor. In the form (0.0.11) it corresponds to the excision theorem in cohomology with local coefficients, see (1.4.18) and (5.2.4) .

In the commutative diagram

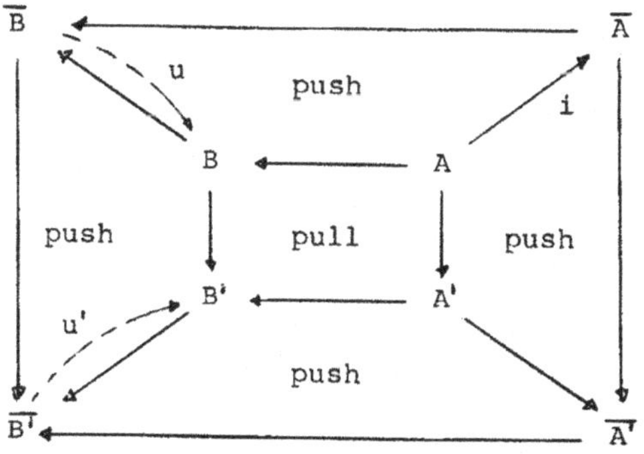

let u be a retraction lifting for u' . Then for the retraction homotopy sets over u and over u'

(0.0.12) <u>Corollary</u>: $\langle \bar{A}, A \rangle_u = \langle \bar{A}', A' \rangle_{u'}$.

If i is the inclusion $A \subset \bar{A} = A \vee U$, (0.0.12) is identical with (0.0.7) .

<u>Proof of</u> (0.0.11) <u>and</u> (0.0.12) :

A proof of (0.0.11) consists in comparing the diagram

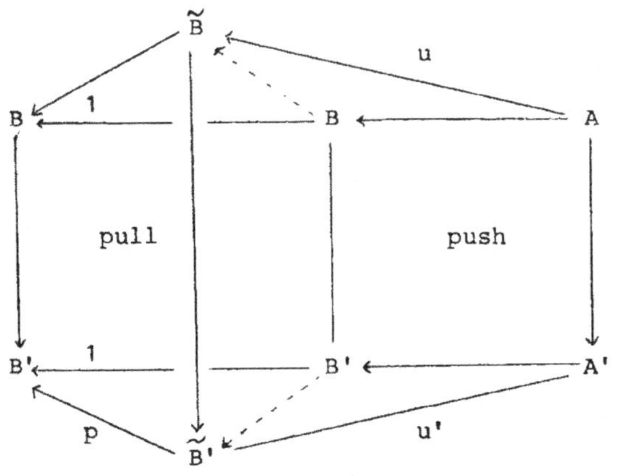

with the diagram in (0.0.10).

A proof of (0.0.12) consists in comparing the diagram

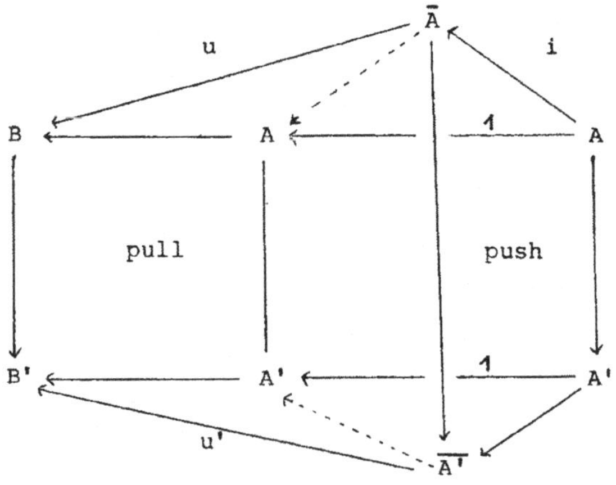

again with the diagram in (0.0.10). Notice that u and u' are given in accordance with (0.0.5) and (0.0.4) respectively. □

(0.1) Cofibrations and fibrations

We here compile some well-known properties of fibrations and cofibrations that we will use in the course of this book. Our source for most of them is the excellent book Homotopietheorie by tom Dieck, Kamps and D. Puppe [24] .

Definition: A map $i: A \to X$ is a cofibration when every commutative diagram in Top

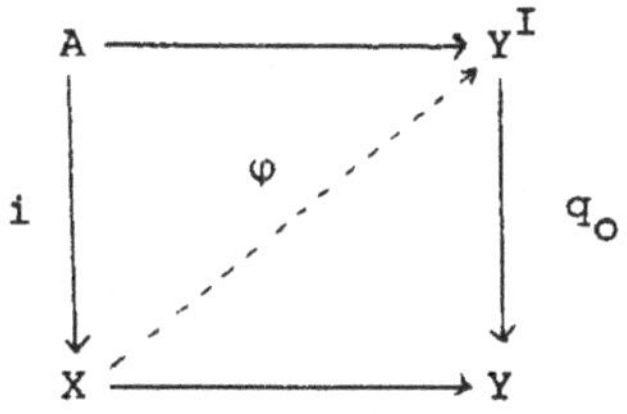

has a homotopy φ extending it. Here $q_o(\sigma) = \sigma(0)$ for $\sigma \in Y^I$, see 5.20 on p.100 of [24] . The space $X/i(A)$ is called the cofiber of i.

Definition: A map $p: Y \to B$ is a fibration when every commutative diagram in Top

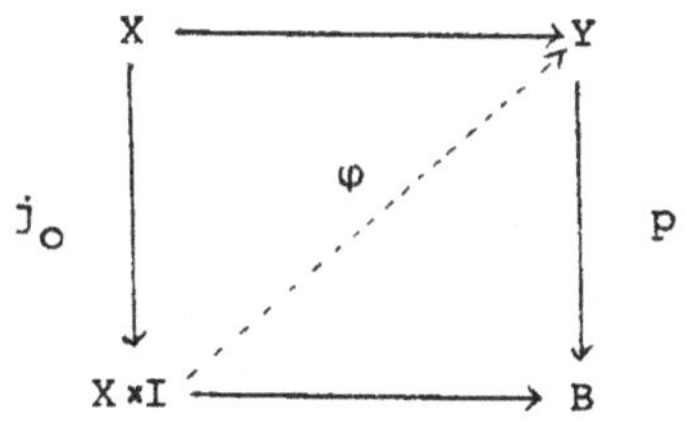

has a homotopy φ extending it. Here $j_o(x) = (x,0)$ for $x \in X$, see 5.5 on p.94 of [24] . The space $F = p^{-1}(*)$ is called the fiber of p.

From now on cofibrations and fibrations will always be pointed maps, although we defined these concepts in Top and not Top^o . If in the above diagrams Top is replaced by Top^o and homotopies are replaced by pointed homotopies, we obtain the concepts of cofibration in Top^o and fibration in Top^o . A cofibration in Top is also a cofibration in Top^o , and a fibration in Top^o is also a fibration in Top , but not conversely. In this book we will use the more prevalent concept of a fibration in Top . This creates some difficulties in general, which we can avoid however by restricting ourselves to well-pointed spaces, see (0.1.2) and (0.1.5) .

Every cofibration $i: A \to X$ is an embedding, since $i(A)$ is closed when X is hausdorff, but not every fibration is an identification map (see 1.17 on p.28 and 6.18, 2 on p.118 of [24]) . Every inclusion of a subcomplex in a CW-complex is a cofibration, and every locally trivial map over a paracompact base space is a fibration, see 5.14 on p.97 of [24] .

(0.1.1) In the commutative diagrams

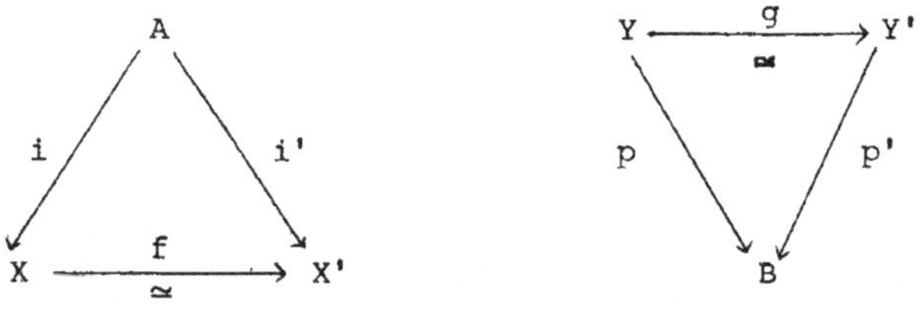

let i and i' be cofibrations and p and p' fibrations, and let f and g be homotopy equivalences in Top . Then f and g are homotopy equivalences under A and over B respectively, see 2.18 and 6.21 of [24] . We say that the map f is a <u>homotopy equivalence under</u> A when there exists a homotopy inverse f' and homotopies $H: f \circ f' \simeq 1$ and $H: f' \circ f \simeq 1$ such that $f'i' = i$ and $H_t \circ i' = i'$ and $\overline{H}_t \circ i = i$ for $t \in I$. Similarly, the map g is a <u>homotopy equivalence over</u> B when there exists a homotopy inverse g' and homotopies $G: g \circ g' \simeq 1$ and $\overline{G}: g' \circ g \simeq 1$ such that $pg' = p'$ and $p'G_t = p'$ and $pG_t = p$ for $t \in I$.

(0.1.2) We call a space X <u>well-pointed</u> when the inclusion $* \subset X$ is a closed cofibration in Top . Let X and X' be well-pointed spaces and $f: X \to X'$ a pointed map that is also a homotopy equivalence in Top . Then f is also a homotopy equivalence in Top° because of (0.1.1). As remarked at the outset, from Chapter 1 onwards we require that all spaces be well-pointed.

(0.1.3) In the cocartesian and cartesian diagrams respectively

$$\begin{array}{ccc} A & \xrightarrow{\bar{\alpha}} & A' \\ {\scriptstyle i}\downarrow & \text{push} & \downarrow{\scriptstyle i'} \\ X & \xrightarrow[\alpha]{} & X' \end{array} \qquad \begin{array}{ccc} Y' & \xrightarrow{\bar{\beta}} & Y \\ \downarrow{\scriptstyle p'} & \text{pull} & \downarrow{\scriptstyle p} \\ B' & \xrightarrow[\beta]{} & B \end{array}$$

let i be a cofibration and p a fibration. Then i' is a cofibration and p' is a fibration. i' and p' are called the <u>induced</u> cofibration and fibration respectively. If $\bar{\alpha}$ or β is a homotopy equivalence in Top , α or $\bar{\beta}$ respectively is also a homotopy equivalence in Top . A corresponding statement in Top° is true for a fibration p in Top°, see 7.38 on p.140, 7.9 on p.127, 7.30 on p.137 and 7.43 on p.142 of [24] .

In particular, it follows that if $i: A \subset X$ is a cofibration and $A \xrightarrow{\simeq} *$ then the projection $X \to X/A$ is a homotopy equivalence in Top. Similarly, if $p: Y \to B$ is a fibration (or a fibration in Top°) with fiber F and if $* \xrightarrow{\simeq} B$, then the inclusion $F \subset Y$ is a homotopy equivalence in Top (in Top°) . This is because the diagrams

$$\begin{array}{ccc} A & \xrightarrow{\simeq} & * \\ \downarrow & \text{push} & \downarrow \\ X & \longrightarrow & X/A \end{array} \qquad \begin{array}{ccc} F & \longrightarrow & Y \\ \downarrow & \text{pull} & \downarrow \\ * & \xrightarrow{\simeq} & B \end{array}$$

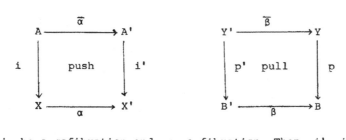

are cocartesian and cartesian respectively.

(0.1.4) Let $p\colon Y \to B$ be a fibration and let $s_1\colon B \to Y$ be a <u>section up to homotopy</u>, that is $ps_1 \simeq 1_Y$. Then s_1 is homotopic to a section s with $ps = 1_Y$. Let $i\colon A \to X$ be a cofibration and let $r_1\colon X \to A$ be a <u>retraction up to homotopy</u> or weak retraction, that is $r_1 i \simeq 1_A$. Then r_1 is homotopic to a retraction r with $ri = 1_A$, see 6.10 on p.112 and 2.27 on p.60 of [24] .

(0.1.5) A fibration $p\colon Y \to B$ has the following <u>relative lifting property</u>. Let $A \subset X$ be a closed cofibration. Then any commutative diagram in Top

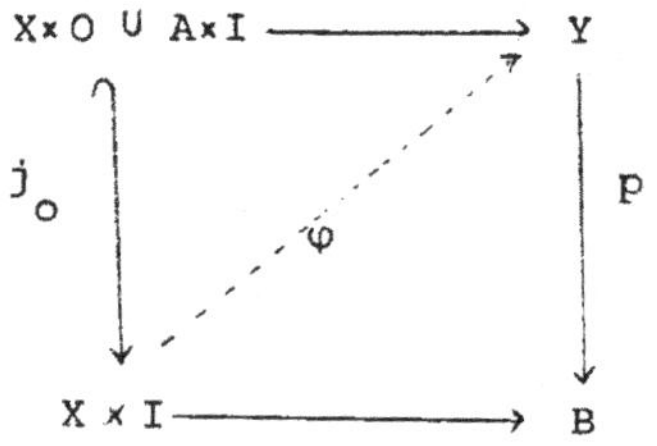

can be extended by a homotopy φ , see p.209 of [24] . We will be using this property frequently. One implication of it is that a fibration p has the pointed lifting property for well-pointed spaces X .

(0.1.6) Let Z be a topological space and $i\colon A \subset X$ a cofibration. Then $i \times 1_Z : A \times Z \subset X \times Z$ is also a cofibration (if Z is locally compact hausdorff this follows easily from the exponential law, see p.97 of [24]). Let $i\colon A \subset X$ be a cofibration, with both A and X locally compact. Then $Z^i\colon Z^X \to Z^A$ is a fibration (this too follows easily from the exponential law (0.0.1)).

(0.1.7) <u>Union and lifting of cofibrations</u>

(A) Let $A_1, A_2 \subset X$ and $A_1 \cap A_2 \subset X$ be closed cofibrations.

Then $A_1 \cup A_2 \subset X$ is also a closed cofibration, see [69] .

(B) Let $p: \tilde{X} \longrightarrow X$ be a fibration and $A \subset X$ a cofibration. Then the inclusion $\tilde{A} = p^{-1}(A) \subset \tilde{X}$ is also a cofibration, see Strøm [113] . This generalizes the first statement of (0.1.6).

(0.1.8) <u>The $\wedge$-product and the $\wedge$-mapping space</u>

For the pointed spaces A and B we define the $\wedge$-product (smash product)

$$A \wedge B = A \times B / A \vee B \quad .$$

We will write (a,b), where $a \in A$ and $b \in B$, for a point in $A \times B$ or $A \wedge B$. The corresponding n-fold products will be denoted by $A^n = A \times \ldots \times A$ and $A^{(n)} = A \wedge \ldots \wedge A$. If A and B are well-pointed then it follows from (0.1.7) and (0.1.3) that $A \wedge B$ is also well-pointed.

We have

$$(A \vee B) \wedge X = (A \wedge X) \vee (B \wedge X)$$

where the $\wedge$-products are to be bracketed first and then the $\vee$-sums. Furthermore, when X and Y are compact and X is hausdorff, or else X and Z are locally compact,

$$(X \wedge Y) \wedge Z = X \wedge (Y \wedge Z) \quad .$$

We now define for pointed spaces A and B the $\wedge$-mapping space $A^{\wedge B}$ to be the subspace of pointed maps in A^B . There is the following exponential law for pointed spaces, see 6.2.38 of [78] .

Let K be a pointed locally compact space. Then the map ϑ in (0.0.1) determines a bijection

$$\vartheta : A^{\wedge(K \wedge B)} \longrightarrow (A^{\wedge K})^{\wedge B}$$

which is a homeomorphism when K and B are compact hausdorff. Furthermore, when B and C are hausdorff

$$A^{\wedge(B \vee C)} = A^{\wedge B} \times A^{\wedge C} \qquad ,$$

see 6.2.32 in [78] .

(0.1.8) Let A be a well-pointed space. Then any (pointed) map $f\colon A \to X$ can be replaced, up to a homotopy equivalence in Top^{o}, by a closed cofibration. To see this we define the mapping cylinder Z_f by means of the cocartesian diagram at the left

$$\begin{array}{ccc} A & \xrightarrow{f} & X \\ j_o \downarrow & \text{push} & \downarrow j_o \\ I^+ \wedge A & \longrightarrow & Z_f \end{array}$$

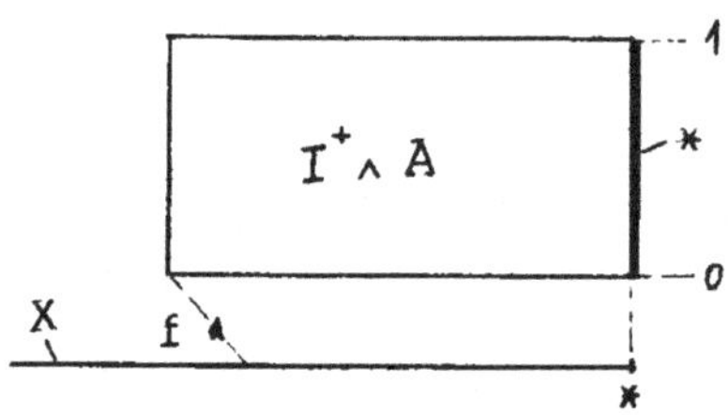

Here $I^+ = \{*\} + I$, so that $I^+ \wedge A = I \times A \,/\, I \times \{*\}$. We have inclusions $j_t\colon A \subset I^+ \wedge A$ defined by $j_t(a) = (t,a)$. These give us the cofibration $(j_o,j_1) : A \vee A \subset I^+ \wedge A$ and the induced cofibration $(j_o,j_1) : (X \vee A) \subset Z_f$ induced by $f \vee 1_A$. The projection $I^+ \wedge A \to A$ with $(t,a) \mapsto a$ determines a map q such that $f = qj_1\colon A \to Z_f \xrightarrow{q} X$. The map q is a homotopy equivalence in Top^{o} with $qj_o = 1_X$ and $j_o q \simeq 1_{Z_f}$ rel $j_o(X)$. The cofiber $C_f = Z_f \,/\, j_1(A)$ of j_1 is called the mapping cone of f . C_f can also be obtained as a pushout

$$\begin{array}{ccc} A & \xrightarrow{f} & X \\ \downarrow & \text{push} & \downarrow \\ CA & \longrightarrow & C_f = CA \cup_f X, \end{array}$$

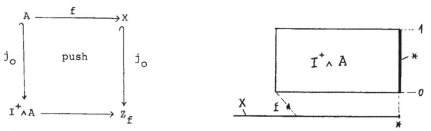

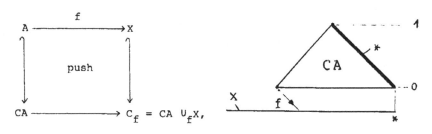

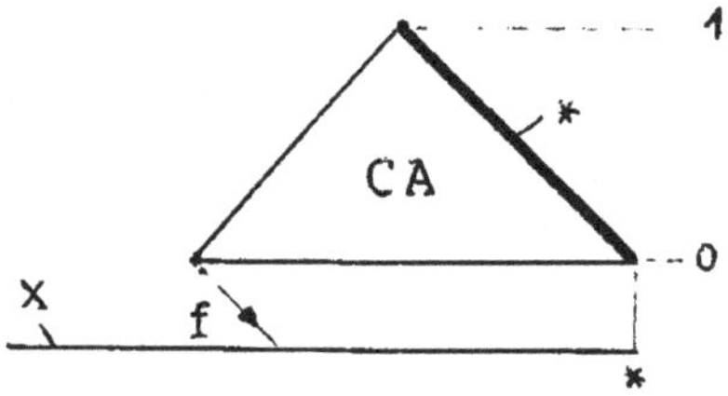

where $CA = I \times A / (I \times \{*\} \cup 1 \times A)$, the cone on A. Since A is well-pointed, $A \subset CA$ and $X \subset C_f$ are cofibrations with cofiber $SA = CA/A = S^1 \wedge A$ (where $S^1 = I/\partial I$). If A and X are well-pointed, so are C_f and Z_f. By 5.11 of [70], Z_f and C_f are CW-complexes when $f: A \to X$ is a cellular map between CW-complexes.

We say $A \xrightarrow{f} X \xrightarrow{g} F$ is a <u>cofiber sequence</u> if there exists a homotopy equivalence h such that $X \subset C_f \xrightarrow[\simeq]{h} F$ is homotopic to g.

(0.1.10) <u>Every pointed map</u> $g: Y \to B$ <u>can be replaced, up to a pointed homotopy equivalence, by a fibration in</u> Top^o. To see this we define the <u>free mapping path space</u> W_g by means of the cartesian diagram in Top^o

$$\begin{array}{ccc} W_g & \longrightarrow & B^I \\ {\scriptstyle q_o}\big\downarrow & \text{pull} & \big\downarrow{\scriptstyle q_o} \\ Y & \xrightarrow{\quad g \quad} & B \end{array}$$

We have maps $q_t: B^I \to B$ defined by $q_t(\sigma) = \sigma(t)$ for $t \in I$, which give us the Top^o fibration $(q_o,q_1): B^I \to B \times B$ and the induced Top^o fibration $(q_o,q_1): W_g \to Y \times B$ induced by $g \times 1_B$. Let $j: Y \to W_g$ be given by $j(y) = (y, g(y))$ where $g(y)$ is the constant path $I \to B$ that takes every $t \in I$ to $g(y)$. Then $g = q_1 j: Y \xrightarrow[\simeq]{j} W_g \longrightarrow B$. The map j is a homotopy equivalence in Top^o with $q_o j = 1$ and $j q_o \simeq 1_{W_g}$ over Y. The fiber $P_g = q_1^{-1}(*)$ of q_1 will be called the <u>mapping path space</u> of g or the <u>fiber of</u> g. P_g can also be obtained as a pullback in

$$\begin{array}{ccc} P_g & \longrightarrow & PB \\ {\scriptstyle q_o}\big\downarrow & \text{pull} & \big\downarrow{\scriptstyle q_o} \\ Y & \xrightarrow{\quad g \quad} & B \end{array}$$

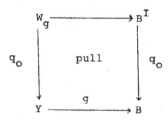

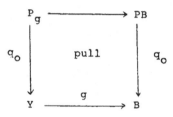

Here $PB = \{\sigma \in B^I \mid \sigma(1) = *\}$ is a subspace of B^I and $q_o(\sigma) = \sigma(0)$ as before. The trivial path $I \to * \in X$ is the base point of PB. The map q_o is again a fibration in Top^o with the loop space $\Omega B = B^{\wedge S^1}$ as fiber. If Y and B are well-pointed, so are ΩB, P_g, W_g (see (0.1.7)(B) and p.180 of [24]).

We say $F \xrightarrow{f} Y \xrightarrow{g} B$ is a <u>fiber sequence</u> if there exists a homotopy equivalence h such that $F \xrightarrow[\simeq]{h} P_g \longrightarrow Y$ is homotopic to f.

(0.1.11) Maps $X \to PB$ are adjoint to maps $CX \to B$, so the homotopy set $[X,P_g]$ can be characterized as the set of homotopy classes of pairs (α,β) with $g\alpha = \beta|_X$ as in

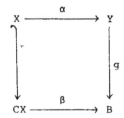

$$\begin{array}{ccc} X & \xrightarrow{\alpha} & Y \\ \big\downarrow & & \big\downarrow g \\ CX & \xrightarrow{\beta} & B \end{array}$$

If $g: Y \subset B$ is an inclusion we write $[X,P_g] = \pi_1^X(g) = \pi_1^X(B,Y)$, see (0.2). Note that CX and PB are contractible in Top^o. Furthermore, P_g and ΩB are CW-spaces when Y and B are CW-spaces ([86], [117]). A CW-space is a space that is homotopy equivalent to a CW-complex.

(0.1.12) Let $p: E \to X$ be a fibration and let $f: A \to X$ be a map. Let A and X and the fiber of p be well-pointed. Then the map p_F induced between the fibers in the diagram

$$\begin{array}{ccccc} P_{\bar f} & \longrightarrow & E_A & \xrightarrow{\bar f} & E \\ {\scriptstyle p_F}\big\downarrow & & {\scriptstyle p}\big\downarrow & \text{pull} & \big\downarrow{\scriptstyle p} \\ P_f & \longrightarrow & A & \xrightarrow{f} & X \end{array}$$

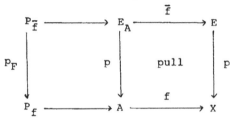

is a homotopy equivalence (over A), see 14.7 on p.204 of [24] .

(0.1.13) The dual statement to (0.2.2) is also true. Let $X \subset Y$ be a cofibration and let $f: X \to A$ be a map. Then the map i_C induced between the cofibers in the diagram

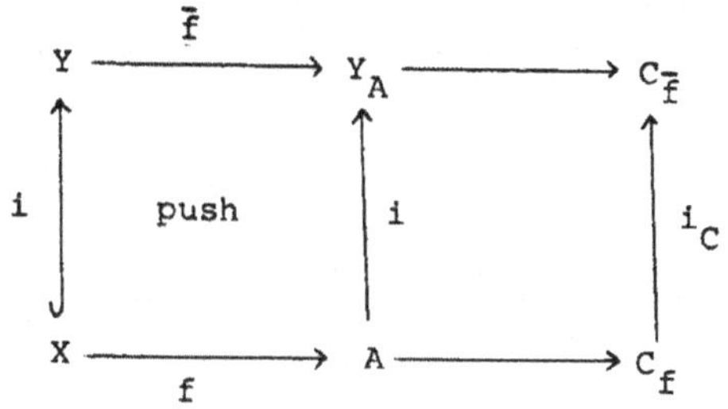

is a homotopy equivalence (under A).

Proof: By (0.1.1), the inclusion $CX \subset CY$ is a homotopy equivalence under X since $X \subset Y \subset CY$ is a cofibration in Top, see 2.24 on p.59 of [24] . Therefore the inclusion $C_f = CX \cup_f A \to CY \cup_{\bar f} A \to C_{\bar f}$ of the corresponding adjunction spaces is a homotopy equivalence under A . Here $CY \cup_f A$ is homeomorphic to $C_{\bar f}$ since $Y_A = (Y + A)/\sim$ is a pushout. ☐

(0.2) Homotopy groups

From the unit interval I we define the 1-sphere $S^1 = I/\partial I$, where $\partial I = I^{\bullet} = \{0,1\}$. We have the comultiplication (see §12 of [24])

$$\mu: S^1 \to S^1 \vee S^1$$

with $\mu(t) = (2t,*)$ for $0 \le t \le 1/2$ and $\mu(t) = (*,2t-1)$ for $1/2 \le t \le 1$. $SA = CA/A = S^1 \wedge A$ is called the suspension of A and $\Omega A = A^{\wedge S^1}$ is called the loop space of A . If A is well-pointed then so are SA and ΩA , see p.180 of [24] . The map μ determines a comultiplication

$$\mu \wedge 1_A : SA \to SA \vee SA$$

and a multiplication

$$1_A^\mu : \Omega A \times \Omega A \longrightarrow \Omega A \qquad ,$$

see (0.1.8). The maps $\mu \wedge 1_A$ and 1_A^μ induce group multiplications

$$+ : [SA,X] \times [SA,X] \to [SA,X]$$

$$\text{and} \quad + : [X,\Omega A] \times [X,\Omega A] \to [X,\Omega A]$$

which we denote by $+$. Inverses in these groups are defined by means of the map $-1 : S^1 \to S^1$ with $(-1)(t) = 1-t$.

The exponential law tells us that

$$[SA,X] = [A,\Omega X] .$$

For $n \geqslant 0$ let $S^n Z$ and $\Omega^n Z$ be the n-fold suspension and n-fold loop space respectively of Z. We define the nth homotopy group

$$\pi_n^Z(X) = [S^n Z, X] = [Z, \Omega^n X].$$

For a pair of spaces (X,A) we define the $(n+1)^{st}$ relative homotopy group $\pi_{n+1}^Z(X,A) = \pi_n^Z(P_i)$, to be the set of homotopy classes of basepoint-preserving maps $(CS^nY, S^nY) \longrightarrow (X,A)$. For these sets and groups we have the exact sequence $(n \geqslant 0)$

$$(0.2.1) \longrightarrow \pi_{n+1}^Z(A) \xrightarrow{i} \pi_{n+1}^Z(X) \xrightarrow{j} \pi_{n+1}^Z(X,A) \xrightarrow{\partial} \pi_n^Z(A) \longrightarrow$$

where i is induced by the inclusion $i : A \subset X$ and j is induced by the identification map $\pi_0 : (CS^nZ, S^nZ) \to (S^{n+1}),*)$, so that $j(f) = f \circ \pi_0$. The map ∂ is defined by restriction. For $n \geqslant 1$ this is a sequence of groups and group homomorphisms, and for $n > 1$ the groups are abelian, see p.194 of [24].

If we take for Z the 0-sphere $S^0 = \{0,1\}$ we obtain the usual homotopy groups

$$\pi_n(X) = \pi_n^{S^0}(X) = [S^n, X]$$

$$\pi_{n+1}(X, A) = \pi_{n+1}^{S^0}(X,A) = [(E^{n+1}, S^n); (X,A)]$$

where $S^n = S^n S^0$ is the n-sphere and $E^{n+1} = CS^n$ is the n-disc. We have $S^{n+m} = S^n \wedge S^m$ and $S^n = S^1 \wedge \ldots \wedge S^1$ by means of the map $(t_1 \ldots t_n) \mapsto (t_1 \ldots t_n)$, and of course $S^n A = S^n \wedge A$ and $\Omega^n A = A^{\wedge S^n}$.

(0.2.1) Definition: A space X together with a map $\mu: X \to X \vee X$ is called a co-H-space when the composition $X \xrightarrow{\mu} X \vee X \subset X \times X$ is homotopic to the diagonal map. A space Y together with a map $\mu: Y \times Y \to Y$ is called an H-space when the composition $Y \vee Y \subset Y \times Y \xrightarrow{\mu} Y$ is homotopic to the folding map.

The induced maps μ^* and μ_* will be denoted again by

$$+ : [X,Z] \times [X,Z] \longrightarrow [X,Z]$$

and $$+ : [Z,Y] \times [Z,Y] \longrightarrow [Z,Y] .$$

If X is a co-H-space, or if Y is an H-space, then $[SX,Z]$ and $[Z,\Omega Y]$ are abelian groups

(0.2.3) Definition: A co-H-space (X,μ) is called a co-H-group when a map $-1: X \to X$ is given such that $-1 + 1 = 1 + (-1) = 0$ in $[X,X]$ and $(\mu \vee 1)\mu \simeq (1 \vee \mu)\mu: X \to X \vee X \vee X$.
An H-space (Y,μ) is called an H-group when a map $-1: Y \to Y$ is given such that $-1 + 1 = 1 + (-1) = 0$ in $[Y,Y]$ and $\mu(\mu \times 1) \simeq \mu(1 \times \mu) : Y \times Y \times Y \to Y$.

For example, the suspension SA is a co-H-group and the loop space ΩA is an H-group. Naturally every topological group is an H-group.

(0.3) Whitehead products

Let A and B be well-pointed spaces. The exact sequence

$$0 \longrightarrow [S(A \wedge B),X] \xrightarrow{\sigma} [S(A \times B),X] \xrightarrow{\rho} [SA,X] \times [SB,X] \longrightarrow 0$$

is part of the cofiber sequence of $A \vee B \subset A \times B$, see [100] or (2.4) below. σ is induced by the identification map $A \times B \twoheadrightarrow A \wedge B$ and ρ is induced by the inclusion $A \vee B \rightarrowtail A \times B$. Let $\rho_A: [SA,X] \longrightarrow [S(A \times B),X]$ and $\rho_B: [SB,X] \longrightarrow [S(A \times B),X]$ be induced by the projections $A \times B \twoheadrightarrow A, B$. With the aid of the exact sequence we define the Whitehead product

$$(0.3.1) \qquad [\ ,\ ] \ : \ \pi_1^A(X) \times \pi_1^B(X) \rightarrow \pi_1^{A \wedge B}(X)$$

with $\quad [\alpha,\beta] = \sigma^{-1}(-\rho_A(\alpha) - \rho_B(\beta) + \rho_A(\alpha) + \rho_B(\beta))$.

The map $[\ ,\ ]$ is bilinear if A and B are co-H-spaces, if they are not then we have additivity rules as described in (0.4). The Whitehead product is anticommutative, that is $[\alpha,\beta] = -(S\tau)^*[\beta,\alpha]$ where $\tau: A \wedge B \twoheadrightarrow B \wedge A$ exchanges A and B. If X is an H-space, all Whitehead products in X vanish. If $X = SA \vee SB$, we have for the inclusions i_{SA}, i_{SB} the Whitehead product map $w = w_{A,B} = [i_{SA}, i_{SB}]: SA \wedge B \twoheadrightarrow SA \vee SB$ for which $[\alpha,\beta] = w^*(\alpha,\beta)$. Let $CA \mathring{\times} CB = CA \times B \cup A \times CB$ be the indicated subspace of $CA \times CB$. The map $w_{A,B}$ can also be obtained as the composition

$$(0.3.2) \qquad w_{A,B} = (\pi_0 \mathring{\times} \pi_0)h \ : \ SA \wedge B \simeq CA \mathring{\times} CB \longrightarrow SA \vee SB \quad .$$

$\pi_o \dot{\times} \pi_o$ is the restriction of $\pi_o \times \pi_o : CA \times CB \longrightarrow SA \times SB$, and h is a homotopy equivalence that we call the <u>join construction</u> for A and B. This result is due to Arkowitz and Barratt. Using (0.3.2) we obtain a homotopy equivalence (see [10])

(0.3.3) $\quad \lambda : (C_w, SA \vee SB) \longrightarrow (SA \times SB, SA \vee SB)$ under $SA \vee SB$.

These statements are clearly true when A and B are 0-spheres.

For the pair $(CSA \vee SB, SA \vee SB)$ we get from the long exact homotopy sequence a short exact sequence

$$0 \to \pi_2^{A \wedge B}(CSA \vee SB, SA \vee SB) \xrightarrow{\partial} \pi_1^{A \wedge B}(SA \vee SB) \xrightarrow{r_2} \pi_1^{A \wedge B}(SB) \longrightarrow 0$$

where $r_{2*}(w_{A,B}) = [0, 1_{SB}] = 0$, that is, $w_{A,B} \in \operatorname{Im} \partial$. We define

$$\tilde{w} = \tilde{w}_{A,B} \in \pi_2^{A \wedge B}(CSA \vee SB, SA \vee SB) \text{ by } \tilde{w}_{A,B} = \partial^{-1} w_{A,B}.$$

This map $\tilde{w}_{A,B}$ defines the relative Whitehead product

$$[\ ,\ ] : \pi_2^A(X,Y) \times \pi_1^B(Y) \longrightarrow \pi_2^{A \wedge B}(X,Y) \quad \text{with} \quad [\tilde{\alpha}, \beta] = \tilde{w}^*_{A,B}(\tilde{\alpha}, \beta)$$

where $\tilde{\alpha} \in \pi_2^A(X,Y)$ and $\beta \in \pi_1^B(Y)$. Again this map is bilinear if A and B are co-H-spaces. In accordance with the definition of $\tilde{w}_{A,B}$ we have $\partial[\tilde{\alpha}, \beta] = [\partial\tilde{\alpha}, \beta]$.

If $A = S^n$ and $B = S^m$ we obtain the Whitehead product $[\ ,\ ] : \pi_{n+1}(X) \times \pi_{m+1}(X) \longrightarrow \pi_{n+m+1}(X)$, and for the pair (X,Y) the relative Whitehead product $[\ ,\ ] : \pi_{n+2}(X,Y) \times \pi_{m+1}(Y) \longrightarrow \pi_{n+m+2}(X,Y)$. These products are bilinear for $m,n \geq 1$. They differ by sign factors from the many Whitehead products used in the literature, see [22] Appendix.

The Whitehead products satisfy the following Jacobi identity: Let X_i $(i = 1,2,3)$ be well-pointed co-H-spaces and let ξ_i : $SX_i \to SX_1 \vee SX_2 \vee SX_3$ denote the inclusions. For a permutation $\sigma \in \Sigma_3$ (Σ_3 is the permutation group on three elements) let $\varepsilon_\sigma \in \{1, -1\}$ be the sign and $v_\sigma : S(X_1 \wedge X_2 \wedge X_3) \to S(X_{\sigma 1} \wedge X_{\sigma 2} \wedge X_{\sigma 3})$ denote the map which permutes coordinates in $X_1 \wedge X_2 \wedge X_3$. Let $\alpha, \beta, \gamma \in \Sigma_3$ be three permutations with the property $\alpha 3 = 1$, $\beta 3 = 2$ and $\gamma 3 = 3$. Then the Jacobi identity

$$0 = \varepsilon_\alpha v_\alpha^*[[\xi_{\alpha 1},\xi_{\alpha 2}],\xi_1] + \varepsilon_\beta v_\beta^*[[\xi_{\beta 1},\xi_{\beta 2}],\xi_2] + \varepsilon_\gamma v_\gamma^*[[\xi_{\gamma 1},\xi_{\gamma 2}],\xi_3]$$

is satisfied. If $X_i = S^{n_i}$ for $i = 1,2,3$ are spheres of dimension $n_i > 1$, then v_σ induces the sign $(-1)^{n(\sigma)}$ where

$$n(\sigma) = \sum_{i<j,\sigma(i)>\sigma(j)} n_{\sigma i}\cdot n_{\sigma j} \ .$$

Thus in the above identity the v_σ^* can be replaced by the signs $(-1)^{n(\sigma)}$, yielding the Jacobi identity for spherical Whitehead products. See in this connection [22] Appendix.

Setting $u_i = n_i + 1$ we get for $\xi_i \in \pi_{u_i}(X)$ the identity

$$0 = [[\xi_1,\xi_2],\xi_3] + (-1)^v [[\xi_1,\xi_3],\xi_2] + (-1)^w [[\xi_2,\xi_3],\xi_1]$$

where $v = u_2 + u_3 + u_2\cdot u_3$ and $w = u_2 + u_3 + u_1\cdot u_2 + u_1\cdot u_3$.

(0.4) Operation of the fundamental group

Let A and B be well-pointed spaces. Using the exact sequence in (0.3) we obtain group operations from the right

$$\pi_1^{A\wedge B}(X) \times \pi_1^{A}(X) \to \pi_1^{A\wedge B}(X) : (\xi,\alpha) \mapsto \xi^{\alpha}$$

$$\pi_1^{A\wedge B}(X) \times \pi_1^{B}(X) \to \pi_1^{A\wedge B}(X) : (\xi,\beta) \mapsto \xi_{\beta}$$

by defining

$$\xi^{\alpha} = \sigma^{-1}(-\rho_A(\alpha) + \sigma(\xi) + \rho_A(\alpha))$$

$$\xi_{\beta} = \sigma^{-1}(-\rho_B(\beta) + \sigma(\xi) + \rho_B(\beta)) .$$

For the exchange map $\tau: A \wedge B \to B \wedge A$ we then have $\tau^*(\xi^{\alpha}) = (\tau^*(\xi))_{\alpha}$. In accordance with Satz 1.4 of [11], $\xi^{\alpha} = \xi$ and $\xi_{\beta} = \xi$ if A or B respectively is a co-H-space, or if X is an H-space.

The Whitehead product satisfies the following additivity rules. If $(\alpha,\beta), (\alpha',\beta') \in [SA,X] \times [SB,X]$ then

$$[\alpha+\alpha',\beta] = [\alpha,\beta]^{\alpha'} + [\alpha',\beta]$$

$$[\alpha,\beta+\beta'] = [\alpha,\beta'] + [\alpha,\beta]_{\beta'}$$

This is easily seen using commutators.

If A and B are spheres, the above operations are nontrivial only when A or B respectively is the 0-sphere. When $B = S^{n-1}$ and $A = S^0$ we recover the usual operation of the fundamental group

$$\pi_n(X) \times \pi_1(X) \longrightarrow \pi_n(X)$$

which we will also denote by $(\xi,\alpha) \mapsto \xi^{\alpha}$ (the case $B = S^0$ and $A = S^{n-1}$ changes nothing since $\xi^{\alpha} = \xi_{\alpha}$).

If we take $X = S(A \wedge B) \vee SA$ or $X = S(A \wedge B) \vee SB$ respectively, we obtain from the inclusions $a: SA \to X$ or $b: SB \to X$ respectively and $i: S(A \wedge B) \to X$ the maps

$$\mu_A = i^a : S(A \wedge B) \to S(A \wedge B) \vee SA$$

$$\mu_B = i_b : S(A \wedge B) \to S(A \wedge B) \vee SB$$

so that we have $\xi^\alpha = \mu_A^*(\xi,\alpha)$, $\xi_\beta = \mu_B^*(\xi,\beta)$. In particular, the operation of the fundamental group is induced by the map $S^n \to S^n \vee S^1$.

Just as with the Whitehead products in (0.3), we can define operations on relative homotopy groups. For the pair $(CSA \wedge B \vee SA, SA \wedge B \vee SA)$ the boundary operator

$$\partial : \pi_2^{A \wedge B}(CSA \wedge B \vee SA, SA \wedge B \vee SA) \to \pi_1^{A \wedge B}(SA \wedge B \vee SA)$$

is injective and $\mu_A \in \operatorname{Im} \partial$, so we can set $\hat{\mu}_A = \partial^{-1}\mu_A$. This map $\tilde{\mu}_A$ induces for a pair (X,Y) the operation

$$\pi_2^{A \wedge B}(X,Y) \times \pi_1^A(Y) \to \pi_2^{A \wedge B}(X,Y) : (\tilde{\xi},\alpha) \to \tilde{\xi}^\alpha$$

where $\tilde{\xi}^\alpha = \tilde{\mu}_A^*(\tilde{\xi},\alpha)$. Naturally we have again for $\partial : \pi_2^{A \wedge B}(X,Y) \to \pi_1^{A \wedge B}(Y)$ the equation $\partial(\tilde{\xi}^\alpha) = (\partial\tilde{\xi})^\alpha$.

In particular, when $A = S^0$ and $B = S^{n-1}$ we recover the operation of the fundamental group

$$\pi_{n+1}(X,Y) \times \pi_1(Y) \to \pi_{n+1}(X,Y)$$

again with $\partial(\tilde{\xi}^\alpha) = (\partial\tilde{\xi})^\alpha$ that is, ∂ is an equivariant homomorphism with respect to the operation of the fundamental group.

The operation of the fundamental group can also be described by using covering transformations of universal coverings. Let X be a CW-complex and $Y \subset X$ be a subcomplex such that the inclusion induces an isomorphism $\pi_1(Y) \cong \pi_1(X)$. Let $\hat{X}$ be the universal covering of X with projection $p : \hat{X} \to X$, and let $\hat{Y} = p^{-1}(Y)$. Then the covering transformations $.\alpha : \hat{X} \to \hat{X} \; : \; x \to x\cdot\alpha$ with $\alpha \in \pi_1(X) = \pi_1(Y)$ induce an operation of $\pi_1(Y)$ on the homotopy group $\pi_n(\hat{X},\hat{Y})$ by $\alpha \to (\cdot\alpha)_*$ for which the projection $p_* : \pi_n(\hat{X},\hat{Y}) \cong \pi_n(X,Y)$ is an equivariant isomorphism, that is $p_*(\xi\cdot\alpha) = (p_*\xi)^{\alpha}$. For the concept of universal covering see [116] .

We will be using the following notation. Let G be a group (often abelian) and let π be a group. We say that π <u>operates on</u> G (from the right) when there is a mapping

$$G \times \pi \xrightarrow{\phi} G \qquad \text{with} \qquad (g,\alpha) \mapsto g\cdot\alpha$$

such that for every $\alpha \in \pi$ the map $\cdot\alpha : G \to G$ is an automorphism of G and such that $(g\cdot\alpha)\cdot\alpha' = g\cdot(\alpha\cdot\alpha')$, $g\cdot 1 = g$. Let $\mathrm{Aut}(G)$ be the group of automorphisms of G . Then π operates on G exactly when the adjoint mapping $\bar{\Phi} : \pi \to \mathrm{Aut}(G)$ with $\bar{\Phi}(\alpha)(g) = g\cdot\alpha^{-1}$ is a homomorphism. Conversely, every homomorphism $\pi \to \mathrm{Aut}(G)$ induces an operation of π on G . If $\pi = \pi_1(X,*)$ is the fundamental group of a space X at the point $*$, we call the group G together with the operation of $\pi = \pi_1(X)$ a <u>local group</u> in X . (As described by Steenrod in Th.1 of [113] , such a local group determines a local coefficient system in the path-connected space X up to equivalence. We will not use this notion (see p.58 F of [116]). When we speak of local coefficients G we will mean merely an operation of $\pi_1(X,*)$ on G.) If further π operates on G and a homomorphism $g : \pi \to \pi$ is given, we call a homomorphism $f : G \to G$ <u>equivariant</u> (with respect to g) if

$f(x^{\alpha}) = (fx)^{g\alpha}$ for $x \in G$, $\alpha \in \pi$. For local groups, an equivariant f will also be called a map of local groups. For instance every map $f: X \longrightarrow Y$ induces a map of local groups $f_*: \pi_n(X) \longrightarrow \pi_n(Y)$. Further examples of local groups will be described in (4.1) Appendix and (1.5.9).

(0.5) Homology and cohomology groups

If π is an abelian group, we denote by $H_n(X, A; \pi)$ and $H^n(X, A; \pi)$ the singular homology and cohomology of the pair (X, A) with coefficients in π. Homology $H_n(..; \pi)$, resp. cohomology $H^n(.., \pi)$ is a covariant resp. contravariant functor from the category of topological pairs to the category of abelian groups. Both functors satisfy the Eilenberg-Steenrod axioms, which we do not need to repeat here, see [116]. If $A \subset X$ is a cofibration, the quotient map $X \to X/A$ (where $X/\phi = X^+ = X + \{*\}$) gives us isomorphisms $H_n(X, A; \pi) \cong \tilde{H}_n(X/A; \pi)$, $H^n(X, A; \pi) \cong \tilde{H}^n(X/A; \pi)$. If $Y = \phi$, we take $\tilde{H}_n(Y) = H_n(Y, *)$ and $\tilde{H}^n(Y) = H^n(Y, *)$ to be the reduced homology and cohomology.

In Chapter 4 a procedure will be given for calculating homology and cohomology by means of cellular chain compleses, in fact for general homology and cohomology with local coefficients. Homology and cohomology are related by the

(0.5.1) Universal coefficient theorem: There is an exact sequence

$$0 \to \operatorname{Ext}(H_{n-1}(X, A), \pi) \to H^n(X, A; \pi) \to \operatorname{Hom}(H_n(X, A), \pi) \to 0$$

which is natural in (X, A), where $H_i(X, A) = H_i(X, A; \mathbb{Z})$.

Homology and homotopy groups are related as follows. For spheres

S^n, $n \geq 1$, there is an isomorphism $H_n(S^n) \cong \mathbb{Z}$. Let $e_n \in H_n(S^n)$ be a generator. Then the <u>Hurewicz homomorphism</u>

$$(0.5.2) \qquad \phi = \phi_n : \pi_n(X) \longrightarrow H_n(X; \mathbb{Z})$$

is given by $\phi_n(\alpha) = \alpha_*(e_n)$. If X is (n-1)-connected (and $\pi_1(X)$ is abelian when $n = 1$) then ϕ_n is an isomorphism.

Cohomology groups can be characterized by means of Eilenberg-MacLane spaces.

(0.5.3) <u>Definition</u>: Let π be a group. An <u>Eilenberg-MacLane space</u> $K_n = K(\pi,n)$, $n > 1$ is a CW-space together with isomorphisms $\alpha: \pi_n(K_n) \cong \pi$ and $\pi_i(K_n) = 0$ for $i \neq n$.

For every pair (π,n), with π abelian if $n \geq 2$, there exists an Eilenberg-Mac Lane space $K(\pi, n)$. It is unique up to homotopy equivalence, that is if (K_n, α) and (K_n', α') are $K(\pi, n)$-spaces then there is a homotopy equivalence $h : K_n \to K_n'$ with $\alpha' h_* = \alpha$ that is unique up to homotopy.

Combining the universal coefficient theorem (0.5.1) and the Hurewicz homomorphism, we obtain for an abelian group π an isomorphism $(m \geq 1)$

$$(0.5.4) \qquad H^m(K(\pi,m) ; \pi) \cong \mathrm{Hom}(\pi,\pi) \qquad .$$

We define the <u>fundamental class</u> $i_m \in H^m(K(\pi,m) ; \pi)$ of $K(\pi,m)$ to be the image of the identity under this isomorphism. We then define a map

$$(0.5.5) \qquad \gamma : [X, K(\pi,m)] \longrightarrow H^m(X; \pi) \quad \text{with} \quad \gamma(f) = f^*(i_m)$$

which is a bijection when X is a CW-space. We will prove this in (4.3.13)

If X is a CW-space and π is abelian, γ gives us an abelian group structure on $[X, K(\pi, m)]$. This structure can also be obtained from a homotopy equivalence

$$(0.5.6) \qquad \lambda : \Omega(K(\pi, m+1)) \simeq K(\pi, m)$$

of which there is only one (up to homotopy), arising from the isomorphism

$$\pi_m(\Omega K(\pi, m+1) \underset{\vartheta}{\cong} \pi_{m+1}(K(\pi, m+1)) \underset{\alpha}{\cong} \pi \quad .$$

Thus, when $K(\pi, m+1) = (K_{m+1}, \alpha)$, the loop space $(\Omega K_{m+1}, \vartheta\alpha)$ is a $K(\pi, m)$.

These properties of Eilenberg-MacLane spaces can be found for example in [116] or [38].

CHAPTER 1. PRINCIPAL FIBRATIONS AND COFIBRATIONS, OBSTRUCTIONS AND DIFFERENCES

We make the general requirement from now on that all spaces be well-pointed.

(1.1) Extension and lifting problems

We will describe here an inductive procedure for dealing with extension and lifting problems. The induction is based on principal decomposition of a cofibration or fibration. The examples we are primarily interested in are CW-decompositions and Postnikov decompositions. Finally, we state two theorems on homotopy equivalences.

Let

(1.1.1)

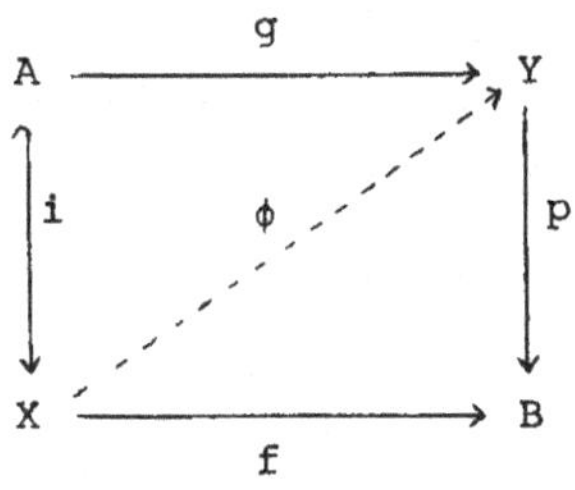

be a commutative diagram in Top^o, with i a closed cofibration and p a fibration.

The two problems in obstruction theory that we want to investigate are

(a) The existence problem: Does a map ϕ exist extending the diagram commutatively?

(b) The classification problem: When a commutative extension exists, how can the set $[X,Y]_B^A$ of homotopy classes under A and over B be characterized?

In considering these problems we will assume that one of the following two conditions holds.

(A) The cofibration $A \subset X$ can be "decomposed" into a filtration

$$A = X_o \subset X_1 \subset \ldots \subset X_n \subset \ldots \subset X$$

where $X_n \subset X_{n+1}$ is a 'principal' cofibration.

(B) The fibration $Y \to B$ can be 'decomposed' into a tower

$$Y \to \ldots \to Y_n \to \ldots \to Y_1 \to Y_o = B$$

where $Y_{n+1} \to Y_n$ is a 'principal fibration.

We will now explain the meanings of 'principal' and 'decompose' in (A) and (B). The exact definitions of 'principal' are given in (1.2), (2.1) and (1.3), (2.2). Assuming condition (A), we want to construct inductively the factoring map ϕ of (1.1.1). That is, we assume that we have a commutative diagram

(1.1.2)

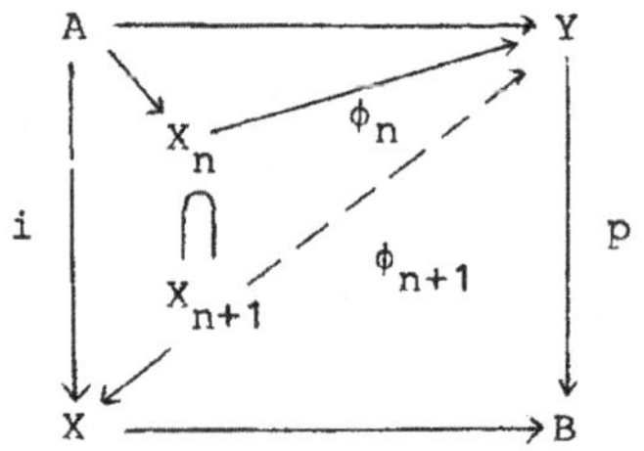

in which the map ϕ_n has already been constructed and where we want to construct the map ϕ_{n+1}. In other words, we are in this situation dealing with the existence and classification problems for the 'principal' cofibration $X_n \subset X_{n+1}$ instead of for $A \subset X$. Thus we need to have a procedure or some criteria for making this inductive step for principal cofibrations. In condition (A), 'decompose' means that a sequence of maps ϕ_n as in (1.1.2) also yields a map ϕ as in (1.1.1). An important example of a decomposition of a cofibration is a CW-decomposition, as described in Theorem (1.4.5). The words 'principal' and 'decompose' have similar meanings in condition (B). Assuming condition (B) holds, we want to construct inductively the factoring map ϕ by means of a sequence of maps ϕ_n for which the diagrams

(1.1.3)

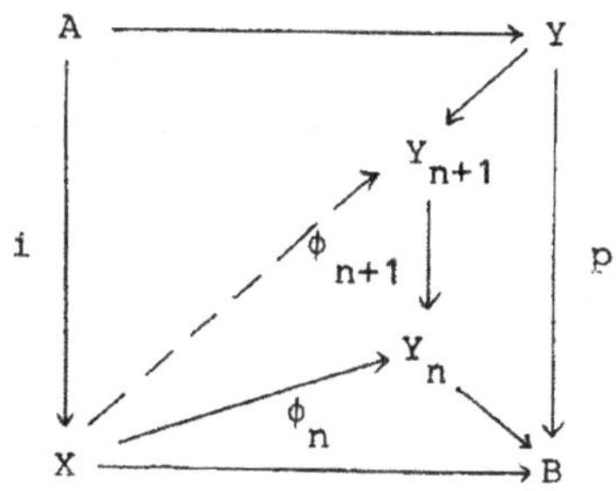

commute. We therefore need to solve the existence and classification problems for the 'principal' fibrations $Y_{n+1} \rightarrow Y_n$. An important example of such a decomposition of a fibration $Y \rightarrow B$ is a Postnikov decomposition, which we describe in (2.3).

The inductive construction of the factoring maps ϕ_n of (1.1.2) and (1.1.3) can be formulated equivalently as follows. By (0.0.9), we can regard (1.1.2) as a problem in the inductive extension of a section.

That is, we have a fibration $\tilde{X} \to X$ $(\tilde{X} = f^{*}Y)$ and induced fibrations $\tilde{X}_n \to X_n$, and we want inductively to construct section extensions u_n of a given section u_o as in the diagram

(1.1.4)

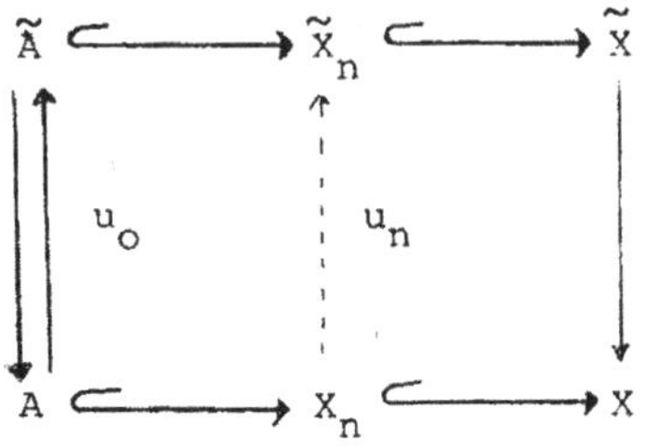

When condition (A) holds, it is often easier to visualize the existence and classification problems of (1.1.1) in this equivalent form as section extension problems. We then first discuss the case of a trivial fibration $\tilde{X} = X \times U$. In this case, (1.1.4) corresponds to the diagram for the construction of extensions u_n of a map u_o

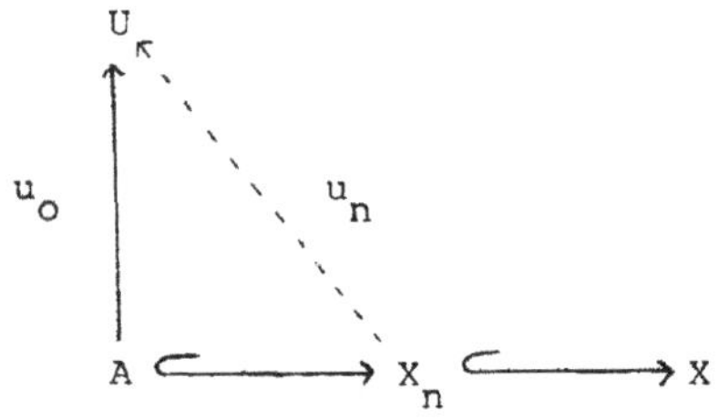

Similarly, (0.0.9) allows us to regard (1.1.3) as a problem in constructing inductively a retraction lifting.

We have a cofibration $Y \subset \bar{Y}$ $(\bar{Y} = g_{*}X)$ and induced cofibrations $Y_n \subset \bar{Y}_n$, and we want inductively to construct retraction liftings u_n of a given retraction u_o as in the diagram

(1.1.5)

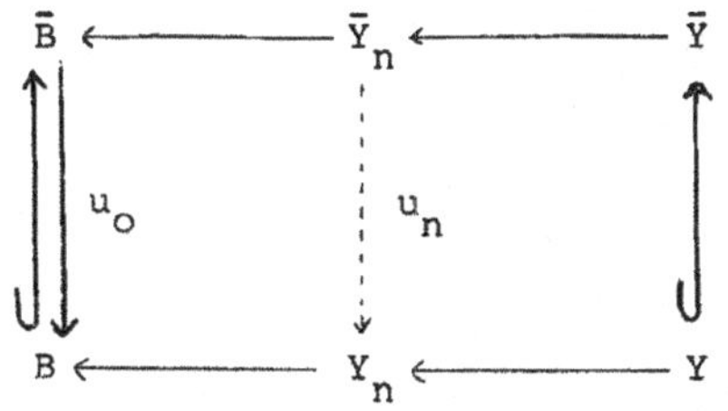

When condition (B) holds, we will deal with the existence and classification problems of (1.1.1) in this equivalent formulation as retraction lifting problems. Let us first look at the case of the trivial cofibration $Y \subset Y \vee U$. In this case (1.1.5) corresponds to a diagram for the inductive construction of liftings u_n of a map u_o

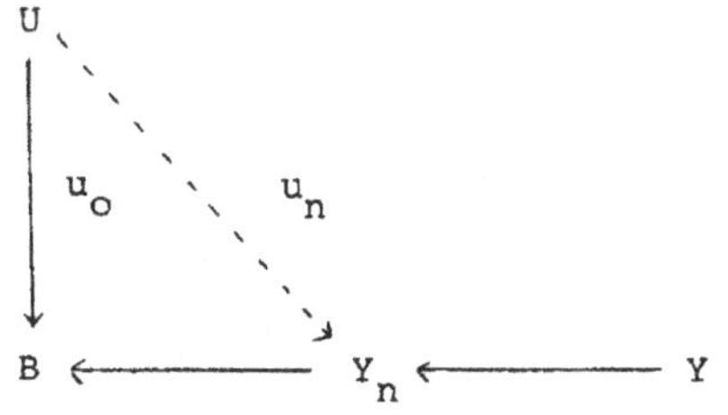

We will now discuss the effect of homotopy equivalences on section homotopy sets and retraction homotopy sets. We will see that it is possible, in the context of the existence and classification problems of (1.1.2), to replace the fibration $Y \to B$ or the cofibration $A \subset X$ by homotopy equivalent fibrations or cofibrations.

(1.1.6) <u>Theorem:</u> <u>In the commutative diagram</u>

(1.1.7)

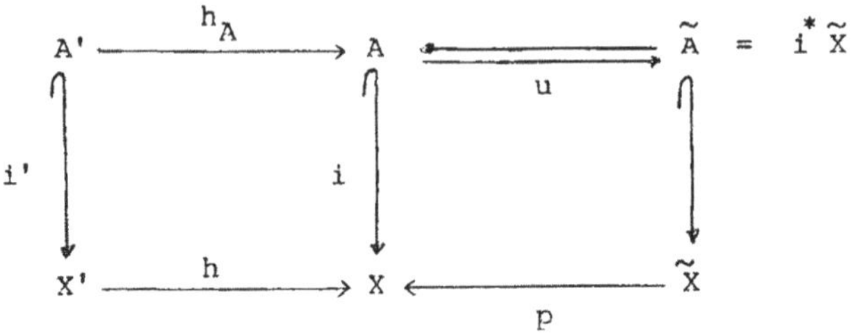

let i and i' be closed cofibrations and p a fibration. If h and h_A are homotopy equivalences, then h induces a bijection

$$h^* : \langle X, \tilde{X}\rangle^u \approx \langle X', h^* \tilde{X}\rangle^{u'}$$

of section homotopy sets relative u and u'. u denotes a section over A and $u' = h_A^* u : A' \to h_A^* \tilde{A}$ the induced section.

In (1.4.3) we obtain this bijection h^* for CW-spaces under weaker assumptions. Dual to (1.1.6) is the

(1.1.7) Theorem: In the commutative diagram

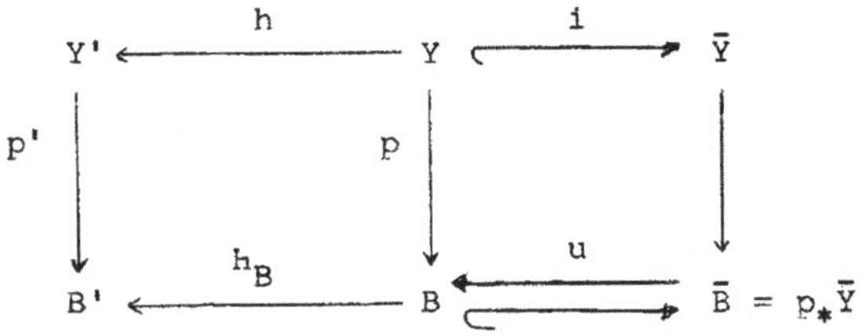

let p and p' be fibrations and i a closed cofibration. If h and h_B are homotopy equivalences, then h induces a bijection

$$h_* : \langle \bar{Y}, Y\rangle_u \approx \langle h_*\bar{Y}, Y'\rangle_{u'}$$

of retraction homotopy sets over u *and* u'. u *denotes a retraction and* $u' = h_{B*}u : h_{B*}\bar{B} \to B'$ *the induced retraction.*

In (1.5.6) we obtain this bijection h_* for Postnikov spaces under weaker assumptions.

Trivial cofibrations and fibrations are special cases to which we can apply the above theorems. With the assumptions of (1.1.6) we have a bijection

$$h^* \; : \; [X, U]^u \approx [X', U]^{u'}$$

where $u' = u \circ h_A$. With the assumptions of (1.1.7), we have a bijection

$$h_* \; : \; [U, Y]_u \approx [U, Y']_{u'}$$

where $u' = h_B \circ u$.

It should be noticed that we have not required the pairs (h, h_A) and (h, h_B) to be homotopy equivalences in the category of pairs. For homotopy sets 'under' and 'over' we have the following corollary.

(1.1.8) *Corollary: In the commutative diagram*

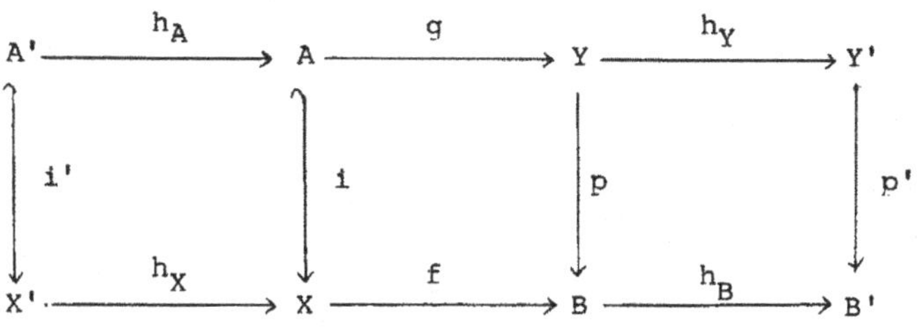

let i *and* i' *be closed cofibrations and let* p *and* p' *be fibrations. Let* h_A, h_X, h_Y, h_B *be homotopy equivalences. Then we have bijections*

$$[X', Y]_B^{A'} \approx [X, Y]_B^A \approx [X, Y']_{B'}^A,$$

induced by h_X *and* h_Y *respectively.*

Making the usual identifications as in (1.1.4) and (1.1.5), one can readily deduce (1.1.8) from the above theorems.

Proof of (1.1.6): Consider the commutative diagram

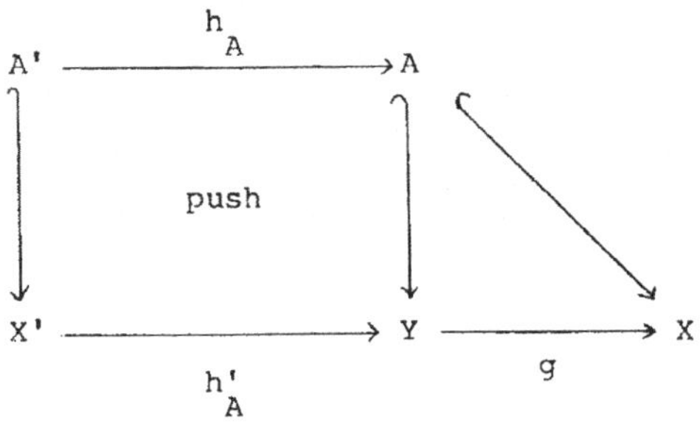

where $h = gh'_A$. Since h_A is a homotopy equivalence, so is h'_A and therefore g as well, cf. (0.1.3). g is in fact a homotopy equivalence under A, cf. (0.1.1). By (0.0.11) and excision, we have a bijection

$$\langle X', \tilde{X}'\rangle^{u'} \approx \langle Y, g^*\tilde{X}\rangle^{u} .$$

We will now show that g induces a bijection

$$g^* : \langle X, \tilde{X}\rangle^{u} \approx \langle Y, g^*\tilde{X}\rangle^{u} ,$$

from which (1.1.6) follows. Let $\bar{g} : X \to Y$ be a homotopy inverse to g and let $H : g\bar{g} \simeq 1$ be the corresponding homotopy relative A. From a section extension $v_o : Y \to g^*\tilde{X}$ of u we obtain the lifting v in the commutative diagram

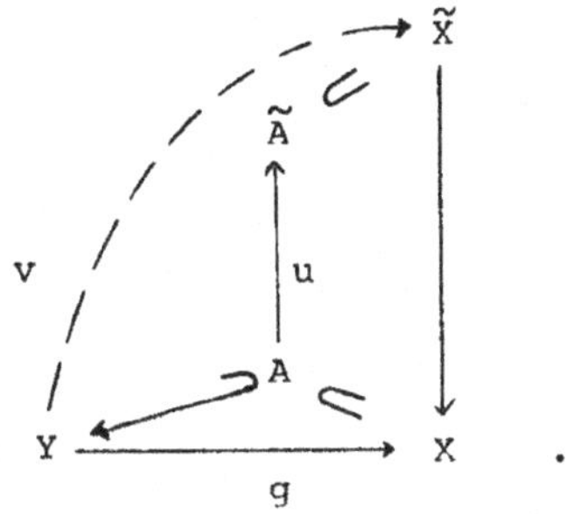

We then have the commutative diagram

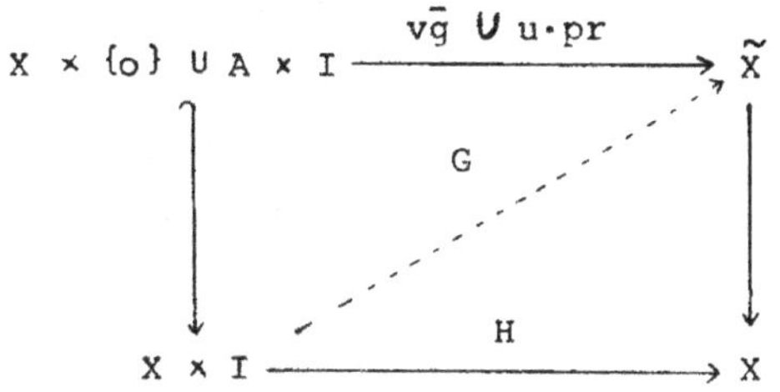

and from it the relative lifting G. Then $G_1 : X \to \tilde{X}$ is a section extension of u and the map $[v_o] \mapsto [G_1]$ is inverse to g^*. □

(1.1.9) <u>Lemma</u>: Let

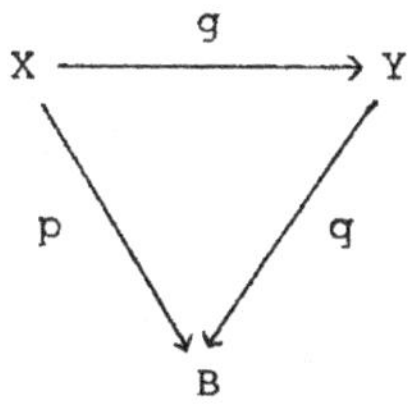

be a commutative diagram in Top^o, where p and q are fibrations and X and Y are well-pointed. If g is a homotopy equivalence in Top, then g is a pointed homotopy equivalence over B in ${Top^o}_B$.

<u>Proof of</u> (1.1.9): By (0.1.1), g has a homotopy inverse $g_1 : Y \to X$ in Top_B. $g_1(x)$ can be connected by a path in the fiber F_q to $* \in F_q \subset Y$. Since Y is well-pointed, this path can be extended to a homotopy $g_1 \simeq \bar{g}$ over B by means of a relative lifting. $\bar{g}$ is then a pointed map, and we have homotopies $H : \bar{g}g \simeq 1$ and $G : g\bar{g} \simeq 1$ in Top_B. Again, a relative lifting can be performed to move these homotopies into homotopies $H' : \bar{g}g \simeq 1$ and $G' : g\bar{g} \simeq 1$ in ${Top^o}_B$. ▭

<u>Proof of</u> (1.1.7): The proof proceeds similarly to that of (1.1.6). We consider the commutative diagram

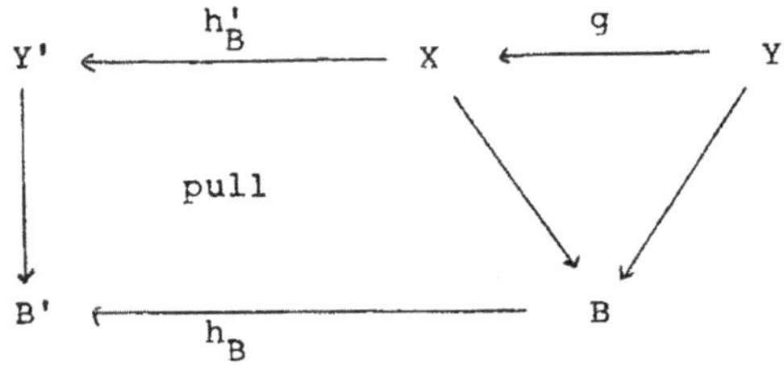

in Top^o with $h = h'_B g$. By (0.1.3), h'_B is a homotopy equivalence in Top^o, and therefore g is one as well. By the preceding lemma, g is in fact a homotopy equivalence in ${Top^o}_B$. In virtue of (0.0.12), we have now only to show that g induces a bijection

$$g_* : \langle \bar{Y}, Y \rangle_u \approx \langle g_* \bar{Y}, X \rangle_u .$$

Let $\bar{g} : X \to Y$ be a homotopy inverse to g and let $H : \bar{g}g \simeq 1$ be a homotopy over B in Top_B^o. From a retraction lifting $v_o : g_*\bar{Y} \to X$ of u we obtain the map v in the commutative diagram

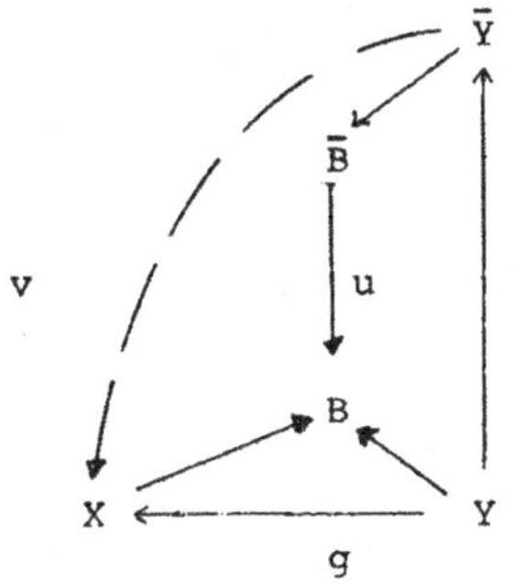

We then have the commutative diagram

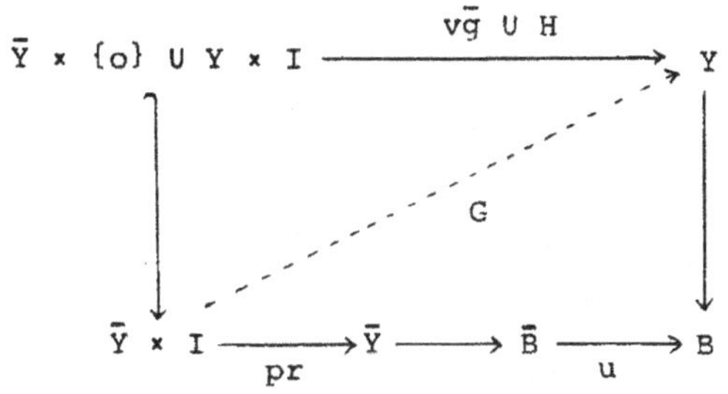

and in it the relative lifting G. Now $G_1 : \bar{Y} \to Y$ is a retraction lifting of u, and the map $[v_o] \to [G_1]$ is inverse to g_*. □

(1.2) Principal cofibrations and extension of maps and sections

We now define principal cofibrations, and show how primary obstructions and differences can be used to solve existence and classification problems for extensions. In other words, we can in fact perform the inductive step described in condition (A) of (1.1.2). First we consider extensions of maps, then we show that extensions of sections in a fibration behave in almost the same way.

(1.2.1) Definition: For a map $f: A \to X$ we say that the cofibration $X \subset C_f$ into the mapping cone is principal. More generally, we call a closed cofibration $X \subset Y$ a principal cofibration when there is a map $f: A \to X$ and a homotopy equivalence $C_f \simeq Y$ under X. The map f is called the coclassifying or attaching map of the principal cofibration $X \subset Y$. (Remember that A and X are assumed to be well-pointed, and that C_f is the reduced mapping cone, see (0.1.9))

We now discuss some properties of cofibrations $X \subset C_f$ which, by (1.1.6), hold also for principal cofibrations $X \subset Y$. A map $A \to U$ is null-homotopic exactly when it can be extended over the cone CA, so a mapping cone C_f has the following characteristic property.

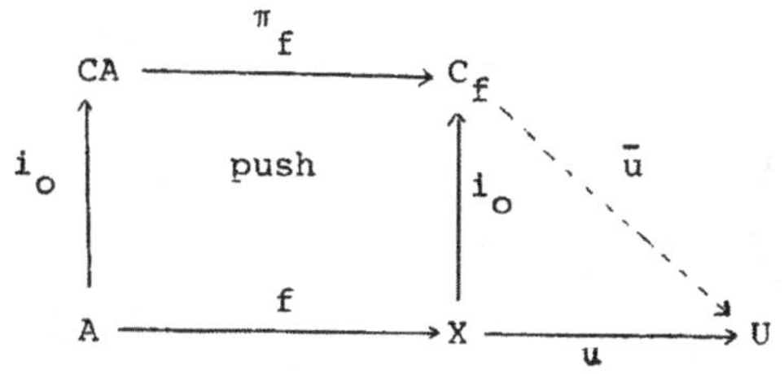

The map u can be extended over C_f (that is, a map $\bar{u}$ with $\bar{u} i_o = u$ exists) exactly when the composition $u \circ f$ is null-homotopic. For this reason we call the homotopy class

$$(1.2.2) \qquad f^*(u) = u \circ f \in [A, U]$$

the (primary) <u>obstruction to extending</u> u. Obstructions of higher order are constructed from such primary obstructions, see (1.2.25).

Let I be the unit interval and let $A \rtimes B = A \times B / A \times \{*\}$. Then the inclusion

$$I \dot{\rtimes} C_f = \{o\} \times C_f \cup I \rtimes X \cup \{1\} \times C_f \subset I \rtimes C_f$$

is a principal cofibration with attaching map

$$(1.2.3) \qquad w_f : SA \longrightarrow I \dot{\rtimes} C_f$$

where

$$w_f(t, a) = \begin{cases} (0, 1 - 3t, a) \in \{o\} \times C_f & \text{for } 0 \leqslant t \leqslant 1/3 \\ (3t - 1, f(a)) \in I \rtimes X & \text{for } 1/3 \leqslant t \leqslant 2/3 \\ (1, 3t - 2, a) \in \{1\} \times C_f & \text{for } 2/3 \leqslant t \leqslant 1 . \end{cases}$$

This is indicated in the sketches

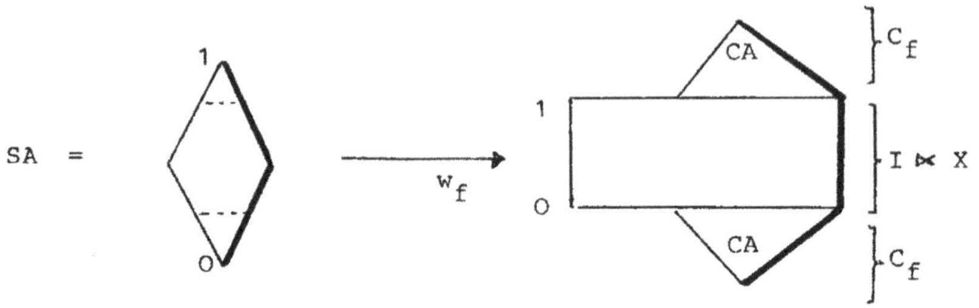

where the heavy lines are identified to the basepoint * . w_f gives rise to the homotopy equivalence

(1.2.4) <u>Lemma</u>: $h : C_{w_f} \simeq I \ltimes C_f$ <u>under</u> $I \dot{\ltimes} C_f$.

<u>Proof</u>: With $f = 1_A = 1$, $w_1 : SA \to I \dot{\ltimes} CA$ is a homeomorphism, see (0.0.3) (I). Since $I \dot{\ltimes} CA \subset I \ltimes CA$ is a cofibration by (0.1.6) and (0.1.7), and $I \ltimes CA$ is contractible, we have a homotopy equivalence

$$h' : C(I \dot{\ltimes} CA) \simeq I \ltimes CA \quad \text{under} \quad I \dot{\ltimes} CA.$$

The composition

$$CSA \xrightarrow{Cw_1} C(I \dot{\ltimes} CA) \underset{h'}{\simeq} I \ltimes CA \xrightarrow{1 \ltimes \pi_f} I \ltimes C_f$$

then determines the homotopy equivalence h of (1.2.4). □

Let $u_o, u_1 : C_f \to U$ be maps and $H : u_{o|X} \simeq u_{1|X} : X \to U$ be a (pointed) homotopy of their restrictions. The triple (u_o, H, u_1) then determines a map

$$u_o \cup H \cup u_1 : \{o\} \times C_f \cup I \ltimes X \cup \{1\} \times C_f \longrightarrow U$$

We call

$$(1.2.5) \qquad d(u_o, H, u_1) = w_f^*(u_o \cup H \cup u_1) \in [SA, U]$$

the (primary) <u>difference</u> of (u_o, H, u_1). By (1.1.2) we have $d(u_o, H, u_1) = 0$ exactly when there is a homotopy $H' : u_o \simeq u_1$ extending H. If $u_{o|X} = u_{1|X} = u$ then we take the stationary homotopy $u \circ pr : I \times X \to X \to U$, and define $d(u_o, u_1) = d(u_o, u\,pr, u_1)$.

We will now describe a few simple properties of the difference construction which are all easy consequences of the definition of w_f. Let $u_2 : C_f \to U$ be another map and let $H' : u_{1|X} \simeq u_{2|X}$ be a homotopy. Then we also have the homotopy $H + H' : u_{o|X} \simeq u_{2|X}$, so in $[SA, U]$ we have

$$(1.2.6) \quad \begin{cases} d(u_o, H, u_1) + d(u_1, H', u_2) = d(u_o, H + H', u_2) \\ \\ d(u_o, u_1) + d(u_1, u_2) = d(u_o, u_2) . \end{cases}$$

The suspension SA <u>co-operates from the right on</u> C_f by means of $\mu : C_f \to C_f \vee SA$ with

$$\mu(t, a) = \begin{cases} (2t, a) \in C_f & \text{for } 0 \leqslant t \leqslant 1/2 \\ \\ (2t - 1, \alpha) \in SA & \text{for } 1/2 \leqslant t \leqslant 1 . \end{cases}$$

Let $* \in Y \subset X$ and $v : Y \to U$. The map μ induces group actions

$$[C_f, U]^v \times [SA, U] \xrightarrow{+} [C_f, U]^v ,$$

(1.2.7)

$$[C_f, U] \times [SA, U] \xrightarrow{+} [C_f, U] ,$$

which will be denoted by + , that is, $\mu^*(\xi, \alpha) = \xi + \alpha$. The difference $d(u_o, u_1)$ is not uniquely defined by the homotopy classes of u_o, u_1 in $[C_f, U]$, but it is by the homotopy classes of u_o, u_1 relative $u : X \to U$. Regarding the group action + , we have

(1.2.8) <u>Theorem</u>: <u>Let</u> $u_o \in [C_f, U]^u$. <u>There is a bijection</u>

$$[C_f, U]^u \xrightarrow{\approx} [SA, U]$$

<u>given by</u> $u_1 \mapsto d(u_o, u_1)$ <u>with inverse</u> $\alpha \mapsto u_o + \alpha$.
<u>That is</u>,

$$d(u_o, u_1 + \alpha) = \alpha \quad \underline{\text{in}} \quad [SA, U],$$

<u>and</u>

$$u_o + d(u_o, u_1) = u_1 \ \underline{\text{in}} \ [C_f, U]^u .$$

For the maps

$$[C_f, U]^u \xrightarrow{\Theta_u} [C_f, U] \xrightarrow{j} [X, U]$$

where Θ_u is the canonical map and j is the restriction, we have

$$\text{Im } \Theta_u = j^{-1}(u) .$$

A consequence of (1.2.8) is thus

(1.2.9) <u>Corollary</u>: <u>Let</u> $f : A \to X$ <u>and let</u> $j : [C_f, U] \to [X, U]$ <u>be the restriction.</u> <u>If</u> $u_o \in j^{-1}(u)$ <u>then</u> $j^{-1}(u) = \{u_o + \alpha | \alpha \in [SA, U]\}$.

Analogous statements hold for the following operation. Let $X \subset Z$ and let $H : C_f \simeq Z$ be a homotopy equivalence under X. Then we have

(1.2.10) $[Z, U]^V \times [SA, U] \xrightarrow{+} [Z, U]^V$

given by

$$u + \alpha = H^{*-1}(H^*(u) + \alpha).$$

We now discuss obstructions and differences for sections of fibrations. In the following, let $\tilde{X}$ be the total space of a fibration $p : \tilde{X} \to X$ with base space X and fiber $F = p^{-1}(*)$. The total space has basepoint $* \in F$. If $Y \subset X$ then $\tilde{Y} \to Y$ is the restricted fibration with $\tilde{Y} = p^{-1}(Y)$.

Now let $\tilde{C}_f \to C_f$ be a fibration with fiber F over the mapping cone C_f of $f : A \to X$. When can a section $u : X \to \tilde{X}$ be extended to a section $u' : C_f \to \tilde{C}_f$ such that the diagram

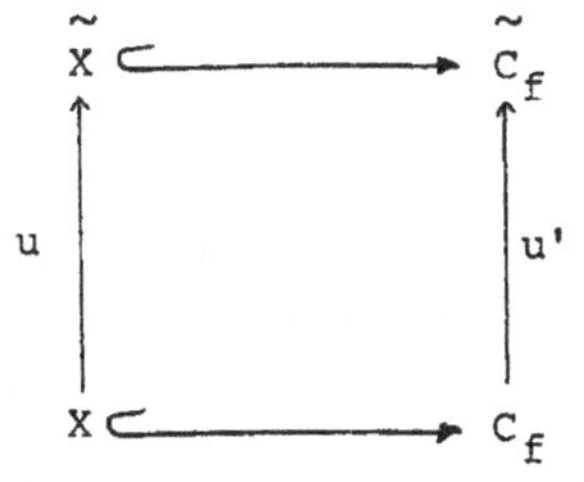

commutes? The obstruction to doing this will be described in (1.2.11), only we must assume additionally that A is a co-H-group. With this assumption, the theory for maps can be applied to sections.

Let A be a co-H-group. Consider the following diagram of groups and group homomorphisms, see [4], [27].

$$\begin{array}{ccc} & & \pi_o^A(F) \\ & & \downarrow i_* \\ \pi_1^A(\tilde{C}_f, \tilde{X}) & \xrightarrow{\partial} & \pi_o^A(\tilde{X}) \\ \cong \downarrow p_* & & p_* \downarrow \uparrow u_* \\ \pi_1^A(C_f, X) & \xrightarrow{\partial} & \pi_o^A(X) \end{array}$$

$\pi_f : (CA, A) \to (C_f, X)$ is an element of $\pi_1^A(C_f, X)$ with $\partial\pi_f = f$. Let $\tilde{\pi}_f = p_*^{-1}(\pi_f)$, see (0.1.12). The exact homotopy sequence of $\tilde{X}$ splits by means of u_*, so i_* is injective. Because $p_*(-\partial\tilde{\pi}_f + u_*(\partial\pi_f)) = 0$, the element

(1.2.11) $\quad f^{\#}(u) = i_*^{-1}(-\partial\tilde{\pi}_f + u_*(\partial\pi_f)) \in \pi_o^A(F)$

is now well-defined.

(1.2.12) <u>Theorem:</u> $f^{\#}(u) = 0$ <u>exactly when the section</u> u <u>over</u> X <u>can be extended to a section</u> u' <u>over</u> C_f.

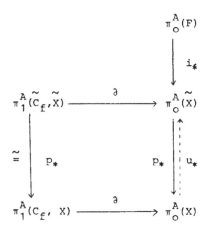

We call $f^{\#}(u)$ the (primary) <u>obstruction to extending the section</u> u. Obstructions of higher order are constructed from such primary ones, see (1.2.28) .

<u>Proof</u>: If u' exists then $\tilde{\pi}_f = u'_*(\pi_f)$ and so $f^{\#}(u) = 0$. Conversely, let $f^{\#}(u) = 0$. The class $\tilde{\pi}_f$ can be represented in such a way that the diagram of pointed pairs

$$\begin{array}{ccc} & & (\tilde{C}_f, \tilde{X}) \\ & \overset{\tilde{\pi}_f}{\nearrow} & \downarrow \\ (CA, A) & \xrightarrow[\pi_f]{} & (C_f, X) \end{array}$$

is commutative. From $f^{\#}(u) = 0$ it follows that $\partial\tilde{\pi}_f = u_* f$. Thus there exists a homotopy $H : \tilde{\pi}_{f|A} \simeq u \circ f : A \to \tilde{X}$. This homotopy gives us a map

$$u' : C_f = CA \cup_f X \longrightarrow \tilde{C}_f$$

with

$$u'(x) = u(x) \quad \text{for} \quad x \in X$$

and

$$u'(t, a) = \begin{cases} H(2t, a) & \text{for} \quad 0 \leqslant t \leqslant 1/2 \\ \\ \tilde{\pi}_f(2t - 1, a) & \text{for} \quad 1/2 \leqslant t \leqslant 1 \, . \end{cases}$$

Since $pu = 1_X$, there is a homotopy $H_o : pu' \simeq 1_{C_f}$ rel X . For H_o we have then the commutative diagram

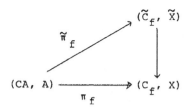

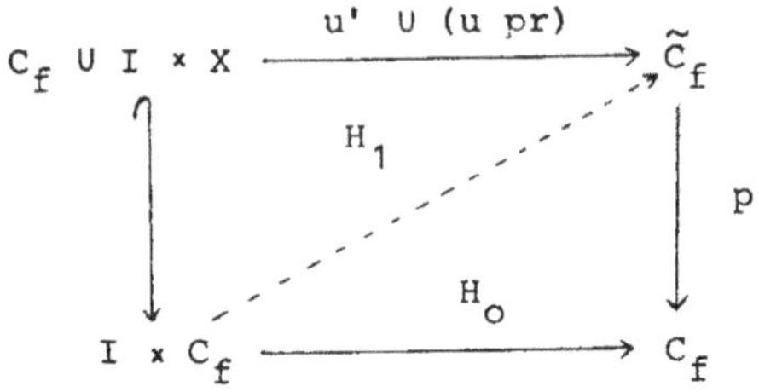

In accordance with (0.1.5), there exists a lifting H_1 that extends the diagram commutatively. Then $H_1 : u' \simeq u''$ is a homotopy with $pu'' = 1_{C_f}$ and $u''_{|X} = u$, so u'' is an extension of u. □

The definition of primary obstructions is natural in the following sense. Let A and B be co-H-groups, and let there be for the maps

(1.2.13)

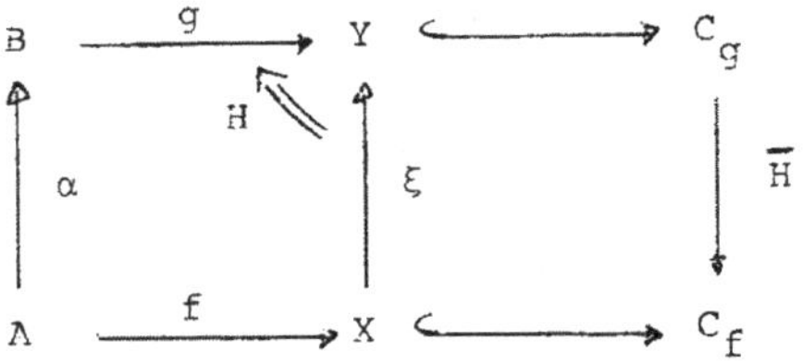

a homotopy $H : \xi f \simeq g\alpha$. Then H defines a map $\bar{H}$ under ξ by

$$\bar{H}(x) = \xi(x) \in Y \qquad \text{for } x \in X \subset C_f$$

$$\bar{H}(t, a) = \begin{cases} H(2t, a) \in Y & , 0 \leqslant t \leqslant 1/2 \\ (2t-1, \alpha(a)) \in C_g & , 1/2 \leqslant t \leqslant 1 . \end{cases}$$

Let $\tilde{C}_g \to C_g$ be a fibration with fiber F and let $\tilde{C}_f \to C_f$ be the fibration induced by $\bar{H}$. Then $\tilde{X} \to X$ is the fibration induced from

$\tilde{Y} \to Y$ by ξ. Thus a section $u : Y \to \tilde{Y}$ determines a section $\xi^* u : X \to \tilde{X}$, and we have

$$f^{\#}(\xi^*(u)) = \alpha^* g^{\#}(u) \tag{1.2.14}$$

for $\alpha^* : [B, F] \to [A, F]$. This is so because the diagram used to define $f^{\#}$ and $g^{\#}$ respectively in (1.2.11) is natural in this sense.

As in (0.0.5), we let $\langle X, \tilde{X} \rangle$ denote the set of section homotopy classes. For the maps

$$\langle X , \tilde{X} \rangle \xrightarrow{\Theta} [X, \tilde{X}] \xrightarrow{p_*} [X, X] ,$$

where Θ takes the section homotopy class of u to the homotopy class of u, we have by (0.1.4)

$$\operatorname{Im} \Theta = p_*^{-1}(1_X) . \tag{1.2.15}$$

Θ is not injective in general. Notice that the primary obstruction $f^{\#}(u)$ of (1.2.11) depends only on the homotopy class Θu of the section u. Given the trivial fibration $\tilde{X} = X \times U \to X$ with fiber U we have once more $\langle X, X \times U \rangle = [X, U]$, and the obstruction $f^{\#}(u)$ of (1.2.11) is equal to the obstruction $f^*(u)$ of (1.2.2).

Similarly to (1.2.5), we reduce the homotopy problem for sections to an extension problem. Given the fibration $p : \tilde{X} \to X$ with fiber F, $1_I \times p : \widetilde{I \times X} = I \times \tilde{X} \to I \times X$ is also a fibration with fiber F. Let $u_0, u_1 : X \to \tilde{X}$ be sections and $H : u_0 \simeq u_1$ a section homotopy. H gives us a section

$$\begin{array}{ccccc}
\widetilde{\partial I \ltimes X} & \hookrightarrow & \widetilde{I \ltimes X} & \longrightarrow & \tilde{X} \\
\uparrow u_o \cup u_1 & & H \, \vdots\uparrow \; \downarrow & \text{pull} & \downarrow \\
\partial I \ltimes X & \hookrightarrow & I \ltimes X & \xrightarrow{\text{pr}} & X
\end{array}$$

where pr is the projection and $\partial I = \{0, 1\}$. Conversely, every section H extending $u_1 \cup u_1$ gives us a section homotopy. As in (1.2.5), we can define the difference of sections. Let $u_o, u_1 :$ $C_f \to \tilde{C}_f$ be sections and $H : u_{o|X} \simeq u_{1|X} : X \to \tilde{X}$ a section homotopy of their restrictions. The triple (u_o, H, u_1) defines a section

$$u_o \cup \bar{H} \cup u_1 : I \dot{\ltimes} C_f \longrightarrow \widetilde{I \dot{\ltimes} C_f} \, .$$

Using the map $w_f : SA \to I \dot{\ltimes} C_f$ from (1.2.3) and also (1.2.11), we define

$$(1.2.16) \qquad d(u_o, H, u_1) = w_f^{\#}(u_o \cup \bar{H} \cup u_1) \in [SA, F].$$

Here A can be any space. H can be extended to a section homotopy $H' : u_o \simeq u_1$ exactly when $d(u_o, H, u_1) = 0$. If $u_{o|X} = u_{1|X} = u$ then we take the constant section homotopy $H = u\,pr$ and define $d(u_o, u_1) = d(u_o, upr, u_1)$. The element $d(u_o, u_1)$ is not well-defined by the section homotopy classes of u_o, u_1 in $\langle C_f, \tilde{C}_f \rangle$, but it is by the section homotopy classes of u_o, u_1 relative $u : X \to \tilde{X}$. A generalization of (1.2.6) is the

(1.2.17) <u>Theorem</u>: <u>If</u> A <u>is a co-H-space, then in</u> [SA, F] <u>we have</u>

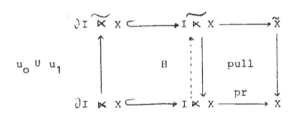

$$d(u_o, H, u_1) + d(u_1, H', u_2) = d(u_1, H + H', u_2)$$

$$d(u_o, u_1) + d(u_1, u_2) = d(u_o, u_2) \ .$$

<u>Proof</u>: Let $J = (I, \{0, 1/2, 1\})$ with basepoint 1. We consider the diagram

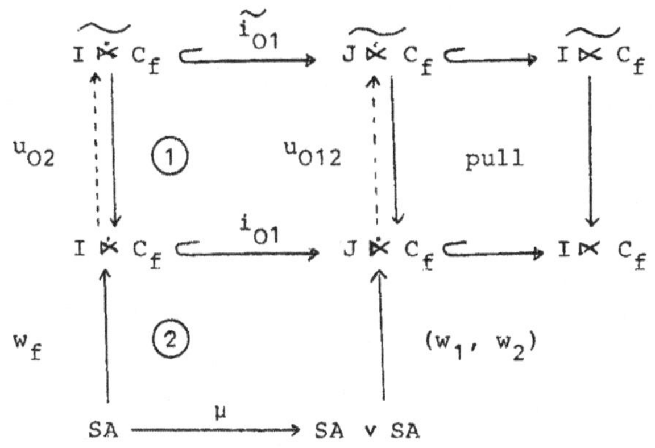

where i_{01} is the inclusion. Here $u_{02} = u_o \cup (H + H') \cup u_2$ is a section, and $u_{012} = u_o \cup H \cup u_1 \cup H' \cup u_2$ is also a section, extending u_{02}. Let $w = w_f$ as in (1.2.3), and let $w_i = (j_i \dot{\ltimes} 1) \circ w : SA \to I \dot{\ltimes} C_f \to J \dot{\ltimes} C_f$ where $J \dot{\ltimes} C_f = \{0, 1/2, 1\} \ltimes C_f \cup I \ltimes C_f$ and where $j_i : I \to J$ is given by $j_o(t) = t/2$ and $j_1(t) = \frac{1+t}{2}$.

It follows immediately from the the definition of w_f that diagram (2) is homotopy commutative in Top^o. From the homotopy equivalence h of (1.2.4) and the homomorphism

$$i_{01*} \ : \ \pi_2^A(I \ltimes C_f, I \dot{\ltimes} C_f) \to \pi_2^A(I \ltimes C_f, J \dot{\ltimes} C_f)$$

we have, because (2) is homotopy commutative,

(1.2.18)

$$i_{01*}(h\pi_w) = \pi_{w_o} + \pi_{w_1}, \quad \text{where} \quad \pi_{w_i} = (j_i \ltimes 1)_*(h\pi_w), \; i = 0, 1.$$

This is proved by first setting $f = 1_A$, the identity, and then using naturality as in the proof of (1.2.4). Thus we also have, in $\pi_2^A(\widetilde{I \ltimes C_f}, \widetilde{J \ltimes C_f})$,

$$i_{01*}(\widetilde{h\pi_w}) = \tilde{\pi}_{w_o} + \tilde{\pi}_{w_1} .$$

In accordance with (1.2.11) and (1.2.16) we have, in $[SA, F]$,

$$d(u_o, H + H', u_1) = i_*^{-1}(-\partial\widetilde{h\pi}_w + u_{02*}(w))$$

$$= i_*^{-1}(-\partial\tilde{\pi}_{w_o} - \partial\tilde{\pi}_{w_o} + u_{012*}(w_o + w_1))$$

$$\overset{(*)}{=} i_*^{-1}(-\partial\tilde{\pi}_{w_o} + u_{01*}(w_o)) + i_*^{-1}(-\partial\tilde{\pi}_{w_1} + u_{12*}(w_1))$$

$$= d(u_o, H, u_1) + d(u_1, H', u_2).$$

The equality (*) holds because $u_{012*}(w_o + w_1) = u_{01*}(w_o) + u_{12*}(w_1)$ where $u_{01} = u_o \cup \bar{H} \cup u_1$ and $u_{12} = u_1 \cup \bar{H'} \cup u_2$ from diagram ②. For this equality, the summands must be reordered: this is possible because we have assumed that A is a co-H-space. □

Let $* \in Y \subset X$ and $v : Y \to \tilde{Y}$ be a section. Generalizing (1.2.7), we define mappings

(1.2.18)

$$\langle C_f, \tilde{C}_f \rangle^v \times [SA, F] \xrightarrow{+} \langle C_f, \tilde{C}_f \rangle^v$$

$$\langle C_f, \tilde{C}_f \rangle \times [SA, F] \xrightarrow{+} \langle C_f, \tilde{C}_f \rangle$$

which we denote by + . These are operations of groups if A is a co-H-space. Let $i : F \subset \tilde{C}_f$ be the inclusion of the fiber and let $\bar{u} \in \langle \bar{u} \rangle \in \langle C_f, \tilde{C}_f \rangle$ and $a \in \alpha \in [SA, F]$. Then the co-operation μ from (1.2.7) gives us the map

$$(1.2.19) \qquad u_a : C_f \xrightarrow{\mu} C_f \vee SA \xrightarrow{(\bar{u},\ ia)} \tilde{C}_f \ .$$

There is a canonical homotopy

$$H^o : p \circ u_a \simeq 1_{C_f} \text{ rel } X \text{ where } pu_a = (1, 0) \circ \mu.$$

The homotopy H^o can be lifted to a homotopy $H^1 : u_a \simeq u_a^1$ rel u, that is, the diagram

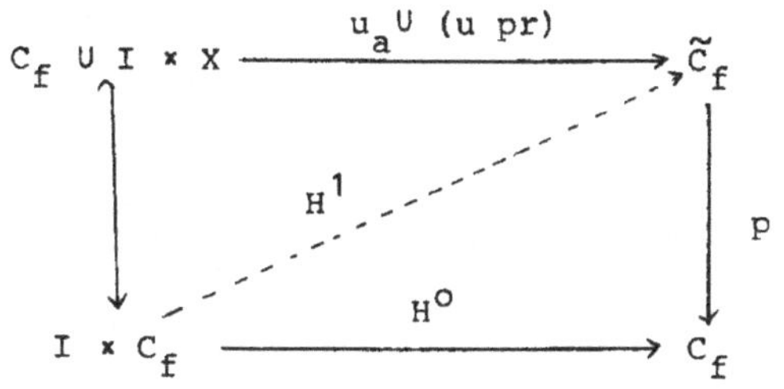

commutes, where $pr : I \times X \to X$ is the projection. Thus $u_a^1 : C_f \to \tilde{C}_f$ is a section extending u.

(1.2.20) Lemma: <u>The section homotopy class</u> $\langle u_a^1 \rangle \in \langle C_f, \tilde{C}_f \rangle^u$ <u>depends only on</u> $\langle \bar{u} \rangle$ <u>and</u> α.

Therefore we can define (1.2.18) by $\langle \bar{u} \rangle + \alpha = \langle u_a^1 \rangle$. We will show in the sequel to Theorem (1.2.21) that this yields a group operation. With Θ as in (1.2.15) we have the equation in $[C_f, \tilde{C}_f]$

$$\Theta(\langle\bar{u}\rangle + \alpha) = \Theta(\langle\bar{u}\rangle) + i_*\alpha$$

because of (1.2.19). Here the operation $+$ on the right is defined as in (1.2.7).

<u>Proof of</u> (1.2.20): Let $a, a' \in \alpha$ and let $A : a \simeq a'$ be a homotopy. Let $\bar{u}, \bar{u}' \in \langle\bar{u}\rangle$ and let $H : \bar{u} \equiv \overline{u'}$ be a section homotopy rel u. With u_a and $u'_{a'}$ as in (1.2.19), let liftings $H^1 : u_a \simeq u_a^1$ and $H^{1'} : u'_{a'} \simeq u'^{1}_{a'}$ be given as in the diagram following (1.2.19). We must show that $u_a^1 \equiv u'^{1}_{a'}$ are section homotopic relative u.

Let $\hat{C}_f = C_f \times I$ and let $\hat{H} = (H \cup A) \circ (\mu \times 1_I) : \hat{C}_f \to \tilde{C}_f$ be the homotopy $u_a \simeq u'_a$ given by H and A. Let $Y = C_f \dot{\times} I \subset \hat{C}_f$. Then we have a map

$$\bar{\bar{H}} = (H_1 \cup H_1') \cup u\,pr : Y \times I = (C_f \times S^0 \times I) \cup X \times I \times I \to \tilde{C}_f$$

where $pr : X \times I \times I \to X$ is the projection. The diagram

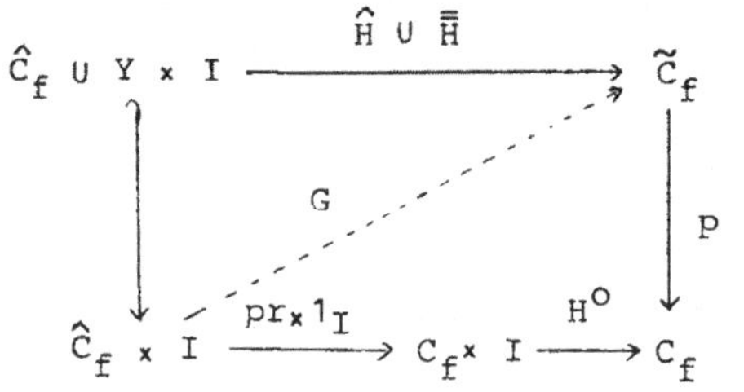

commutes. Let G be a lifting which extends the diagram commutatively. Then $G_{|C_f \times I \times \{1\}} : u_a^1 \equiv u'^{1}_{a'}$ rel u is the desired section homotopy. ☐

We generalize (1.2.8) to obtain the

(1.2.21) Theorem: Let A be a co-H-space. Let $u_o \in \langle C_f, \tilde{C}_f\rangle^u$ where $u : X \to \tilde{X}$ is a section. Then there is a bijection

$$\langle C_f, \tilde{C}_f\rangle^u \approx [SA, F]$$

defined by $u_1 \mapsto d(u_o, u_1)$ with inverse $\alpha \mapsto u_o + \alpha$.

We use this theorem to show that (1.2.18) defines a group operation.

(1.2.22) Corollary: Let A be a co-H-space. Given $u_o \in \langle C_f, \tilde{C}_f\rangle^u$ and $\alpha, \beta \in [SA, F]$, we have, in $\langle C_f, \tilde{C}_f\rangle^u$,

$$u_o + 0 = u_o$$

$$(u_o + \alpha) + \beta = u_o + (\alpha + \beta).$$

Proof of (1.2.22): Because of (1.2.21),

$$\begin{cases} d(u_o, u_o + \alpha) = \alpha \\ d(u_o + \alpha, (u_o + \alpha) + \beta) = \beta \end{cases}$$

and by (1.2.17)

$$d(u_o, (u_o + \alpha) + \beta) = \alpha + \beta.$$

On the other hand,

$$d(u_o, u_o + (\alpha + \beta)) = \alpha + \beta.$$

Thus the injectivity of the map $u_1 \mapsto d(u_o, u_1)$ gives us the equation

$$u_o + (\alpha + \beta) = (u_o + \alpha) + \beta. \qquad \square$$

For the maps

$$\langle C_f, \tilde{C}_f \rangle^u \xrightarrow{\Theta_u} \langle C_f, \tilde{C}_f \rangle \xrightarrow{j} \langle X, \tilde{X} \rangle$$

where Θ_u is the canonical map and j is the restriction, we have $\mathrm{Im}\ \Theta_u = j^{-1}(u)$. This follows from the relative lifting property (0.1.5). Thus (1.2.21) gives us, in analogy with (1.2.9), the

(1.2.23) Corollary: Let A be a co-H-space. If $u_o \in \langle C_f, \tilde{C}_f \rangle$ and $j(u_o) = u$ then $j^{-1}(u) = \{u_o + \alpha \mid \alpha \in [SA, F]\}$.

Proof of (1.2.21): We show first that $d(u_o, u_o + \alpha) = \alpha$. Let $f_1 : A \to \{0\} \times C_f \cup I \ltimes X$ be given by $f_1(a) = (1, f(a))$. Then $I \dot{\ltimes} C_f = C_{f_1}$ and for the section

$$\bar{u} = u_o \cup \overline{upr} \cup u_o : C_f \dot{\ltimes} I \to \widetilde{C_f \dot{\ltimes} I}$$

we have $w_f^{\#}(\bar{u}) = d(u_o, u_o) = 0$. On the other hand, for the section $\bar{u} + \alpha = u_o \cup \overline{upr} \cup (u_o + \alpha)$ we have the equation

$$w_f{}^*(\bar{u} + \alpha) = w_f{}^*(\bar{u}) + i_*\alpha$$

in $[SA, \widetilde{C_f \dot{\ltimes} I}]$. This follows directly from the definition of w_f. Thus it follows from the definition of $w_f^{\#}$ in (1.2.11) (see the following lemma (1.2.24)) that

$$w_f^{\#}(\bar{u} + \alpha) = w_f^{\#}(\bar{u}) + \alpha = \alpha.$$

By definition of the difference, we have $d(u_o, u_o + \alpha) = w_f^{\#}(\bar{u} + \alpha)$, so $d(u_o, u_o + \alpha) = \alpha$. This means that the mapping $u_1 \longmapsto d(u_o, u_1)$

is surjective. We will now show that the mapping $u_1 \mapsto d(u_o, u_1)$ is injective also. To this end let u_1 and u_1' be such that

$$d(u_o, u_1') = d(u_o, u_1).$$

Then by (1.2.17)

$$d(u_1, u_1') = d(u_1, u_o) + d(u_o, u_1') = -d(u_o, u_1) + d(u_o, u_1') = 0$$

Thus by the obstruction property $u_1 \simeq u_1'$ rel u. □

In proving (1.2.21) we used the following lemma, which will also be of importance later.

(1.2.24) Lemma: Let $f : A \to X$ and $g : B \to C_f$ where B is a suspension. Let $\tilde{C}_g \to C_g$ be a fibration with fiber F. Let $u, u + \alpha \in \langle C_f, \tilde{C}_f \rangle$ with $\alpha \in [SA, F]$. If in $[B, \tilde{C}_f]$

$$g^*(u + i_*\alpha) = g^*(u) + i_*(\beta)$$

for some $\beta \in [B, F]$, then in $[B, F]$

$$g^{\#}(u + \alpha) = g^{\#}(u) + \beta.$$

Proof: By definition in (1.2.11) we have

$$g^{\#}(u) = i_*^{-1}(-\partial\tilde{\pi}_g + u_*(g))$$

where $u_*(g) = g^*(u)$. Accordingly,

$$g^{\#}(u + \alpha) = i_*^{-1}(-\partial\tilde{\pi}_g + g^*(u + i_*\alpha))$$

$$= i_*^{-1}(-\partial\tilde{\pi}_g + g^*(u) + i_*\beta)$$

$$= g^{\#}(u) + \beta \ . \quad \Box$$

We will be using the following simple kind of direct limit.

(1.2.25) <u>Definition</u>: Let $\mathcal{K}$ be a category and

$$X_o \to .. \to X_n \xrightarrow{i_n} X_{n+1} \to \ldots$$

a sequence of morphisms in $\mathcal{K}$. An object X_∞ is called a <u>direct limit</u> (written $\varinjlim X_n$) of this sequence when there are morphisms $j_n : X_n \to X_\infty$ with $j_{n+1} \circ i_n = j_n$ such that X_∞ has the universal property: to each X together with a system of morphisms g_n : $X_n \to X$ such that $g_{n+1} \circ i_n = g_n$, there exists exactly one morphism $g : X_\infty \to X$ such that $g \circ j_n = g_n$. X_∞ is uniquely determined up to isomorphism by this property.

In the categories of sets and of topological spaces, direct limits always exist. A commutative diagram

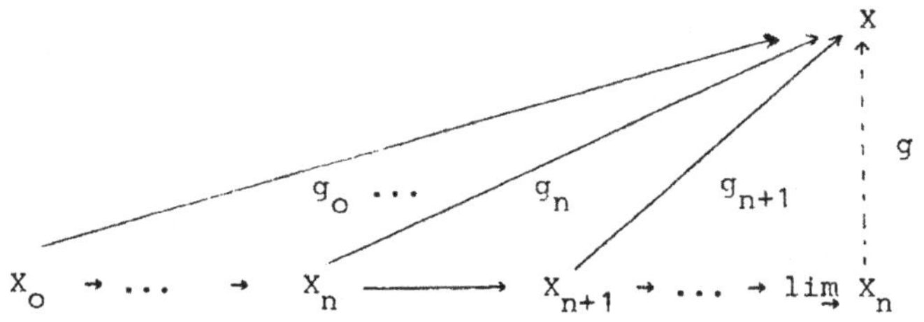

thus defines a map from the direct limit.

If all $X_n \longrightarrow X_{n+1}$ are closed cofibrations, $X_o \to \varinjlim X_n$ is also a closed cofibration

If X is a space with a filtration of principal cofibrations

$$X_o \subset X_1 \subset \ldots \subset X = \varinjlim X_i$$

we call X an <u>iterated principal cofibration</u>, or an iterated mapping cone. Let $f_i : A_i \to X_{i-1}$ be the attaching maps of X and let $H_i : X_{i+1} \simeq C_{f_i}$ be homotopy equivalences under X_i.

Given a map $u : X_o \to U$, we consider the extension problem $(i \geqslant 1)$

$$\begin{array}{ccccc} & & X & & \\ & & \cup & & \\ & & X_i & & \\ & & \cup & & \\ A_i & \xrightarrow{f_i} & X_{i-1} & \overset{\bar{u}}{\dashrightarrow} & \\ & & \cup & & \\ & & X_o & \xrightarrow{u} & U \end{array}$$

By (1.2.2), the map u can be extended over X_i exactly when the subset

(1.2.26) $$\mathcal{H}_i(u) = \{f_i^*(\bar{u}) \mid \bar{u} \text{ extends } u\} \subset [A_i, U]$$

contains the trivial element $0 \in \mathcal{H}_i(u)$. We call $\mathcal{H}_i(u)$ the <u>i-th order obstruction</u> to extending u over X.

<u>Warning</u>: If u can be extended over $X = \varinjlim X_i$, then $0 \in \mathcal{H}_i(u)$

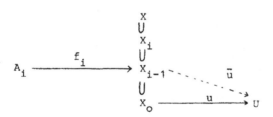

for all i , but the converse is not true. If $0 \in \mathcal{H}_i(u)$ for all i , then there exists an extension $u_i : X_i \to U$ for every i , but these extensions may not be compatible with each other, that is, in general $u_{i|X_{j-1}}$ will not be equal to u_{i-1} . An example due to Gray in which this happens will be described in (1.2.30).

If the following very restrictive condition is satisfied, it is possible to extend over the direct limit.

(1.2.27) <u>Lemma:</u> <u>If</u> $0 \in \mathcal{H}_n(u)$ <u>and</u> $[A_i, U] = 0$ <u>for all</u> $i > n$, <u>then</u> u <u>can be extended over</u> $X = \varinjlim X_i$.

The obstructions of higher order $i \geq 2$ can be empty. An i-th order obstruction is defined only when all obstructions of order $k < i$ vanish.

Since $X_o \subset X_{i-1}$ is a cofibration, the i-th order obstruction depends only on the homotopy class $u \in [X_o, U]$. Naturally the definition (1.2.26) allows us to write

$$\mathcal{H}_{i+j}(u) = \bigcup \{ \mathcal{H}_i(\bar{u}) \mid \bar{u} \text{ extends } u \text{ over } X_j \} .$$

$\mathcal{H}_1(u) = f_1^*(u)$ is the primary obstruction. $\mathcal{H}_2(u)$ and $\mathcal{H}_3(u)$ are called the secondary and tertiary obstructions, respectively, to extending u .

Given an iterated mapping cone X with filtration as above, we define the <u>associated iterated mapping cone</u> Y with filtration

$$Y_o \subset Y_1 \subset \dots \subset Y$$

to be

$$(1.2.28) \quad \begin{cases} Y = I \ltimes X \\ Y_i = \{0\} \times X \cup I \ltimes X_i \cup \{1\} \times X , \quad i \geq 0 . \end{cases}$$

The attaching maps are ($i \geq 1$)

$$w_i = (1 \ltimes H_i)\ w_{f_i} : SA_i \longrightarrow I \ltimes C_{f_{i-1}} \longrightarrow I \ltimes (X_{i-1}, X_{i-2}) \subset Y_{i-1} \quad .$$

Let $u_o, u_1 : X \to U$ be maps and let $H : u_o|X_o \simeq u_1|X_o$ be a homotopy. The triple (u_o, H, u_1) defines a map

$$u_o \cup H \cup u_1 : Y_o = \{0\} \times X \cup I \times X_o \cup \{1\} \times X \longrightarrow U \quad .$$

The i-th order obstruction to extending this map over Y_i is called the <u>i-th order difference</u>

$$(1.2.29) \qquad D_i(u_o, H, u_1) = \mathcal{H}_i(u_o \cup H \cup u_1) \subset [SA_i, U] \quad .$$

Writing $u_o^{\ i} = u_o|X_i$ and $u_1^{\ i} = u_1|X_i$ we have

$$D_i(u_o, H, u_1) = \left\{ d(u_o^{\ i}, H', u_1^{\ i}) \mid H' : u_o^{\ i-1} \simeq u_1^{\ i-1} \text{ extends } H \right\}$$

The homotopy $H : u_o|X_o \simeq u_1|X_o$ however can be extended to a homotopy $u_o|X_i \simeq u_1|X_i$ exactly when $0 \in D_i(u_o, H, u_1)$. We have thus reduced the homotopy problem for maps to an extension problem. If $u_o|X_o = u_1|X_o = u$, then we set $D_i(u_o, u_1) = D_i(u_o, u\ pr, u_1)$.

It can be seen from (1.2.6) that the i-th order difference $D_i(u_o, H, u_1)$ is either empty or else is a coset of the subgroup $D_i(u_o, u_o) \subset [SA_i, U]$. In (3.2.14) we will characterize the i-th order difference as an element in the E_i –term of a spectral sequence. We will see further on that the obstruction sets $\mathcal{H}_i(u)$ are also cosets of subgroups when certain stability conditions are satisfied.

<u>Warning</u>: If $0 \in D_i(u_o, u_1)$ for all i, it does not follow that $u_o \simeq u_1$ rel u. Let $\mathbb{C}P_\infty$ be complex projective space. Then $\mathbb{C}P_\infty = \varinjlim \mathbb{C}P_n$ is an iterated principal cofibration. Gray [36] showed that there is a map $f: \mathbb{C}P_\infty \to S^3$ which is not null-homotopic, although the restriction $f|\mathbb{C}P_n$ is null-homotopic for all n.

(1.3) Principal fibrations and lifting of maps and retractions

The ensuing discussion of principal fibrations is a rather detailed dualization of (1.2) for principal cofibrations that should be of some help to readers unacquainted with duality techniques. We show how primary obstructions and differences for principal fibrations can be used to solve existence and classification problems for liftings. In other words, we can in fact perform the inductive step described in condition (B) of (1.1.3). We start by considering the familiar situation of map liftings, then go on to show that almost identical considerations apply to retraction liftings.

(1.3.1) Definition: For a map $f: X \to A$ we say that the fibration $P_f \to X$ of the mapping path space is principal, see (0.1.10). More generally, we call a fibration $Y \to X$ a principal fibration when there is a map $f: X \to A$ and a homotopy equivalence $P_f \simeq Y$ over X. The map f is called the classifying map of the principal fibration.

Example: Let G be a topological group with classifying G-principal bundle $E_G \to B_G$. A G-principal bundle $Y \to X$ is induced by its classifying map $f: X \to B_G$. $Y \to X$ is thus a principal fibration in the sense of (1.3.1) with classifying map f, since $Y \simeq P_f$ over X. There is a homotopy equivalence of the fibers $\Omega B_G \simeq G$, see [108].

We now present some properties of fibrations $P_f \to X$. Using (1.1.7), it can be seen that they hold for principal fibrations $Y \to X$ as well. A map $f: U \to A$ is null-homotopic exactly when it can be lifted to a map into the path space PA , see (0.1.10). This property of a path space corresponds to the following characteristic property of a mapping path space P_f .

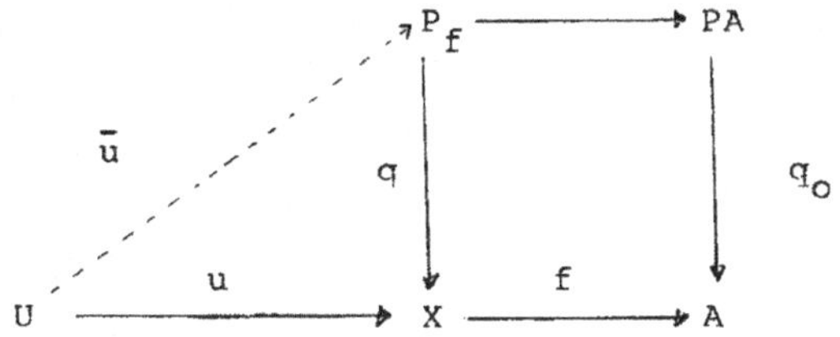

The map u can be lifted to P_f, that is there exists a pointed map $\bar{u}$ with $q\bar{u} = u$, exactly when the composition f o u is nullhomotopic. For this reason, we call the homotopy class

(1.3.2) $\qquad f_*(u) = f \circ u \in [U, A]$

the (primary) <u>obstruction to lifting</u> u. Such primary obstructions are used to define higher-order obstructions to lifting a map.

Dual to the principal cofibration $I \dot{\rtimes} C_f \subset I \rtimes C_f$ of (1.2.4) is the principal fibration p defined by

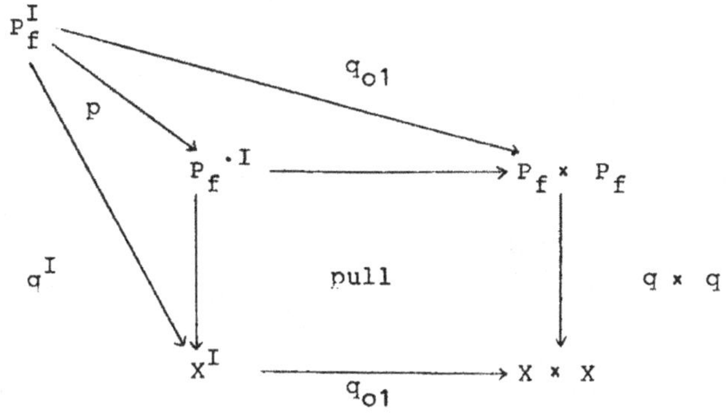

Here $q_{01}(\sigma) = (\sigma(0),\sigma(1))$. It is easy to see that p is a fibration. The classifying map for p is

(1.3.3) $$w_f : P_f^{\cdot I} \longrightarrow \Omega A$$

which is defined on an element

$$(\sigma,(x_0, \tau_0),(x_1, \tau_1)) \in P_f^{\cdot I} \subset X^I \times P_f \times P_f$$

to be the loop

$$w_f(\sigma, (x_0, \tau_0), (x_1, \tau_1)) = - \tau_0 + f \circ \sigma + \tau_1$$

in ΩA. This loop is the sum of paths as indicated in

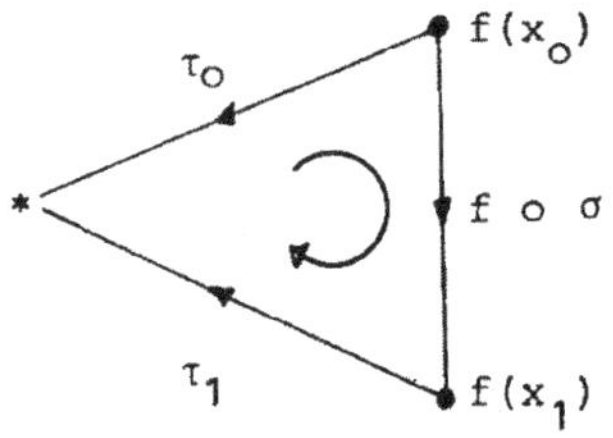

.

There is also a homotopy equivalence dual to that of (1.2.4).

(1.3.4) <u>Lemma</u>: $jh : P_f^I \simeq P_{w_f}$ <u>over</u> $P_f^{\cdot I}$.

<u>Proof</u>: For $f = 1_A = 1$ we see that $w_1 : PA^{\cdot I} \to \Omega A$ is a homeomorphism, cf. (0.0.3) (II). Since $p : PA^I \to PA^{\cdot I}$ is a fibration and PA^I is contractible, we have a homotopy equivalence

(see (1.1.9))

$$h' : PA^I \simeq P(PA^{\cdot I}) \quad \text{over} \quad PA^{\cdot I} .$$

The composition

$$P_f^I \xrightarrow{\pi_f^I} PA^I \xrightarrow{h'} P(PA^{\cdot I}) \xrightarrow{P_{w_1}} P\Omega A$$

then determines the homotopy equivalence h of (1.3.4). □

Let $u_o, u_1 : U \to P_f$ and let $H : qu_o \simeq qu_1 : U \to X$ be a homotopy. The triple (u_o, H, u_1) defines a map $\bar{H} \succeq (u_o, u_1) : U \to P_f^{\cdot I}$ since $P_f^{\cdot I}$ is a pull-back, see (0.0) . As in (1.2.5) we call

(1.3.5) $$d(u_o, H, u_1) = w_{f*}(\bar{H} \succeq (u_o, u_1)) \in [U, \Omega A]$$

the primary difference of (u_o, H, u_1). By (1.3.4), $d(u_o, H, u_1) \equiv 0$ exactly when there is a homotopy $H' : u_o \simeq u_1$ lifting H, that is, $(q^I) \circ H' = H$. If $qu_o = qu_1 = u$ then we take the homotopy $u\,pr : I \times U \to X$ and define $d(u_o, u_1) = d(u_o, u\,pr, u_1)$.

We will now describe a few simple properties of the difference construction which are all easy consequences of the definition. Let $u_2 : U \to P_f$ be a second map and let $H' : qu_1 \simeq qu_2$ be a homotopy. Then we have

(1.3.6) $$d(u_o, H, u_1) + d(u_1, H', u_2) = d(u_o, H + H', u_2) \quad .$$

The loop space ΩA <u>operates from the right on</u> P_f by μ : $P_f \times \Omega A \to P_f$ with $\mu(x, \tau, \sigma) = (x, \tau + \sigma)$, where $\tau + \sigma$ is the addition of paths. The map μ induces group actions

$$[U, P_f]_v \times [U, \Omega A] \xrightarrow{+} [U, P_f]_v$$

(1.3.7)

$$[U, P_f] \times [U, \Omega A] \xrightarrow{+} [U, P_f]$$

which we denote by + , so that $\mu_*(\xi, \alpha) = \xi + \alpha$. Here $[U, P_f]_v$ is the homotopy set of liftings over $v : U \to Y$ (0.0.5), $P_f \to X \to Y$ is a space over Y. The element $d(u_o, u_1)$ is not uniquely defined by the homotopy classes of u_o, u_1 in $[U, P_f]$, but it is by the homotopy classes of u_o, u_1 relative $u : U \to X$. For the above group actions + we have, dually to (1.2.8), the

(1.3.8) <u>Theorem:</u> <u>Let</u> $u_o \in [U, P_f]_u$. <u>Then there is a bijection</u>

$$[U, P_f]_u \xrightarrow{\approx} [U, \Omega A]$$

<u>given by</u> $u_1 \mapsto d(u_o, u_1)$ <u>with inverse</u> $\alpha \mapsto u_o + \alpha$.

For the maps

$$[U, P_f]_u \xrightarrow{\Theta_u} [U, P_f] \xrightarrow{q_*} [U, X]$$

where Θ_u is the canonical map, we have $\operatorname{Im} \Theta_u = q_*^{-1}(u)$ (since $q : P_f \to X$ is actually a fibration in Top^o). Thus (1.3.8) implies the

(1.3.9) <u>Corollary</u>: <u>Let</u> $f : X \rightarrow A$ <u>and let</u> $q_* : [U, P_f] \rightarrow [U, X]$ <u>be induced by the projection</u>. <u>If</u> $u_o \in q_*^{-1}(u)$ <u>then</u> $q_*^{-1}(u) = \{u_o+\alpha \mid \alpha \in [U, \Omega A]\}$.

There are similar statements for the following operation. Let $q : Z \rightarrow X$ be a fibration and let $h : P_f \simeq Z$ be a (pointed) homotopy equivalence over X . Then we have the operation

$$(1.3.10) \qquad [U, Z]_V \times [U, \Omega A] \xrightarrow{+} [U, Z]_V \quad , \quad u_o+\alpha = h_*(h_*^{-1}(u_o)+\alpha)$$

We saw in (1.2) that the theory for extending maps could be formulated more generally for sections of fibrations. We will now show that the properties of liftings of maps discussed above hold, more generally, for liftings of retractions. In the following, let $\overline{X}$ be the total space of a cofibration $X \subset \overline{X}$ with cofiber $F = \overline{X}/X$. Given a map $p : X \quad Y$ let $Y \subset \overline{Y}$ be the cofibration, with cofiber F , induced by p .

Let a map $f : X \rightarrow A$ and a closed cofibration $P_f \subset \overline{P_f}$ be given and let $X \subset \overline{X}$ be induced by $q : P_f \rightarrow X$. When can a retraction $u : \overline{X} \rightarrow X$ be lifted to a retraction $u' : \overline{P}_f \rightarrow P_f$ such that the diagram

$$(1.3.11) \qquad \begin{array}{ccc} \overline{P}_f & \xrightarrow{u'} & P_f \\ \downarrow{\scriptstyle \overline{q}} & & \downarrow{\scriptstyle q} \\ \overline{X} & \xrightarrow{u} & X \end{array}$$

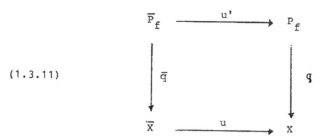

commutes? The obstruction to doing this can be described, dually to (1.2.11), if we assume that A is an H-group. We consider the diagram of groups and group homomorphisms

$$\begin{array}{ccc}
 & & [F, A] \\
 & & \downarrow j^* \\
[C_{\bar q}, A] & \xrightarrow{\delta} & [\bar X, A] \\
\cong \downarrow i_C^* & & i^* \downarrow \uparrow u^* \\
[C_q, A] & \xrightarrow{\delta} & [X, A]
\end{array}$$

The map δ is the restriction to the mapping cones C_q, $C_{\bar q}$. By (0.1.13), the induced map between the cofibers $i_C : C_q \to C_{\bar q}$ is a homotopy equivalence, so i_C^* is an isomorphism. In $[C_q, A]$ there is an element π_f with $\delta\pi_f = f$ defined

$$\pi_f : C_q = CP_f \cup_q X \to A$$

$$\begin{cases} \pi_f(x) = f(x) & \text{for } x \in X \\ \pi_f(t, (x, \tau)) = \tau(t) & \text{for } x \in X,\ \tau \in PA,\ \tau(0) = f(x),\ t \in I. \end{cases}$$

Let $\tilde\pi_f = i_C^{*-1}(\pi_f) \in [C_q, A]$. The long exact sequence for the cofibration $X \xrightarrow{i} \bar X \xrightarrow{j} F = \bar X/X$ splits by u^*, so i^* is injective (A was assumed to be an H-group). On the other hand $i^*(-\delta\tilde\pi_f + u^*(\delta\pi_f)) = 0$ since $i^*u^* = 1$, so

$$(1.3.12) \quad f_{\#}(u) = j^{*-1}(-\delta\tilde\pi_f + u^*(\delta\pi_f)) \in [F, A]$$

is well-defined.

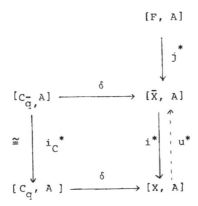

(1.3.13) <u>Theorem:</u> $f_{\#}(u) = 0$ <u>exactly when the retraction</u> $u : \bar{X} \to X$ <u>can be lifted to a retraction</u> $u' : \bar{P}_f \to P_f$.

We call $f_{\#}(u)$ the (primary) <u>obstruction to lifting the retraction</u> u. The corresponding higher-order obstructions are defined from such primary obstructions.

<u>Proof</u>: If u' exists, then $(u', u) : \bar{q} \to q$ determines a map $u'_C : C_{\bar{q}} \to C_q$ with $u'_C i_C = 1$. But we have $i_C^{*-1}(\pi_f) = \tilde{\pi}_f = u'_C{}^*(\pi_F)$, and so

$$-\delta\tilde{\pi}_f + u^*(\delta\pi_f) = -\delta u'_C{}^*(\pi_f) + u^*(\delta\pi_f)$$

$$= - u^* \delta\pi_f + u^* (\delta\pi_f) = 0 .$$

Thus $f_{\#}(u) = 0$ as well.

Now suppose $f_{\#}(u) = 0$. We consider the diagram

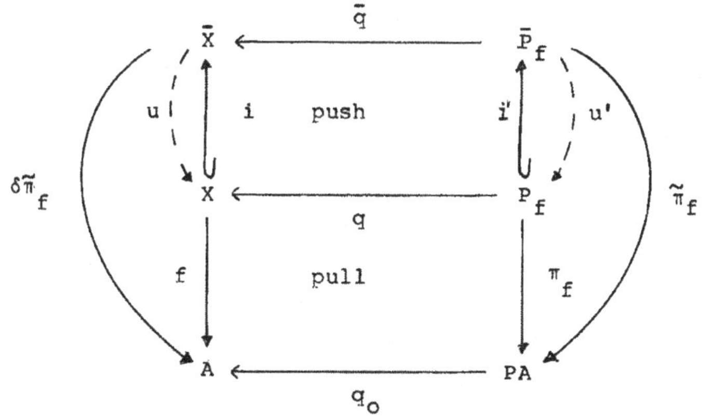

Any extension $\tilde{\pi}_f$ of π_f in this diagram represents the adjoint of the class $\tilde{\pi}_f \in [C_q, A]$ above. Therefore $\delta\tilde{\pi}_f$ is represented on the pushout $\bar{X} = (X \cup \bar{P}_f)/\sim$ by $\delta\tilde{\pi}_f = f \cup q_o \tilde{\pi}_f$. From $f_{\#}(u) = 0$ it follows that $\delta(\tilde{\pi}_f) = f \circ u \in [\bar{X}, A]$ and therefore there is a homotopy

$$H : f \circ u \simeq \delta \tilde{\pi}_f \ .$$

Using H, we define a map u' over u in the above diagram by the formulas

$$\begin{cases} qu' = u\bar{q} \\ (\pi_f u')(x)(t) = \begin{cases} H(\bar{q}x, 2t) & \text{for } 0 \leqslant t \leqslant 1/2 \\ \tilde{\pi}_f(x)(2t-1) & \text{for } 1/2 \leqslant t \leqslant 1 \end{cases} \end{cases}$$

where $x \in \bar{P}_f$. From u' we obtain a homotopy H_o : $u' i' \simeq 1_{P_f}$ over X.

This homotopy H_o allows us to construct the commutative diagram

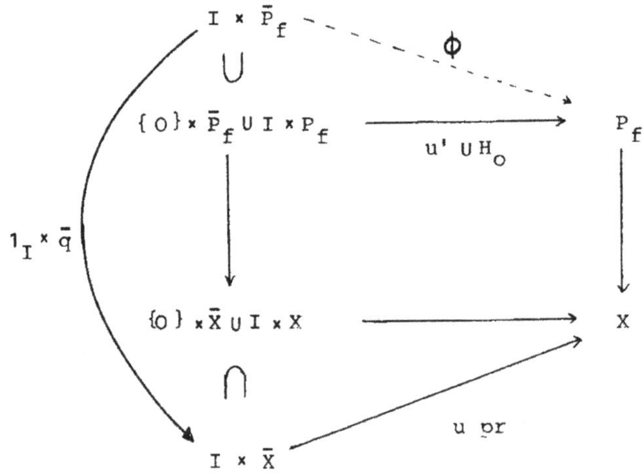

Because of the relative lifting property (0.1.5), there exists a homotopy ϕ extending the diagram commutatively. ϕ_1 is then a retraction of i' that lifts u, that is $q\,\phi_1 = uq$. $\square$

The primary obstruction is natural in the following sense. Given maps

(1.3.14)
$$\begin{array}{ccccc} P_g & \longrightarrow & Y & \xrightarrow{\;g\;} & B \\ \uparrow{\scriptstyle \overline{H}} & & \uparrow{\scriptstyle \xi} & \overset{H}{\Rightarrow} & \uparrow{\scriptstyle \alpha} \\ P_f & \longrightarrow & X & \xrightarrow{\;f\;} & A \end{array}$$

we assume there is a homotopy $H : g\xi \simeq \alpha f$. Then H determines the map $\overline{H}$ over ξ with $\overline{H}(x,\tau) = (\xi(x),\sigma)$ for $\tau \in PA$ where $\sigma \in PB$,

$$\sigma(t) = \begin{cases} H(2t,\, x) & \text{for } 0 \leq t \leq 1/2 \\ \alpha\tau(2t-1) & \text{for } 1/2 \leq t \leq 1 \end{cases} .$$

Let $P_f \subset \overline{P}_f$ be a cofibration and let $P_g \subset \overline{P}_g$ be the cofibration induced by H. Then $Y \subset \overline{Y}$ is induced from $X \subset \overline{X}$ by ξ.

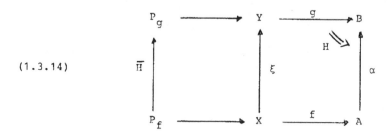

Therefore a retraction $u : \bar{X} \to X$ also determines a retraction $\xi_*(u) : \bar{Y} \to Y$. If F is the cofiber and $\alpha_* : [F, A] \to [F, B]$ then

(1.3.15) $$\alpha_* f_\#(u) = g_\# \xi_*(u) \quad .$$

This follows because the diagram from which the definition of f and g in (1.3.12) derives is natural in this sense. As in (0.0.4) , let $\langle \bar{X}, X \rangle$ denote the set of retraction homotopy classes. For the maps

$$\langle \bar{X}, X \rangle \xrightarrow{\Theta} [\bar{X}, X] \xrightarrow{i^*} [X, X]$$

where Θ associates to the retraction homotopy class of u its homotopy class, we have

(1.3.16) $$\operatorname{Im} \Theta = i^{*-1}(\ _X) \quad .$$

Notice that the primary obstruction $f_\#(u)$ of (1.3.12) depends only on the homotopy class Θu of the retraction u . For the trivial cofibration $X \subset X \vee U$ with cofiber U we again have $\langle X \vee U, X \rangle = [U, X]$, see (0.0). In this case the obstruction $f_\#(u)$ of (1.3.12) is equal to the obstruction $f_*(u)$ of (1.3.2). $\square$

In analogy with (1.3.5) , we now reduce the <u>homotopy problem</u> for retractions to a lifting problem. Let $X \subset \bar{X}$ be a cofibration and let $c : X \to X^I$ be the map to constant paths $c(x)(t) = x$ and let $X^I \subset \overline{X^I}$ be the induced cofibration.

Let $u_o, u_1 : \bar{X} \to X$ be retractions and let $H : u_o \equiv u_1$ be a retraction homotopy. Then H gives us a map $\bar{H} : \bar{X} \to X^I$ extending c, so from $\bar{H}$ we get a retraction $\bar{H} : \overline{X^I} \to X^I$ lifting (u_o, u_1) over q_{o1}.

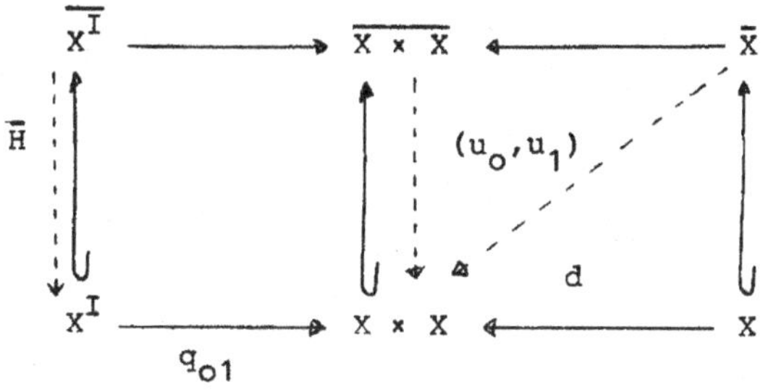

Here $X \times X \subset \overline{X \times X}$ is the cofibration induced by the diagonal d. Conversely, every such retraction $\bar{H}$ lifting (u_o, u_1) gives us a retraction homotopy $u_o \equiv u_1$.

As in (1.3.5), we define the difference of retractions. Let $P_f \subset \overline{P_f}$ be a cofibration with cofiber F, let $u_\tau' : \bar{P}_f \to P_f$ with $\tau = 0, 1$ be retractions lifting the retractions $u_\tau : \bar{X} \to X$, and suppose $H : u_o \equiv u_1$ is a retraction homotopy. The triple (u_o', H, u_1') defines a retraction $\bar{H} \underline{\times} (u_o', u_1')$.

(1.3.17)

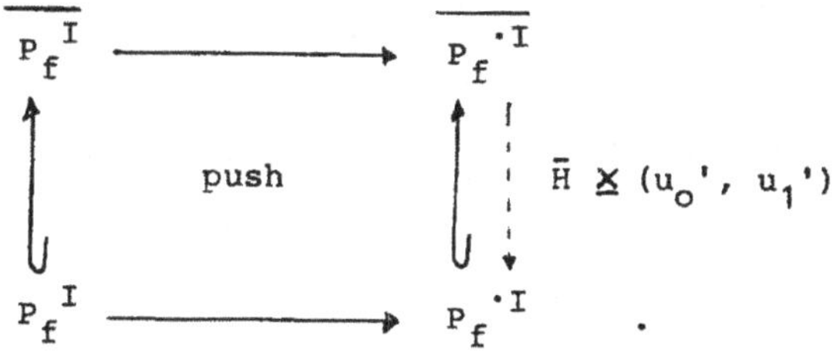

Generalizing (1.3.5), we define

(1.3.18) $$d(u_o', H, u_1') = w_{f\#}(\bar{H} \underline{\times} (u_o', u_1')) \in [F, \Omega A].$$

Here w_f is given as in (1.3.3) and $w_{f\#}$ is defined by (1.3.12). Thus A can be arbitrary. Because of (1.1.7) we see that the retraction homotopy $H : u_o \simeq u_1 : \bar{X} \to X$ can be lifted to a retraction homotopy $H' : u_o' \simeq u_1' : \bar{P}_f \to P_f$ (that is, such that $qH_t' = H_t q$) exactly when $d(u_o', H, u_1') = 0$.

We call $d(u_o', H, u_1')$ the <u>primary difference</u> of (u_o', H, u_1'). If $u_o = u_1 = u$ then we take the constant retraction homotopy $H = u \circ pr : u_o \equiv u_1$, and define $d(u_o', u_1') = d(u_o', u\, pr, u_1)$. The difference $d(u_o', u_1')$ is not well-defined by the retraction homotopy classes of u_o' and u_1' in $\langle \bar{P}_f, P_f \rangle$, but it is by the retraction homotopy classes of u_o' and u_1' over $u : \bar{X} \to X$ see the definitions in (0.0.8) and (0.0.9) .

As a generalization of (1.3.6), and dual to (1.2.17),

(1.3.19) <u>Theorem: Let</u> A <u>be an H-space. Then in</u> $[F, \Omega A]$.

$$d(u_o, H, u_1) + d(u_1, H', u_2) = d(u_o, H + h', u_1)$$

$$d(u_o, u_1) + d(u_1, u_2) = d(u_o, u_2) \; .$$

<u>Proof:</u> Let $J = (I, \{0, 1/2, 1\})$. Consider the diagram

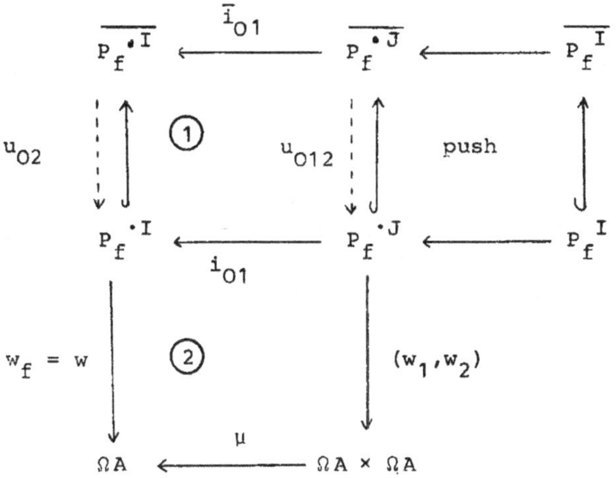

where $P_f^{\cdot J}$ is defined for the pair J as $P_f^{\cdot I}$ is for the pair $(I, I^{\cdot})$. i_{01} is induced by the inclusion $I = (I, I^{\cdot}) \subset J$ and $u_{02} = \overline{(H + H')} \veebar (u_o, u_1)$ is defined as in (1.3.17). The map $u_{012} = \overline{(H + H')} \veebar (u_o, u_1, u_2)$ is defined correspondingly, so that u_{012} lifts u_{01} in ①, that is

$$(1) \qquad u_{02} \circ \overline{i}_{01} = i_{01} \circ u_{012}.$$

Further, let $j : I \to J$ $(\tau = u, 1)$ denote the maps with $j_o(t) = t/2$ and $j_1(t) = (1 + t)/2$. These induce maps $w_\tau = w_f \circ \gamma_\tau^{\cdot}$: $P_f^{\cdot J} \to P_f^{\cdot I} \to \Omega A$, where $\gamma_\tau^{\cdot} = 1^{\cdot j_\tau}$. It then follows easily from the definition of w_f that the diagram ② is homotopy commutative in Top^o, where μ is the addition of loops. We can now continue the proof along dual lines to that of (1.2.17). Using the homotopy equivalence $h : P_f^I \to P_w$ over $P_f^{\cdot I}$ from (1.3.4), we construct the commutative diagram

$$\begin{array}{ccccc}
C_q & \hookleftarrow & P_f^{\cdot I} & \xleftarrow{q} & P_w \\
\uparrow \beta & & \uparrow i_{01} & & \uparrow h \\
C_{q'} & \hookleftarrow & P_f^{\cdot J} & \xleftarrow{q'} & P_f^{I} \\
\downarrow \gamma_\tau & & \downarrow \gamma_\tau^{\bullet} & & \downarrow h \circ (1^{j_\tau}) \\
C_q & \hookleftarrow & P_f^{\cdot I} & \xleftarrow{q} & P_w
\end{array}$$

where β and γ_τ $(\tau = 0, 1)$ are the induced maps on the mapping cones. For the element $\pi_w \in [C_q, \Omega A]$ from (1.3.12) with $w = w_f$, we have in $[C_q, \Omega A]$ the equation

$$(2) \qquad \beta^*(\pi_w) = \gamma_0{}^*(\pi_w) + \gamma_1{}^*(\pi_w).$$

This follows from the homotopy commutative diagram ② by first setting $f = 1_A$ and then using the naturality of the construction. Because of (0.1.13), equation (2) implies a corresponding equation for the total spaces of the cofibrations. That is, with $\bar{q}' : \overline{P_f{}^I} \to \overline{P_f{}^{\bullet J}}$, we have in $[C_{\bar{q}'}, \Omega A]$ the equation

$$(3) \qquad \bar{\beta}^*(\tilde{\pi}_w) = \bar{\gamma}_0{}^*(\tilde{\pi}_w) + \bar{\gamma}_1{}^*(\tilde{\pi}_w).$$

Thus we have in $[\overline{P_f{}^{\cdot J}}, \Omega A]$, for the cofiber map $j : \overline{P_f{}^{\cdot J}} \to F$, the equations

$$j^* d(u_0, H + H', u_1) = i_{01}{}^*(-\delta\tilde{\pi}_w + u_{02}^*(w)), \qquad \text{see (1.3.18)},$$

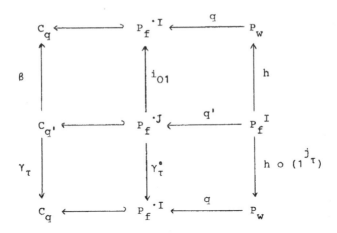

$$= - \delta\bar{\beta}^*\tilde{\pi}_w + u_{012}{}^* i_{01}{}^*(w), \qquad \text{see (1)},$$

$$= -\delta\bar{\gamma}_o{}^*(\tilde{\pi}_w) - \delta\bar{\gamma}_1{}^*(\tilde{\pi}_w) + u_{012}{}^* i_{01}{}^*(w) , \qquad \text{see (3)},$$

$$= -\bar{\gamma}_o^{\cdot}{}^*\delta(\tilde{\pi}_w) - \bar{\gamma}_1^{\cdot *}\delta(\tilde{\pi}_w) + u_{012}^*(w_1 + w_2), \qquad \text{see (2)},$$

$$\overset{(*)}{=} \bar{\gamma}_o^{\cdot *}(-\delta\tilde{\pi}_w + u_{01}{}^*(w)) + \bar{\gamma}_1^{\cdot *}(-\delta\tilde{\pi}_w + u_{12}{}^*(w)),$$

$$= j^* d(u_o, H, u_1) + j^* d(u_1, H', u_2) , \qquad \text{see (1.3.18)}.$$

Equation (*) follows from the definition of u_{012} , with $u_{01} = H \underline{\times} (u_o, u_1)$ and $u_{12} = H' \underline{\times} (u_1, u_2)$. Since j^* is injective, we have proved the proposition. □

Generalizing (1.3.7) and dualizing (1.2.18), we define

$$(1.3.20) \qquad \begin{array}{l} \langle \bar{P}_f, P_f \rangle_v \times [F, \Omega A] \xrightarrow{+} \langle \bar{P}_f, P_f \rangle_v \\ \langle \bar{P}_f, P_f \rangle \times [F, \Omega A] \xrightarrow{+} \langle \bar{P}_f, P_f \rangle \end{array}$$

which are operations of groups if A is an H-space. Let $j : \bar{P}_f \to F$ be the projection onto the cofiber, and let $\bar{u} \in \langle \bar{u} \rangle \in \langle \bar{P}_f, P_f \rangle_u$ and $a \in \alpha \in [F, \Omega A]$. We then obtain, by means of the operation μ from (1.3.7), the map

$$(1.3.21) \qquad u_a : \bar{P}_f \xrightarrow{(\bar{u}, aj)} P_f \times \Omega A \xrightarrow{\mu} P_f .$$

For the inclusion $i : P_f \subset \bar{P}_f$ we then have a (canonical) homotopy

$$H^{\circ} : u_a i \simeq 1_{P_f} \quad \text{over} \quad X, \quad (\text{since} \quad u_a i = \mu'(1, 0)).$$

Furthermore, the diagram

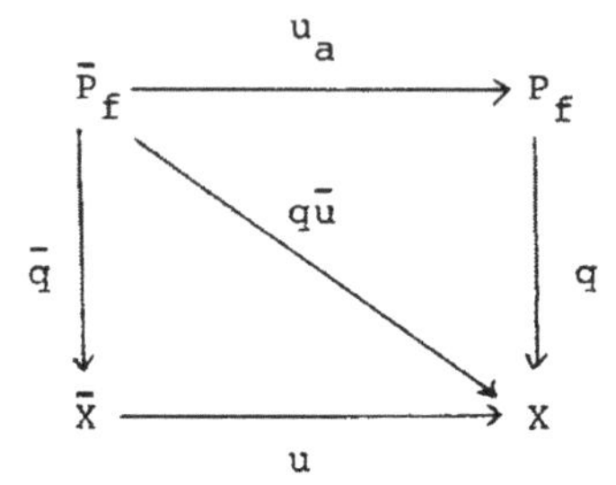

is commutative. Since i is a cofibration, H° has the relative lifting H^1 in

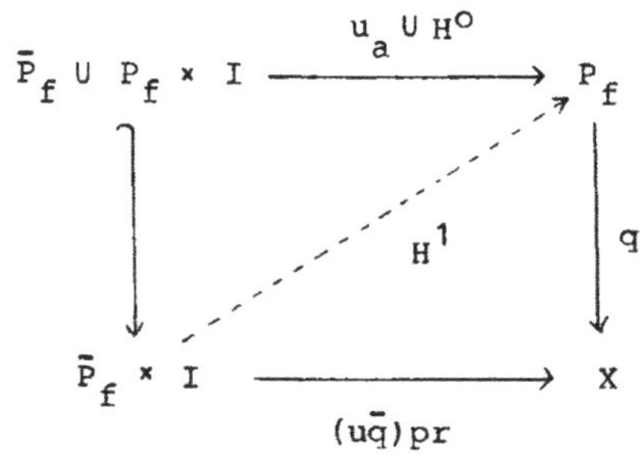

Here $u_a^{\ 1} = H_1^{\ 1} : \bar{P}_f \to P_f$ is a retraction that lifts $u : \bar{X} \to X$.

(1.3.22) <u>Lemma:</u> <u>The retraction homotopy class</u> $\langle u_a^{\ 1} \rangle \in \langle \bar{P}_f, P_f \rangle_u$ <u>depends only on</u> $\langle u \rangle$ <u>and</u> α.

For this reason we can define (1.3.20) by $\langle u \rangle + \alpha = \langle u_a^{\ 1} \rangle$. That this yields a group operation is the statement of the next theorem (1.3.24). With θ as in (1.3.16), we have in $[\bar{P}_f, P_f]$ the equation

$$\Theta(\langle \bar{u} \rangle + \alpha) = (\Theta \ \langle \bar{u} \rangle) + j^{*}\alpha$$

in virtue of (1.3.21). The operation $+$ on the righthand side of the equation is defined as in (1.3.7).

<u>Proof of</u> (1.3.22): Let $a, a' \in \alpha$ and let $A : a \simeq a'$ be a homotopy. Let $\bar{u}, \bar{u}' \in \ \bar{u}\rangle$ and let $H : \bar{u} \equiv \bar{u}'$ be a retraction homotopy over u. Let u_a and $u_a{}'$ be defined from $(\bar{u}, a)$ and $(\bar{u}', a')$ as in (1.3.21), and let the relative liftings $H^1 : u_a \simeq u_a{}^1$ and $H^{1'} : u_a{}' \simeq u_a{}^{1'}$ be defined as in the diagram following (1.3.21). We must show that $u_a{}^1 \equiv u_a{}^{1'}$ are retraction homotopic over u.
Now we have the homotopy

$$\hat{H} : \bar{P}_f \times I \xrightarrow{(H,\ A \circ (j \times 1_I))} P_f \times \Omega A \xrightarrow{\mu'} P_f$$

where $q\hat{H} = (u\bar{q})\,pr$. Therefore the diagram

$$\begin{array}{ccccc}
\bar{P}_f \times I \times 0 \cup \bar{P}_f \times I^{\cdot} \times I & & \xrightarrow{\hat{H} \cup (H^1,\ H^{1'})} & & P_f \\
\big\downarrow & & \phi \ \nearrow & & \big\downarrow q \\
\bar{P}_f \times I \times I & \xrightarrow[pr]{} & \bar{P}_f & \xrightarrow[u\bar{q}]{} & X
\end{array}$$

is commutative. Since $\bar{P}_f \times I^{\cdot} \subset \bar{P}_f \times I$ is a cofibration there exists a relative lifting ϕ , so $\phi|\bar{P}_f \times I \times \{1\} : u_a{}^1 \equiv u_a{}^{1'}$ is a retraction homotopy over u. $\square$

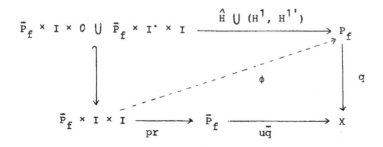

A generalization of (1.3.8), dual to (1.2.21), is the

(1.3.23) Theorem: Let A be an H-space and let $u_o \in \langle \bar{P}_f, P_f \rangle_u$ where $u : \bar{X} \to X$ is a retraction. Then there is a bijection

$$\langle \bar{P}_f, P_f \rangle_u \xrightarrow{\approx} [F, \Omega A]$$

given by $u_1 \mapsto d(u_o, u_1)$ with inverse $\alpha \mapsto u_o + \alpha$.

This theorem implies that the operation (1.3.20) is a group operation.

(1.3.24) Corollary: Let A be an H-space and let $u_o \in \langle \bar{P}_f, P_f \rangle_u$ and $\alpha, \beta \in [F, \Omega A]$. Then in $\langle \bar{P}_f, P_f \rangle_u$

$$u_o + 0 = u_o \quad ,$$

$$(u_o + \alpha) + \beta = u_o + (\alpha + \beta) \quad .$$

This is proved exactly as was (1.2.22). For the maps

$$\langle \bar{P}_f, P_f \rangle_u \xrightarrow{\Theta_u} \langle \bar{P}_f, P_f \rangle \xrightarrow{q} \langle \bar{X}, X \rangle \quad ,$$

where Θ_u is the canonical map and q takes the retraction $\bar{u}$ to the retraction $q \circ \bar{u}$ (see (0.0.4)), we have $\operatorname{Im} \Theta_u = q^{-1}(u)$. Therefore

(1.3.25) Corollary: Let A be an H-space and let $u_o \in \langle \bar{P}_f, P_f \rangle$ with $qu_o = u$. Then $q^{-1}(u) = \{u_o + \alpha \mid \alpha \in [F, \Omega A]\}$.

<u>Proof of</u> (1.3.23): (Dual to the proof of (1.2.21)). We first show that $d(u_o, u_o + \alpha) = \alpha$. Let $w = w_f$. For the retraction

$$\bar{u} = (u \circ pr) \underline{\times} (u_o, u_o) : \overline{P_f{}^{\cdot I}} \to P_f{}^{\cdot I} ,$$

see (1.3.17) , we have $w_{\#}(\bar{u}) = d(u_o, u_o) = 0$. On the other hand, for the retraction

$$\bar{u} + \alpha = (u \circ pr) \underline{\times} (u_o, u_o + \alpha) : \overline{P_f{}^{\cdot I}} \to P_f{}^{\cdot I}$$

we have the equation

$$w_*(\bar{u} + \alpha) = w_*(\bar{u}) + j^*(\alpha)$$

in $[\overline{P_f{}^{\cdot I}}, \Omega A]$. Here the operation + from the right is defined, since $P_f{}^{\cdot I}$ is a mapping path space (in the diagram

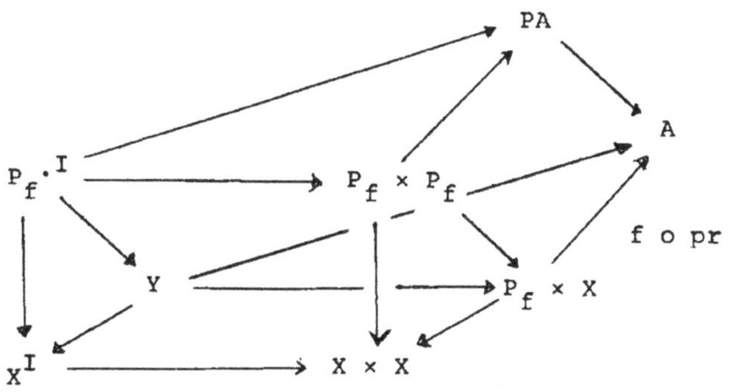

Y is a pull-back and all the squares are cartesian).
In accordance with the following lemma, we then have

$$w_{\#}(\bar{u} + \alpha) = w_{\#}(\bar{u}) + \alpha = \alpha$$

as well. Since $d(u_o, u_o + \alpha) = w_{\#}(\bar{u} + \alpha)$, also $d(u_o, u_o + \alpha) = \alpha$. We now show that the mapping $u_1 \mapsto d(u_o, u_1)$ is injective. Let u_1, u_1' be such that $d(u_o, u_1') = d(u_o, u_1)$. Then by (1.3.19) we have

$$d(u_1, u_1') = d(u_1, u_o) + d(u_o, u_1') = -d(u_o, u_1) + d(u_o, u_1') = 0$$

so it follows from (1.3.18) that $u_1 \equiv u_1'$ over u. $\square$

(1.3.26) <u>Lemma</u>: <u>Let</u> $f : X \to A$ <u>and</u> $g : P_f \to B$, <u>where</u> B <u>is a loop space.</u> <u>Let</u> $P_g \subset \overline{P_g}$ <u>be a cofibration with cofiber</u> $j : \bar{P}_g \to F$. <u>Let</u> $u, u + \alpha \in \langle \bar{P}_f, P_f \rangle$ <u>with</u> $\alpha \in [F, \Omega A]$. <u>If in</u> $[\bar{P}_f, B]$

$$g_*(u + j^*(\alpha)) = g_*(u) + j^*(\beta)$$

<u>for some</u> $\beta \in [F, B]$, <u>then in</u> $[F, B]$

$$g_{\#}(u + \alpha) = g_{\#}(u) + \beta .$$

<u>Proof:</u> By definition

$$g_{\#}(u) = j^{*-1}(-\delta\tilde{\pi}_g + u^*(g))$$

where $u^*(g) = g_*(u)$. Thus we also have

$$\begin{aligned} g_{\#}(u + \alpha) &= j^{*-1}(-\delta\tilde{\pi}_g + (u + \alpha)^*(g)), \\ &= j^{*-1}(-\delta\tilde{\pi}_g + g_*(u + j^*(\alpha))), \\ &= j^{*-1}(-\delta\tilde{\pi}_g + g_*(u) + j^*(\beta)) = g_{\#}(u) + \beta \quad . \end{aligned}$$ $\square$

We will be using the following simple kind of inverse limit.

(1.3.27) Definition: Let $\mathcal{K}$ be a category and let

$$\to Y_n \xrightarrow{p_n} Y_{n-1} \to \cdots \to Y_o$$

be a sequence of morphisms in $\mathcal{K}$. An object Y_∞ is called an inverse limit (written $\varprojlim Y_n$) of the sequence when there are morphisms $q_n : Y_\infty \to Y_n$ with $p_n \circ q_n = q_{n-1}$ and such that the following universal property holds. For each Y together with a system of morphisms $g_n : Y \to Y_n$ such that $g \circ q_n = g_n$ there exists exactly one morphism $g : Y \to Y_\infty$ such that $g \circ q_n = g_n$. Y_∞ is uniquely determined up to isomorphism by this property.

In the categories of sets and of topological spaces, inverse limits always exist. A commutative diagram

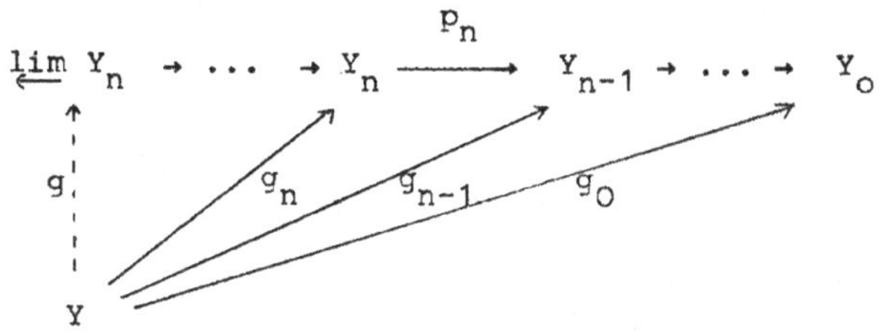

thus defines a map g into the inverse limit. If all the p_n are fibrations (in Top^o), then $\varprojlim Y_n \to Y_o$ is a fibration (in Top^o).

A space X together with a tower of principal fibrations

$$X = \varprojlim X_i \to \dots \to X_i \to X_{i-1} \to \dots \to X_o$$

is called an <u>iterated principal fibration</u> or iterated mapping path space.

Let $f_i : X_{i-1} \to A_i$ be the classifying maps for X and let $H_i : P_{f_i} \simeq X_i$ be homotopy equivalences over X_i. Given a map $u : U \to X_0$, we consider the lifting problem $(i \geq 1)$

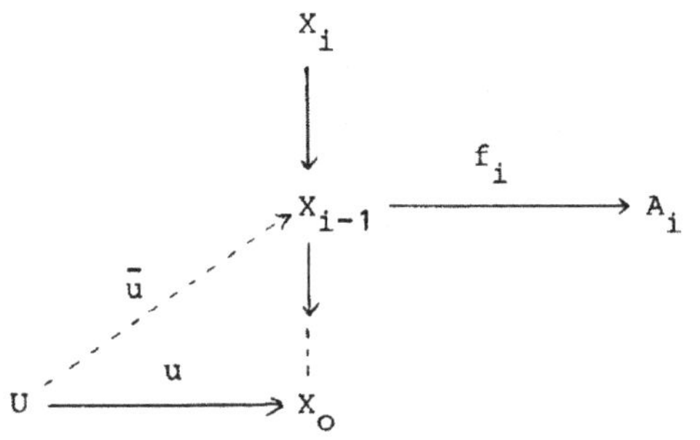

By (1.3.2), the map u can be lifted to X_i exactly when the subset

(1.3.28) $$\mathcal{H}^i(u) = \{f_{i*}(\bar{u}) \mid \bar{u} \text{ lifts } u\}$$

of $[U, A_i]$ contains the trivial element $0 \in \mathcal{H}^i(u)$. We call $\mathcal{H}^i(u)$ the <u>i-th order obstruction</u> to lifting u to X_i.

Let $X \subset \bar{X}$ be a cofibration with cofiber F and let $X_i \subset \bar{X}_i$ be the induced cofibrations. We consider the obstruction situation

$$\begin{array}{ccccc}
\bar{X}_i & \overset{\bar{\bar{u}}}{\dashrightarrow} & X_i & & \\
\downarrow & & \downarrow & & \\
\bar{X}_{i-1} & \underset{\bar{u}}{\overset{\hookleftarrow}{\longrightarrow}} & X_{i-1} & \xrightarrow{\quad f_i \quad} & A_i \\
\downarrow & & \downarrow & & \\
\bar{X}_o & \underset{u}{\overset{\hookleftarrow}{\longrightarrow}} & X_o & &
\end{array}$$

where each A_i is an H-group. The retraction u can be lifted to a retraction $\bar{\bar{u}}$ exactly when the subset

$$\mathcal{H}^i(u) \;=\; \{f_{i\#}(\bar{u}) \mid \bar{u} \text{ lifts } u\}$$

of $[F, A_i]$ contains the trivial element $0 \in \mathcal{H}^i(u)$. As before, we call $\mathcal{H}^i(u)$ an i-th order obstruction. As in (1.2.27), we must make a

<u>Warning:</u> Even if $0 \in \mathcal{H}^i(u)$ for all i, it does not follow that u can be lifted to $X = \varprojlim X_i$.

We can say the following about inverse limits. Let $Y_n \xrightarrow{p_n} Y_{n-1} \to \dots \to Y_o$ be fibrations and $Y = \varprojlim Y_n$. Let $Y \subset \bar{Y}$ be a closed cofibration with cofiber F and let $Y_n \subset \bar{Y}_n$ be the induced cofibrations. Let $p : Y_o \to D$ and let $u : \bar{D} \to D$ be a retraction of the induced cofibration $D \subset \bar{D}$. The maps p_n

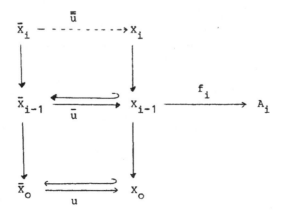

induce maps

$$p_{n*}: \langle \bar{Y}_n, Y_n \rangle_u \longrightarrow \langle \bar{Y}_{n-1}, Y_{n-1} \rangle_u$$

between retraction homotopy sets. We therefore obtain

$$\phi : \langle \bar{Y}, \varprojlim Y_n \rangle_u \longrightarrow \varprojlim \langle \bar{Y}_n, Y_n \rangle_u .$$

(1.3.29) Theorem: ϕ is surjective. If $\varprojlim Y_n \to Y_o$ is an iterated principal fibration with classifying maps $Y_{i-1} \to A_i$ and if there is an $N < \infty$ such that $[F, \Omega A_i] = 0$ for $i > N$, then ϕ is bijective..

Setting $D = *$ and $\bar{Y} = Y \vee U$, we have of course the

(1.3.30) Corollary: If $\varprojlim Y_n \to Y_o$ is an iterated principal fibrations with classifying maps $Y_{i-1} \to A_i$ and if $[U, \Omega A_i] = 0$ for $i > N$, then the canonical map

$$\phi : [U, \varprojlim Y_n] \to \varprojlim [U, Y_n]$$

is bijective.

Remark: The map ϕ is not injective in general. If we take U in (1.3.30) to be the sphere S^n, $n \geqslant 2$, then ϕ is a homomorphism of abelian groups. The kernel of ϕ is $\varprojlim^1 \pi_{n+1}(Y_i)$, see [23] p. 254. It seems probable that the theory for ϕ with maps can be generalized to retractions.

Proof of (1.3.32): Let $\xi_n \in \langle \bar{Y}_n, Y_n \rangle_u$ be a sequence of elements with $p_{n*}\xi_n = \xi_{n-1}$. Such a sequence represents an element of $\varprojlim \langle \bar{Y}_n, Y_n \rangle_u$. It follows from the relative lifting property of the fibrations p_n that the sequence can be represented by elements $u_n \in \xi_n$ such that $p_{n*}u_n = u_{n-1}$. Such a sequence u_n determines an element $\xi \in \langle \bar{Y}, \varprojlim Y_n \rangle^u$ with $\phi(\xi) = \{\xi_n\}$.
Now let $v, w : \bar{Y} \to Y$ be retractions with $\phi(u) = \phi(w)$. Then the induced retractions $v_i, w_i, : \bar{Y}_i \to Y_i$ are retraction homotopic over u. The retraction homotopy $H : v_N \equiv w_N$ can be lifted inductively to a retraction homotopy $v \equiv w$ since $[F, \Omega A_i] = 0$ for $i > N$, by (1.3.18). □

Theorem (1.3.22) has an exact dual for sections and direct limits. Let $X_o \subset \ldots \subset X_n \overset{i_n}{\subset} X_{n+1} \subset \ldots$ be closed cofibrations and let $X = \varinjlim X_i$, see (1.2.25). Let $\tilde{X} \to X$ be a fibration with fiber F and let $\tilde{X}_n \to X_n$ be the induced fibrations. Let $i : D \to X_o$ and let $u : D \to \tilde{D}$ be a section of the induced fibration $\tilde{D} \to D$. The maps i_n induce maps

$$i_n^* : \langle X_{n+1}, \tilde{X}_{n+1} \rangle^u \longrightarrow \langle X_n, \tilde{X}_n \rangle^u$$

of section homotopy sets. We therefore obtain

$$\psi : \langle \varinjlim X_n, \tilde{X} \rangle^u \longrightarrow \varprojlim \langle X_n, \tilde{X}_n \rangle^u.$$

(1.3.34) <u>Theorem</u>: <u>ψ is surjective. If $X_o \subset \varinjlim X_n$ is an iterated principal cofibration with attaching maps $A_i \to X_{i-1}$ and if $[SA_i, F] = 0$ for $i > N$, then ψ is bijective.</u>

(1.4) CW-spaces

We will here discuss CW-decompositions of a closed cofibration $A \subset X$. These decompositions can be used in the inductive procedure for solving the existence and classification problems when condition (A) of (1.1.1) holds. We will use the obstructions and differences defined in (1.2) to prove a relative version of the Whitehead theorem and of the approximation theorem of James-Thomas.

(1.4.1) Definition: A pair (X, A) of spaces, together with a filtration

$$A \subset X_0 \subset X_1 \subset \dots \subset X = \lim_{\rightarrow} X_i$$

is called a relative CW-complex when

(i) X_0 is the disjoint union of A with a discrete set of points, and

(ii) for every $i \geq 0$ there is a (non-pointed) map

$$f_{i+1} : \bigcup_e S_e^i \longrightarrow X_i, \quad e \in Z_{i+1},$$

from a disjoint union of spheres $S_e^i = S^i$ into X_i, such that

$$X_{i+1} = X_i \cup_{f_{i+1}} \bigcup_e E_e^{i+1}$$

is the adjunction space of f_{i+1}, where $E_e^{i+1} = CS_e^i$.

X_i is called the i-skeleton of (X, A) and Z_{i+1} is the set of (i+1)-cells of $X_{i+1} - X_i$. The restriction $f_e = f_{i+1} | S_e^i$ is called the attaching map of the cell e. If $A = \phi$ is empty, X is called a CW-complex. In this case, X^o is a set of points (the 0-cells) with the discrete topology. A CW-space is a space that is homotopy equivalent to a CW-complex. The inclusion $A \subset X$

of a relative CW-complex (X, A) is a closed cofibration. We call a pair (X', A) of spaces a <u>relative CW-space</u> when $A \subset X'$ is a closed cofibration and there exist a relative CW-complex (X, A) and a homotopy equivalence $X \to X'$ under A. If a relative CW-space (X', A) is such that the corresponding CW-complex has only cells of dimension $\leqslant n$, then we write $\dim (X' - A) \leqslant n$. CW-complexes have the following fundamental property, see [116] p. 404.

(1.4.2) <u>Cellular approximation theorem</u>: <u>Let</u> $f : (X, A) \to (X', A')$ <u>be a map between relative CW-complexes. Then there is a map</u> $g : (X, A) \to (X', A')$ <u>with</u> $g(X_n) \subset (X'_n)$ <u>that is homotopic to</u> f <u>relative</u> A.

The map g is called a cellular approximation to f.

Since we have required that the attaching map of a principal cofibration be basepoint preserving, the skeletal filtration of a CW-complex is not an iterated principal cofibration. We therefore define a normalized form of CW-complex.

(1.4.3) <u>Definition</u>: Let (X, A) be a relative CW-complex, $* \in A$. (X, A) is said to be <u>strictly pointed</u> when

(i) $X^o = A$ and $\operatorname{im} f_1 = \{*\}$, and

(ii) all attaching maps f_e of cells in $X - A$ are pointed.

Given a strictly pointed relative CW-complex (X, A), it is clear that $X_1 = A \vee \bigvee_{e \in Z_1} S^1$ and that $X_{i+1} = C_{f_{i+1}}$ is a mapping cone, where $f_{i+1} : \vee S^i_e \to X_i$. Thus $A \subset X$ is an iterated principal cofibration. The following lemma shows that it often suffices to consider only strictly pointed CW-complexes.

(1.4.4) Lemma: Let (X, A) be a relative CW-complex with both X and A path-connected. Then there is a strictly pointed CW-complex $(\bar{X}, A)$ and a homotopy equivalence $\bar{X} \to X$ under A.

Proof: We inductively alter the skeletons X_i of (X, A). Since A is path-connected, there is a homotopy $f_1 \simeq f_1'$ such that $f_1'(f_1^{-1}(A)) = \{*\}$. This gives us a homotopy equivalence $X_1 \simeq X_1' = X_o \cup_{f_1'} \bigcup_e E^1$. Since X is path-connected, X_1'/A is a path-connected 1-dimensional CW-complex. We choose a maximal tree T in X_1'/A. Then we have $X_1' = A \vee X_1'/A \simeq A \vee (X_1'/A)/T = \bar{X}_1$. Now let $h_i : X_i \to \bar{X}_i$ be a homotopy equivalence with $\bar{X}_i$ strictly pointed. The maps $h_i f_e$ for $e \in Z_{i+1}$ are homotopic to pointed maps f_e'. We now define $f_{i+1} = (f_e' | e \in Z_{i+1})$ and $\bar{X}_{i+1} = C_{f_{i+1}}$. □

Thus, if $i \geq 1$, $\bar{X}$ in (1.4.4) has the same set of i-cells as X.

(1.4.5) Theorem (CW-decomposition): Let $A \subset X$ be a closed cofibration with X path-connected. Then there exists a relative CW-complex (X', A) and a map $h' : X' \to X$ under A that is a weak homotopy equivalence, see (1.4.7).

We call such a CW-complex X' together with h' a CW-decomposition or CW-model of (X, A).

(1.4.6) Note: Let $A \subset X$ be a closed cofibration where X is path-connected and $* \in A$. If (X, A) is r-connected with $r \geq 1$, then there exists a CW-decomposition (X', A) of (X, A) that is strictly pointed and is such that $X_r' = A$, that is, $X' - A$ has only cells of dimension $\geq r + 1$.

Proof of (1.4.5) and (1.4.6): For every $\alpha \in \pi_1(X, A)$ we choose a representative $h_\alpha \in \alpha$. Let the map $(h_1, f_1) : (\vee_\alpha E^1, \vee_\alpha S^o) \to (X, A)$

be the union of the f_α. Then h_1 gives us a map $h_1 : X_1' = C_{f_1} \to X$ under A that is 1-connected. We now assume that we have the skeleton X_n' and an n-connected map $h_n : X_n' \to X$ under A, $n \geqslant 1$. Let $Z_{n+1} \subset \pi_{n+1}(Z_{h_n}, X_n')$ be a generating set, and let $r : Z_{h_n} \to X$ be the retraction. From representatives $h_\alpha \in \alpha$ for $\alpha \in Z_{n+1}$ we obtain a map

$$(h_{n+1}, f_{n+1}) : (\underset{\alpha}{\vee} E^{n+1}, \underset{\alpha}{\vee} S^{n+1}) \xrightarrow{h_\alpha} (Z_{h_n}, X_n') \xrightarrow{(r, h_n)} (X, X)$$

This gives us a map $h_{n+1} : X'_{n+1} = C_{f_{n+1}} \to X$ under A that is $(n+1)$-connected. This follows from a diagram chase between the exact sequences of the pairs (X'_{n+1}, X_n') and (Z_{h_n}, X_n').

If the conditions of note (1.4.6) are satisfied, we start the induction with generating elements $\alpha \in \pi_{r+1}(X, A)$ and so arrive at the statement of (1.4.6). □

(1.4.7) <u>(relative) Whitehead theorem: Let (X, A) and (X', A) be relative CW-spaces with X and X' path-connected. Let $f : X \to X'$ be a map under A inducing an isomorphism $f_* : \pi_*(X) \to \pi_*(X')$. Then f is a homotopy equivalence under A.</u>

If we take $A = \{*\}$ to be a single point, we obtain the usual Whitehead theorem. We call a map $f : X \to X'$ a <u>weak homotopy equivalence</u> when X and X' are path-connected and f induces an isomorphism $f_* : \pi_*(X) \to \pi_*(X')$. Thus a weak homotopy equivalence between CW-spaces is also a homotopy equivalence. We now want to prove the Whitehead theorem in a somewhat more general form for iterated relative CW-spaces. Let

$$A = X_0 \subset \ldots \subset X_n \subset \ldots \subset \bar{X} = \varinjlim X_n$$

be a chain of inclusions of relative CW-spaces (X_n, X_{n-1}) where the X_n are path-connected for $n \geq 1$. We call $(\bar{X}, A)$ an <u>iterated relative CW-space</u>. As a matter of fact, CW-spaces are closed under iteration, that is

(1.4.8) <u>Theorem</u>: <u>An iterated relative CW-space</u> $(\bar{X}, A)$ <u>is a relative CW-space</u>.

In proving this theorem we will use

(1.4.9) <u>Lemma (general Whitehead theorem)</u>: <u>Let</u> (X, A), (X', A) <u>be iterated relative CW-spaces</u>. <u>Let</u> $f : X' \to X$ <u>be a map under</u> A <u>inducing isomorphisms</u> $f_* : \pi_*(X') \to \pi_*(X)$. <u>Then</u> f <u>is a homotopy equivalence under</u> A.

By (1.4.8), this lemma is not more general than the Whitehead theorem (1.4.7).

<u>Proof of</u> (1.4.8): $A \subset X$ is a closed cofibration and $\pi_0(X) = \varinjlim \pi_0(X_n) = 0$. By (1.4.3), there exists a relative CW-space (X', A) and a map $h : X' \to X$ under A that is a weak homotopy equivalence. (1.4.9) then tells us that h is a homotopy equivalence under A, and so (X, A) is a relative CW-space.

The proof of the Whitehead theorem, and of the more general (1.4.9), essentially rests on the following lemma, which can be proved easily by induction because the relevant obstructions vanish.

(1.4.10) <u>Section lemma</u>: Let (X, A) be a relative CW-complex with skeletons X_n where X is path-connected. Let $\tilde{X} \to X$ be a fibration with fiber F. Let $\tilde{A} \to A$ be the restricted fibration and $u : A \to \tilde{A}$ a section.

(i) Suppose $X = X_n$. If F is $(n-1)$-connected, u can be extended to a section $X \to \tilde{X}$, and if F is n-connected, then any two section extensions are section homotopic relative A.

(ii) Suppose $X_n = A$ and $\pi_i(F) = 0$ for $i \geq n$. Then u can be extended to a section $X \to \tilde{X}$, and any two section extensions are section homotopic relative A.

<u>Proof</u>: If F is 0-connected, the section $A \to \tilde{A}$ can be extended over X_1. Now replace (X, X_1) by a strictly pointed CW-complex $(\bar{X}, X_1)$ as in (1.4.4), and use (1.1.6). The statement of the lemma then follows from an inductive application of (1.2.12) and (1.2.16). □

<u>Proof of</u> (1.4.9): Consider the following commutative diagram, in which f has been replaced by a fibration q.

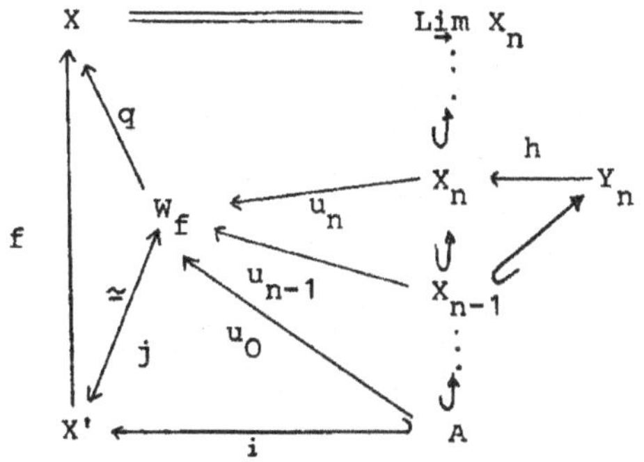

Here (Y_n, X_{n-1}) is a relative CW-complex and h is a homotopy equivalence under X_{n-1}. $u_o = j \circ i$ is a commutative extension of the diagram. We now assume that we have a commutative extension u_{n-1} of the diagram for $n \geq 1$. We construct a commutative extension u_n as follows. We first form the commutative diagram

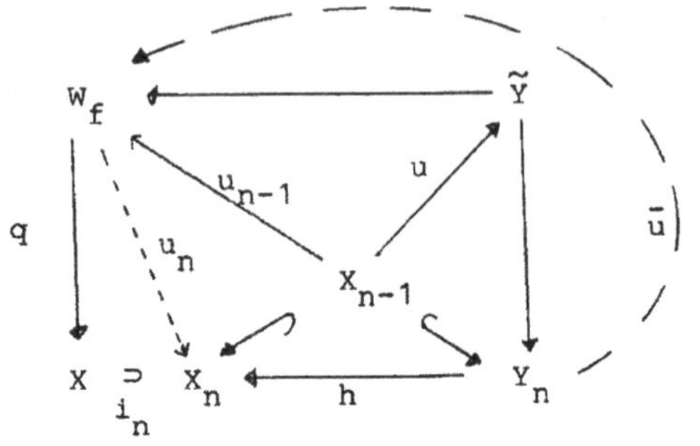

from the above maps. Here $\tilde{Y} \to Y_n$ is the fibration induced by q, that is, the outer rectangle in the diagram is cartesian. The section u is given by u_{n-1}. Denoting the fiber of q by P_f, we have $\pi_i(P_f) = 0$ for all $i \geq 0$. By the previous section lemma, the section u can be extended over Y_n. Such a section extension gives us a map $\bar{u}$ as in the preceding diagram with $\bar{u}|X_{n-1} = u_{n-1}$ and $q\bar{u} = i_n h$. By (1.1.6), $\bar{u}$ then gives us a section u_n extending u_{n-1} such that $u_n h \equiv \bar{u}$ rel X_{n-1}. Taking $u_\infty = \varinjlim u_n$, we obtain a section of q. Let $q_o : W_f \to X'$ be the homotopy inverse to j as in (0.1.10). Since u_∞ extends the section $u_o = ji$, we have a map $\bar{f} = q_o \circ u_\infty : X \to X'$ under A with $f\bar{f} \simeq 1$. This same procedure, applied to $\bar{f}$ instead of f, gives us a map $\bar{\bar{f}}$ with $\bar{f}\,\bar{\bar{f}} \simeq 1$, so $f \simeq f\,\bar{f}\,\bar{\bar{f}} \simeq \bar{\bar{f}}$. Therefore f is a homotopy equivalence, hence a homotopy equivalence under A. □

A map $f : A \to B$ is <u>n-connected</u> if the fiber P_f is $(n-1)$-connected, see (0.2). This is the case exactly when

(1.4.11) $$f_* : \pi_i(A) \to \pi_i(B) \text{ is an } \begin{cases} \text{isomorphism for } i < n \\ \text{epimorphism for } i \leq n. \end{cases}$$

<u>Example</u>: Let (X, A) be a relative CW-complex with skeletons X_n.

Then the inclusion $X_n \subset X$ is n-connected. This follows from the cellular approximation theorem.

The property (1.4.11) of an n-connected map is generalized in the following theorem.

(1.4.12) Theorem: Let $f : A \to B$ be n-connected, $n \leq \infty$. If X is a path-connected CW-complex, then the induced map

$$f_* : [X, A] \to [X, B] \text{ is } \begin{cases} \text{bijective for } \dim X \leq n-1 \\ \text{surjective for } \dim X \leq n \end{cases}.$$

If A and B are path-connected CW-spaces and Y is a space with $\pi_j(Y) = 0$ for $j \geq m_Y$, then the induced map

$$f^* : [B, Y] \to [A, Y] \text{ is } \begin{cases} \text{bijective for } n \geq m_Y \\ \text{injective for } n \geq m_Y - 1. \end{cases}$$

We will prove a generalization of this theorem. To start with, we generalize the first statement of the theorem to apply to sections of fibrations, thus obtaining the following approximation theorem.

(1.4.13) Theorem (Approximation of the total space): In the commutative diagram (in Top^O)

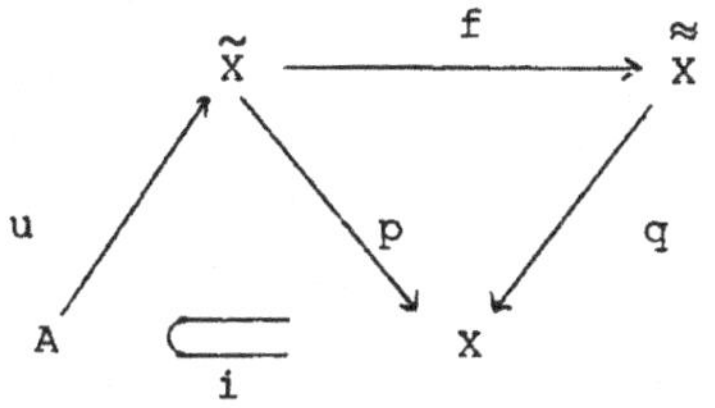

let p *and* q *be fibrations,* u *and* $u' = fu$ *sections. In addition, let* (X, A) *be a relative CW-space with* X *path-connected. If* $\tilde{X} \to \tilde{\tilde{X}}$ *is n-connected,* $n \leq \infty$, *then the induced map*

$$f_* : \langle X, \tilde{X}\rangle^u \longrightarrow \langle X, \tilde{\tilde{X}}\rangle^{u'} \text{ is } \begin{cases} \text{bijective for } \dim (X - A) \leq n - 1 \\ \\ \text{surjective for } \dim (X - A) \leq n . \end{cases}$$

where $f_*(u_o) = f \circ u_o$. *In particular, if* $\dim (X - A) \leq n - 1$, *then* u *can be extended exactly when* u' *can be extended.*

If we take $\tilde{X}$ and $\tilde{\tilde{X}}$ to be trivial fibrations in this approximation theorem and set $A = \{*\}$, we obtain the first statement of (1.4.12). If f is a homotopy equivalence, it follows from (1.1.8) that f_* is bijective.

The approximation theorem is particularly useful in connection with Postnikov decompositions of a fibration. When we take a Postnikov decomposition of a fibration $\tilde{X} \to X$ (see (1.5)), we have fibrations $\tilde{X}_{n-1} \to X$ and n-connected maps $h_{n-1} : \tilde{X} \to \tilde{X}_{n-1}$ over X. Thus if $\dim (X - A) < n$, h_{n-1} induces a bijection $h_{n-1*} : \langle X, \tilde{X}\rangle^u \to \langle X, \tilde{X}_{n-1}\rangle^{u'}$ where $u' = h_{n-1} u$. Because of this bijection, weeneed only (n-1) Postnikov sections in order to solve the existence and classification problems of (1.1.3).

In the preceding theorem, we approximated the total space of a fibration. If we approximate the base space, we get the

(1.4.14) *Theorem* (Approximation of the base space): *In the commutative diagram* (*in* Top^o)

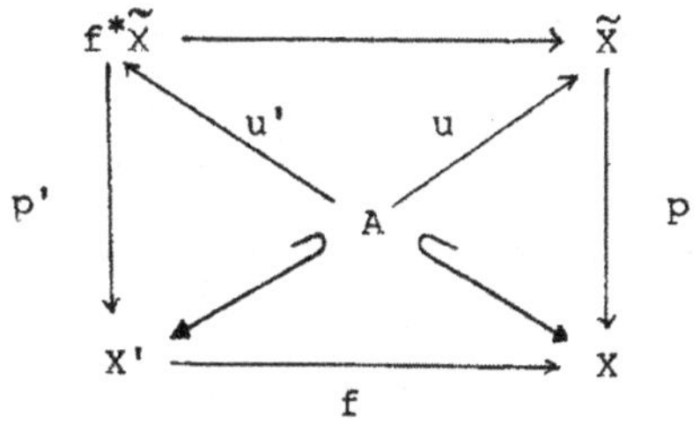

<u>let</u> p <u>be a fibration with fiber</u> F <u>and</u> p' <u>the induced fibration</u>. <u>Let</u> (X, A) <u>and</u> (X', A) <u>be relative CW-spaces with</u> X <u>and</u> X' <u>path-connected</u>. <u>Let</u> $\pi_i(F) = 0$ <u>for</u> $i \geq m_F$. <u>If</u> f <u>is an</u> n-<u>connected map</u>, $n \leq \infty$, <u>then the induced map</u>

$$f^* : \langle X, \tilde{X}\rangle^u \longrightarrow \langle X', f^*\tilde{X}\rangle^{u'} \quad \underline{\text{is}} \quad \begin{cases} \underline{\text{bijective for}} & n \geq m_F \\ \underline{\text{injective for}} & n \geq m_F - 1. \end{cases}$$

If we take $\tilde{X}$ to be the trivial fibration in this theorem and set $A = \{*\}$, we obtain the second statement of (1.4.12). If $n = \infty$, then f is a homotopy equivalence under A by the Whitehead theorem, so (1.4.14) follows from (1.1.6). The proof of the approximation theorems essentially rests on the section lemma (1.4.10).

<u>Proof of the approximation theorems</u> (1.4.13) <u>and</u> (1.4.14): We first prove (1.4.13). By (1.1.6) we can assume that (X, A) is a relative CW-complex. We begin by proving (1.4.13) for the case that f is a fibration.

Let $\dim (X - A) \leq n$ and, in the diagram

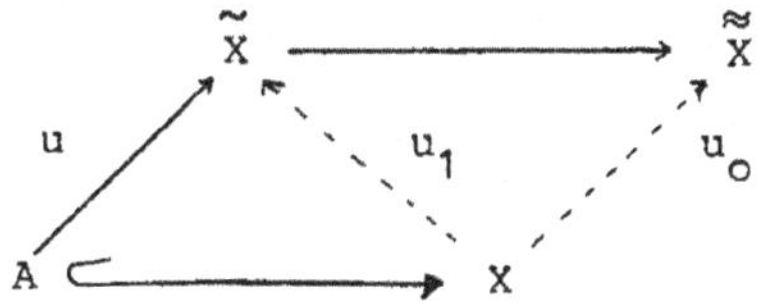

let u_o be a section of q extending $u' = fu$. Then we obtain the diagram

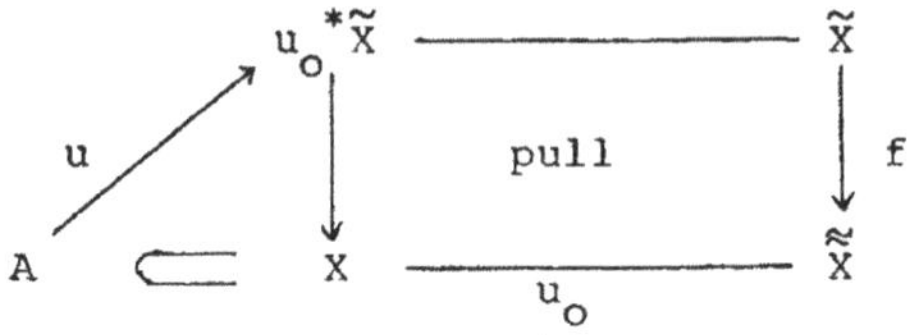

where by (i) the section u can be extended over X. This extension gives us a lifting u_1 of u_o that extends u and is a section of p. Therefore f is surjective. Now let $\dim(X - A) < n$. We consider the diagram

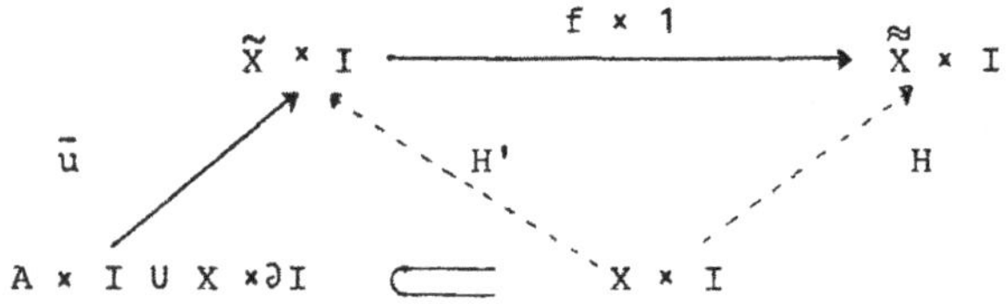

Let $u_o, u_1 : X \to \tilde{X}$ be section extensions of u and let $H : f \circ u_o \simeq f \circ u$ be a section homotopy relative $u' = f \circ u$. Then H gives us a section H of $q \times 1$ extending $(f \times 1)\,\bar{u}$. Since $\dim(X \times I - (A \times I \cup X \times \partial I)) \leqslant n$, we conclude for the same reasons as above that H can be lifted to a section H' which extends $\bar{u}$. H' is therefore a section homotopy $u_o \simeq u_1$ relative $\bar{u}$, so f_* is injective. If f in (1.4.13) is not a fibration, we first replace it by a fibration $\bar{f}$, as described in (0.1.10). We then have the commutative diagram in Top^o

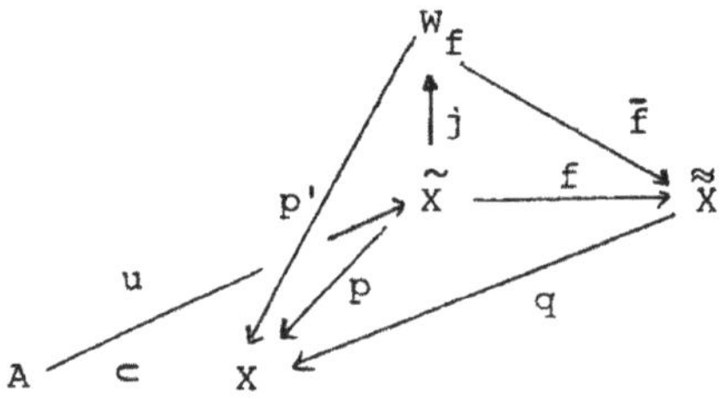

where $p' = q\bar{f}$ is a fibration. Since j is a homotopy equivalence, (1.1.6) tells us that $j_* : \langle X, \tilde{X}\rangle^{u} \approx \langle X, W_f\rangle^{ju}$ is a bijection. Since we know that the theorem is true for the fibration $\bar{f}$, it is also true for f. This proves (1.4.13).

The proof of (1.4.14) is quite analogous. The map f gives us a diagram

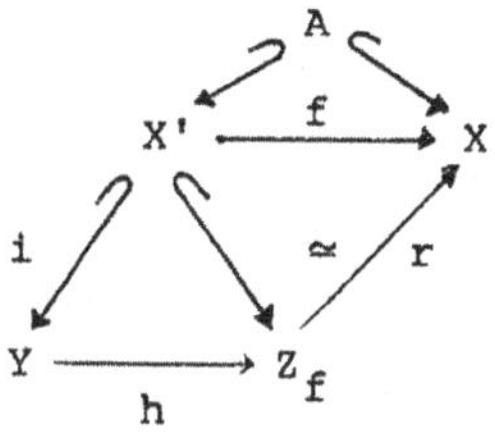

where r is the retraction of the mapping cylinder, and h is a CW-decomposition of (Z_f, X') as in (1.4.6). By the Whitehead theorem, rh is a homotopy equivalence under A. By (1.1.6) we have a bijection $(rh)^* : \langle X, \tilde{X}\rangle^{u} \approx \langle Y, \tilde{Y}\rangle^{w}$ where $\tilde{Y} = (rh)^* \tilde{X}$ and $v = (rh)^* u$. If $n \geq m_F$ then (ii) of the section lemma tells us that $i^* : \langle Y, \tilde{Y}\rangle^{v} \longrightarrow \langle X', f^*\tilde{X}\rangle^{u'}$ is surjective, since $Y_n = X'$. Because $Y \times I - (Y \times \dot{I} \cup X' \times I)$ has only cells of dimension $\geq n + 2$, it follows from (ii) that i^* is also injective for $n + 1 \geq m_F$. It follows that (1.4.14) is true, since $f^* = i^*(rh)^*$.

□

A consequence of (1.4.13) is

(1.4.15) <u>Theorem</u> (Uniqueness of CW-models): <u>Let</u> $A \subset X$ <u>be a closed cofibration with</u> X <u>path-connected.</u> <u>In the commutative diagram</u>

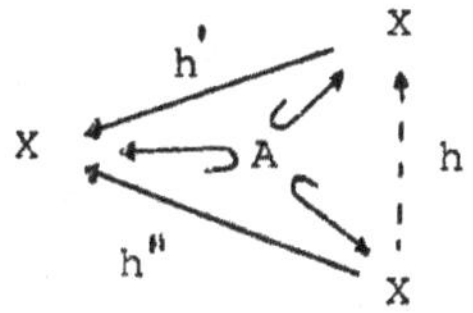

<u>let</u> X', h' <u>and</u> X", h" <u>be CW-decompositions of</u> (X, A). <u>Then there exists a homotopy equivalence</u> h <u>under</u> A <u>such that</u> $h'h \simeq h''$ <u>under</u> A, <u>and</u> h <u>is uniquely determined, up to homotopy under</u> A, <u>by</u> h' <u>and</u> h".

<u>Proof:</u> Taking the relevant maps in (1.4.13) to be trivial fibrations, we see that h' induces a bijection $h_*' : [X'', X']^A \approx [X'', X]^A$. Now let $h = (h_*')^{-1}(h'')$. By the Whitehead theorem, h is a homotopy equivalence under A. ☐

<u>Note:</u> CW-models can be used to define a functor between homotopy categories. Let Top^A be the category of spaces under A and let $Cofiber^A$ be the full subcategory of Top^A whose objects are closed cofibrations $A \subset X$. Let CW^A be the full subcategory of $Cofiber^A$ whose objects are relative CW-spaces (X, A), and let $I : CW^A \to Cofiber^A$ be the inclusion functor. Choosing CW-models gives us a functor

$$\mathcal{S} : Cofiber^A/\simeq \;\to\; CW^A/\simeq$$

between the corresponding homotopy categories as well, see p.11 of [24]. That is, for each object $A \subset X$ we choose a CW-model $\sigma_X : \mathcal{S}X \to X$ under A. By (1.4.13), there is then exactly one map $\mathcal{S} : [X, Y]^A \longrightarrow [\mathcal{S}X, \mathcal{S}Y]^A$ with $\sigma_Y \circ \mathcal{S}(\alpha) = \alpha \circ \sigma_X$ in $Top^A/\simeq$. $\mathcal{S}$ is therefore a functor, and the composition $\mathcal{S} \circ I$ is naturally equivalent to the identity, see 16.21 on p.139 of [38].

Singular homology and cohomology have the following characteristic property with regard to CW-decompositions.

(1.4.16) <u>Theorem:</u> <u>Let</u> $h : X' \to X$ <u>be a CW-decomposition of a path-connected space</u> X, i.e. X' <u>is a CW-complex and</u> h <u>is a weak</u>

homotopy equivalence. Then h induces isomorphisms $H_n(X'; \pi) \cong H_n(X; \pi)$, $H^n(X'; \pi) \cong H^n(X; \pi)$.

This theorem can be proved using the singular polytope of X.

(1.4.17) Corollary: Let $f : X \to Y$ be a weak homotopy equivalence. Then f induces isomorphisms $H_n(X; \pi) \cong H_n(Y, \pi)$ and $H^n(Y; \pi) \cong H^n(X; \pi)$.

Proof: f determines a map $f' : X' \to Y'$ between the CW-decompositions of X and Y. By the Whitehead theorem f' is a homotopy equivalence, and so (1.4.17) follows from (1.4.16). ☐

In the next section we will see that the implication in (1.4.17) can be reversed for Postnikov spaces.

We now look once more at the map γ of (0.5.5), in connection with (1.4.16). For a CW-decomposition $h : X' \to X$, the diagram

(1.4.18)
$$\begin{array}{ccc} [X, K(\pi, m)] & \xrightarrow{\gamma} & H^m(X, \pi) \\ {\scriptstyle h^*}\downarrow & & \downarrow{\scriptstyle h^*}\ \cong \\ [X', K(\pi, m)] & \xrightarrow[\cong]{\gamma} & H^m(X', \pi) \end{array}$$

is commutative. That is, γ is equivalent to h^* up to isomorphisms. The singular cohomology groups can thus be defined alternatively as $H^m(X; \pi) = [X', K(\pi, m)]$ where X' is a CW-decomposition of X (see definition 21.1 in [38]). Because of the uniqueness of CW-decompositions, as expressed in (1.4.15), this definition is independent of the choice of X', h'.

In the next section we will use

(1.4.19) <u>Theorem</u> (CW-model of a fibration) : <u>Let</u> $p : Y \longrightarrow B$ <u>be a fibration with</u> Y <u>path-connected. Then there is a fibration</u> $p' : Y' \to B$ <u>and a map</u> $h : Y' \to Y$ <u>over</u> B, <u>where</u> Y' <u>is a CW-space and</u> h <u>is a weak homotopy equivalence</u>.

<u>Proof:</u> We first take a CW-decomposition $\pi : Y" \to Y$ of Y as in (1.4.5). We then have the commutative diagram

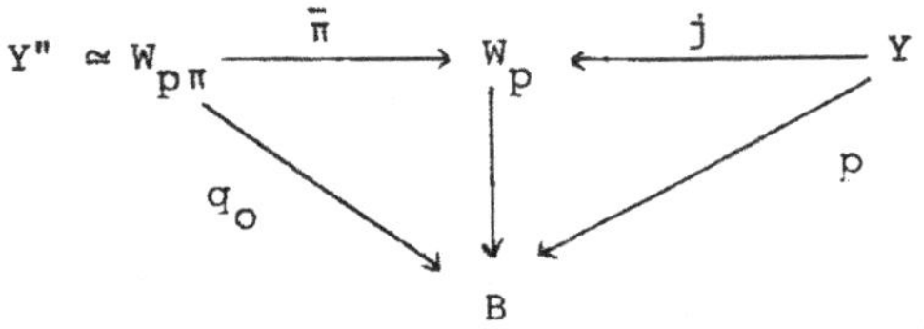

where $\bar{\pi}$ is induced by π. W_p and $W_{p\pi}$ are the path fibrations of (0.1.10). Since j is a homotopy equivalence and therefore a homotopy equivalence over B, there is a $\bar{j}$ inverse to j. We then set $Y' = W_{p\pi}$, $p' = q_o$ and $h = \bar{j} \circ \bar{\pi}$. ☐

The CW-model of a fibration is unique up to homotopy equivalence, that is

(1.4.20) <u>Lemma:</u> <u>Let</u> h' <u>and</u> $h"$ <u>be CW-models of the fibration</u> p <u>in the diagram</u>

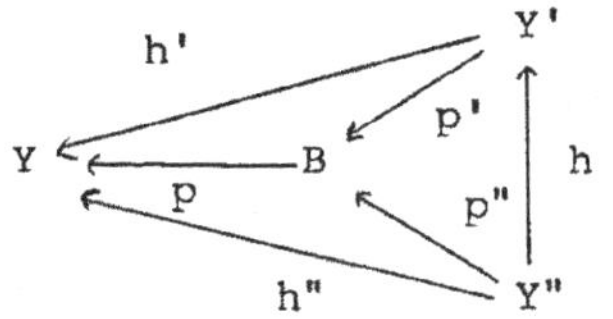

<u>Then there is a homotopy equivalence</u> h <u>over</u> B <u>such that</u> $h'h \simeq h"$ <u>over</u> B, <u>and</u> h <u>is uniquely determined up to homotopy over</u> B.

<u>Proof:</u> h' induces a map $h_1' : p"^*Y' \to p"^*Y$, and $h"$ determines a section $h" : Y" \to p"^*Y$. By (1.4.13), h_1' induces a bijection $h_1'{}_* : \langle Y", p"^*Y'\rangle \approx \langle Y", p"^*Y\rangle$. We now define h to be

$(h_1{}'_*)^{-1}(h")$. □

(1.4.21) Corollary: Let $p : Y \to B$ be a fibration with fiber F. If F and B are CW-spaces, so is Y.

Proof: Let (Y', p', h') be a CW-decomposition of p with fiber F'. Then, by Milnor [86] , F' is a CW-space. Therefore $F' \to F$ is a homotopy equivalence by the Whitehead theorem. Since a CW-space has a countable (numerable) null-homotopic cover, (9.3) of [24] ensures that h' is a homotopy equivalence too. □

This corollary was proved in a different way by J. Stasheff in [117] .

(1.5) Postnikov spaces

We now investigate the concept of Postnikov spaces, which are dual to CW-spaces. Whereas CW-spaces are built up inductively from principal cofibrations of maps $S^n \to X$, Postnikov spaces are constructed inductively from principal fibrations of maps $Y \to K(\pi, n)$. The role of the sphere S^n is taken over in the dual concept by the Eilenberg-MacLane space $K(\pi, n)$, accordingly the role of homotopy groups for CW-spaces is played by cohomology groups for Postnikov spaces. The Whitehead theorem and the approximation theorem of James-Thomas will be seen to have easily formulable duals for Postnikov spaces. Nilpotent and complete spaces are examples of Postnikov spaces, see (1.5.11). We thus immediately recover Dror's generalization of the Whitehead theorem. Finally, we define the Postnikov model of a fibration, an analogue of the CW-model of a cofibration. Localizations of a space are important examples of Postnikov models.

(1.5.1) Definition: Let R be a commutative ring with 1.

(i) Let π be an R-module and let $K(\pi, n)$, $n > 1$, be the Eilenberg-MacLane space for π. Let B be a CW-space. We call a principal fibration $Y \to B$ with classifying map $B \to K(\pi, n)$ an R-Postnikov space over B (of order 0)

(ii) We call a fibration $Y \to B$ an R-Postnikov space over B (of order n, $n \geq 1$). when Y and B are path-connected CW-spaces and there exists a tower

$$\varprojlim Y^i \to \ldots \to Y^i \to Y^{i-1} \to \ldots \to Y^o = B$$

of R-Postnikov spaces Y^i over Y^{i-1} (of order $n - 1$) and a map $h : Y \to \varprojlim Y^i$ over B that is a weak homotopy equivalence.

R-Postnikov spaces have properties which are, in a sense, dual to those of CW-spaces discussed in (1.4). The author does not know if R-Postnikov spaces of order 1 are closed with respect to iteration. This is why the above iterative definition has been made, corresponding to property (1.4.8) of CW-spaces.

(1.5.2) Definition: Let $p : Y \to B$ be a fibration between CW-spaces Y and B. Let $Y' \to B$ be an R-Postnikov space over B and let $h : Y \to Y'$ be a map over B that induces isomorphisms h^* : $H^*(Y', \pi) \cong H^*(Y, \pi)$ for all R-modules π. We call the R-Postnikov space Y' over B together with the map h *an R-Postnikov model of* $Y \to B$.

In (1.4.5) we obtained CW-models of a closed cofibration fairly easily. It is much more complicated to construct an R-Postnikov model of a fibration. We will give examples where this is possible. If an R-Postnikov model of a fibration exists, it is uniquely determined up to equivalence by (1.5.8). We will prove this using the following dual Whitehead theorem.

(1.5.3) Theorem (dual Whitehead theorem): *Let* Y *and* Y' *be R-Postnikov spaces over* B. *Let* $f : Y \to Y'$ *be a map over* B *that induces isomorphisms* $f^* : H^*(Y'; \pi) \to H^*(Y, \pi)$ *for all R-modules* π. *Then* f *is a homotopy equivalence over* B.

The proof of this theorem consists essentially of the following retraction lemma, which is dual to the section lemma (1.4.10). The retraction lemma is easily seen to hold because the relevant obstructions vanish.

(1.5.4) Retraction lemma: Let $Y = \varprojlim Y_{(i)} \to B$ be an iterated principal fibration, thus every $Y_{(n)} \to Y_{(n-1)}$ has a classifying map $Y_{(n-1)} \to K(\pi_n, d_n + 1)$. Let $Y \subset \bar{Y}$ be a closed cofibration with cofiber $F = \bar{Y}/Y$ and let $B \subset \bar{B}$ be the induced cofibration. Let $u : \bar{B} \to B$ be a retraction.

(i) Let $Y = Y_{(n)}$, $n < \infty$. If $[F, K(\pi_i, d_i + 1)] = 0$ for $i \leq n$, then u can be lifted to a retraction $\bar{Y} \to Y$. If in addition $[F, K(\pi_i, d_i)] = 0$ for $i \leq n$, then any two such liftings are retraction homotopic over u.

(ii) Let $B = Y_{(n)}$. If $[F, K(\pi_i, d_i + 1)] = 0$ for $i > n$, then u can be lifted to a retraction $\bar{Y} \to Y$. If in addition $[F, K(\pi_i, d_i)] = 0$ for $i > n$, then any two such liftings are retraction homotopic over u.

This lemma follows from a simple inductive application of the properties of obstructions and differences described in (1.3.13) and (1.3.18), using the homotopy equivalence (0.5.6).

(1.5.5) Remark: It is in general difficult to calculate the homotopy set $[F, K(\pi, d)]$ of a topological space F. If F is a CW-space, the problem is one in singular cohomology, since by (0.5.5) we then have $[F, K(\pi, d)] = \tilde{H}^d(F; \pi)$. This is why we have required Postnikov spaces to be CW-spaces. If in definition (1.5.1) of R-Postnikov space we omit the condition that Y and B be CW-spaces, and instead require that h in (ii) be a homotopy equivalence, we call $Y \to B$ an R-Postnikov tower over B. An obstruction theory for Postnikov towers can be built up completely independently of CW-spaces, but it is necessary to replace singular cohomology by homotopy sets $[F, K(\pi, d)]$. The dual Whitehead theorem then has the following form.

Theorem: Let Y and Y' be R-Postnikov towers over B. Let

$f : Y \to Y'$ be a map over B that induces isomorphisms

$$f^* : [Y', K(\pi, n)] \to [Y, K(\pi, n)]$$

for all $n \geqslant 0$ and all R-modules π. Then f is a homotopy equivalence over B.

This theorem has a proof strictly dual to that of the general Whitehead theorem (1.4.9) and which uses the retraction lemma instead of the section lemma. We leave it to the reader to dualize the other theorems of (1.4) for Postnikov towers, where this is possible.

Proof of (1.5.3): We will prove the statement for Postnikov spaces of order 2. Consider the commutative diagram

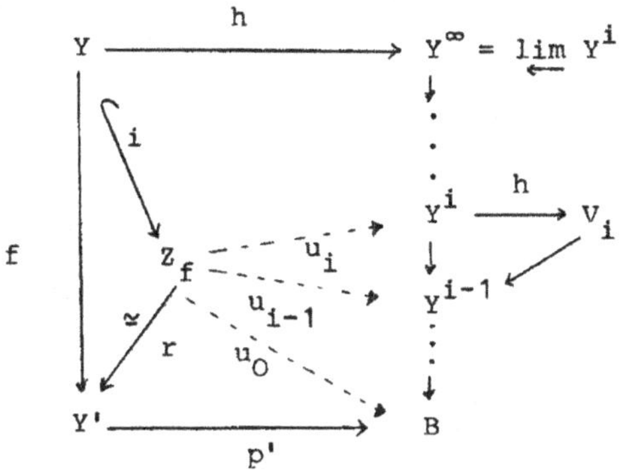

where f is replaced by a cofibration i. The V_i are R-Postnikov spaces of order 1 over Y^{i-1}, and the maps h are weak homotopy equivalences. These stipulations amount to requiring that $Y \to B$ be a Postnikov tower of order 2. $u_o = p'r$ is a commutative extension of the diagram. We now assume that we have a commutative extension u_{i-1}, $i \geqslant 1$. We construct a commutative extension u_i as follows. Using the same maps as in the preceeding diagram, we form the diagram

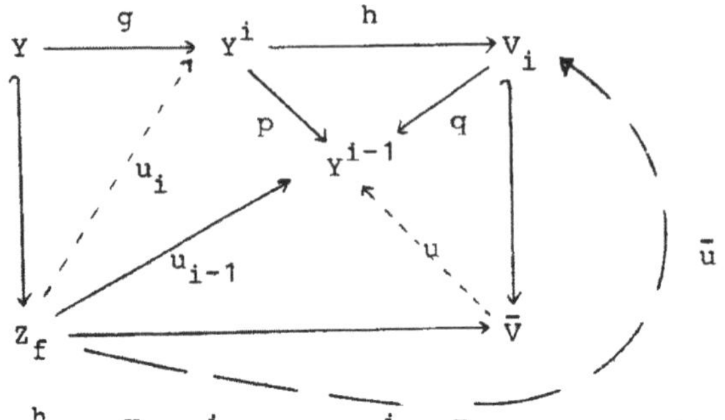

where $g : Y \xrightarrow{h} Y^{\infty} \to Y^i$, and $V^i \subset \bar{V}$ is the cofibration induced by $Y \subset Z_f$. The retraction u is defined by u_{i-1} and q. Since Y and Y' are CW-spaces, so is the cofiber $C_f = Z_f/Y = \bar{V}/V_i$. f induces isomorphisms in cohomology, so $[C_f, K(\pi, i)] = \tilde{H}^i(C_f, \pi) = 0$ for all R-modules π, $i \geqslant 0$. By (ii) of the retraction lemma, u can be lifted to a retraction $\bar{V} \to V_i$. To this retraction corresponds a map $\bar{u}$ as in the diagram with $q\bar{u} = u_{i-1}$ and $hg = \bar{u}i$. By (1.4.13), $\bar{u}$ gives us a map u_i over Y^{i-1} with $g = u_i \circ i$ and $hu_i \simeq \bar{u}$ over Y^{i-1}. In applying (1.4.13), we look at section extensions of g and hg in the pulled-back fibrations $u_{i-1}^{*}(p)$ and $u_{i-1}^{*}(q)$ in order to obtain u_i. In the diagram

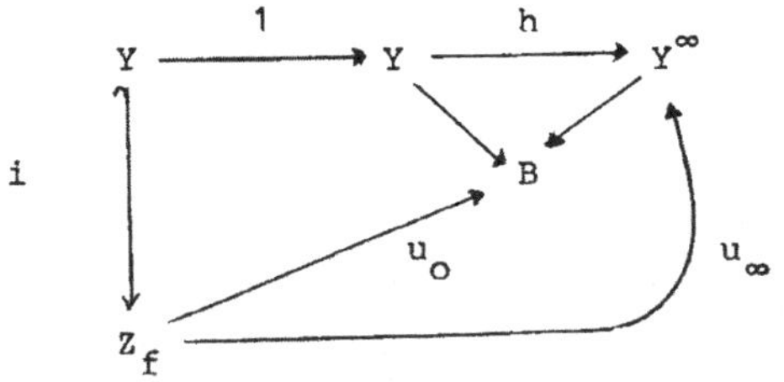

let $u_{\infty} = \varprojlim u_i$. Applying (1.4.13) exactly as above, we obtain from u_{∞} a map $\bar{f}$ over B with $1_Y = \bar{f}i$ and $h\bar{f} \simeq u_{\infty}$ over B. The inclusion $Y' \subset Z_f$ is such that $g = \bar{f}|Y' : Y' \to Y$ is a map over B with $gf \simeq 1_Y$. Thus g also induces isomorphisms in homology. Since Y' is a Postnikov space over B, we can apply the above procedure to g instead of f. We thus obtain a retraction $\bar{g} : Z_g \to Y'$ and we define $k = \bar{g}|Y'$, so now $kg \simeq 1_{Y'}$. Therefore $k \simeq kgf \simeq f$, and so f is a homotopy equivalence and in fact a homotopy equivalence over B. This completes the proof of (1.5.3) for Postnikov spaces of order 2. The full statement for Postnikov spaces of higher order can be proved quite analogously. □

We now dualize the approximation theorem of James-Thomas.

(1.5.6) *Approximation theorem for retractions: In the commutative diagram of CW-spaces*

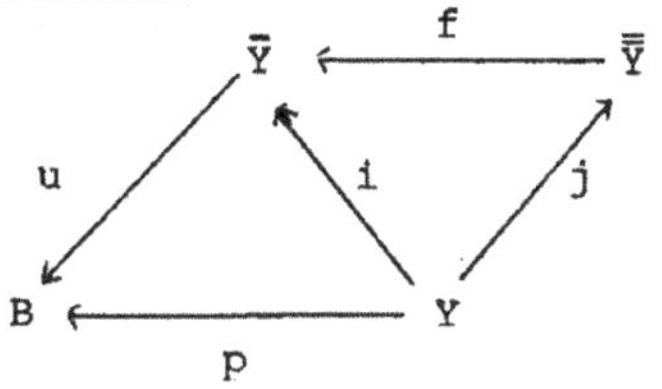

let i *and* j *be closed cofibrations and let* u *and* $u' = u \circ f$ *be retractions, that is,* $u'j = p = ui$. *Further let* $p : Y \to B$ *be an R-Postnikov space over* B. *If* f *induces isomorphisms* $f^* : H^*(\bar{Y}, \pi) \cong H(\bar{\bar{Y}}, \pi)$ *for all R-modules* π, *then* $f^* : \langle \bar{Y}, Y \rangle_u \to \langle \bar{\bar{Y}}, Y \rangle_{u'}$ *is a bijection of retraction homotopy sets, where* $f^*(u_o) = u_o \circ f$.

This theorem dualizes only the case $n = \infty$ of (1.4.13). For Postnikov spaces $Y \to B$ of order 1, the case $n < \infty$ (cf. (1.4.13)) can be dualized with the aid of the retraction lemma. We leave the details to the reader. The proof of the approximation theorem proceeds dually to that of (1.4.13), using an induction as in the proof of the dual Whitehead theorem. If we assume the relevant maps in (1.5.6) to be trivial cofibrations, we get the

(1.5.7) *Corollary: Let* $p : Y \to B$ *be an R-Postnikov space over* B.

Let $f : U' \to U$ *be a map between CW-spaces that induces isomorphisms* $f^* : H^*(U', \pi) \to H^*(U, \pi)$ *for every R-module* π. *Then* $f^* : [U, Y]_u \to [U', Y]_{uf}$ *is a bijection where* $u : U \to B$.

A consequence of this, dual to (1.4.15), is the

(1.5.8) *Theorem* (Uniqueness of R-Postnikov models): *Let* $Y \to B$ *be a fibration of CW-spaces. In the commutative diagram*

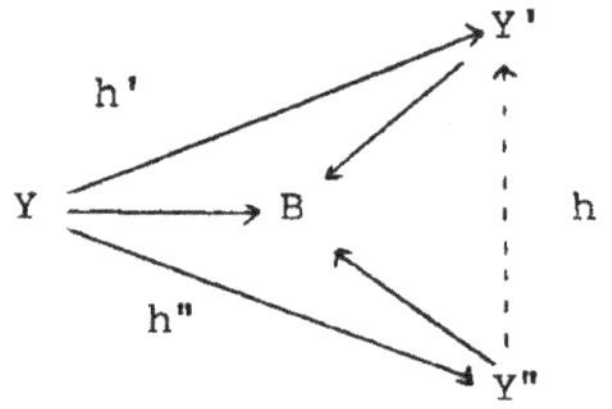

let Y' *and* Y'' *be R-Postnikov models of* p. *Then there is a homotopy equivalence* h *over* B *such that* $hh' \simeq h''$ *over* B *and* h *is uniquely determined up to homotopy over* B *by* h' *and* h''.

Proof: By the above corollary, h' induces a bijection $h'^* : [Y', Y'']_B \longrightarrow [Y, Y'']_B$. With $h = (h'^*)^{-1}(h'')$ the theorem follows from the dual Whitehead theorem. ☐

Note: Just as with CW-spaces, we can define a functor between the relevant homotopy categories by means of Postnikov models. Let Top_B be the category of spaces over B, and $R\text{-}Fibr_B$ be the full subcategory whose objects are fibrations $Y \to B$ possessing an R-Postnikov model. Further let $R\text{-}Postn_B$ be the full subcategory of R-Postnikov spaces over B and let $I : R\text{-}Postn_B \to R\text{-}Fibr_B$ be the inclusion functor. Choosing Postnikov models gives us a functor

$$R : R - Fibr_B / \simeq \longrightarrow R - Postn_B / \simeq$$

between homotopy categories, such that $R \circ I$ is naturally equivalent to the identity. This functor R is defined using (1.5.7) quite similarly to the way the functor S for CW-spaces was defined. That is, given $Y \to B$, we choose an R-Postnikov model r_Y : $Y \to RY$ over B. There is then exactly one map $R : [X, Y]_B \to [RX, RY]_B$ such that $R(\alpha) \circ r_X = r_Y \circ \alpha$ in $Top_B/\simeq$. For a subring $R \subset \mathbb{Q}$, we will characterize localization (1.5.14) as an instance of the functor R.

We now give various examples of R-Postnikov spaces and R-Postnikov models. For this we need the following definitions.

(1.5.9) <u>Definition</u>: Let $p : Y \to B$ be a fibration with fiber F. We have an isomorphism $\pi_n(F) \cong \pi_{n+1}(Z_p, Y)$ where Z_p is the mapping cylinder of p. By (0.4), this isomorphism gives us a group operation $\varphi_n : \pi_n(F) \times \pi_1(Y) \to \pi_n(F)$, $n \geq 1$. and we also write $\varphi_n(\xi, \alpha) = \xi^\alpha$. Then $i_*(\xi^\alpha) = (i_*\xi)^\alpha$ where $i : F \subset Y$ is the inclusion.

A more detailed treatment of this group operation is to be found in the appendix to (4.1).

(1.5.10) <u>Definition</u>: Let π and G be groups (where the multiplication in G is written additively), and let $\phi : G \times \pi \to G$, $(g, \alpha) \mapsto g \cdot \alpha$ be a group operation of π on G. If H is a normal subgroup of G invariant with respect to ϕ ($\phi(H \times \pi) \subset H$) we define

$$\Gamma_\phi H = \text{the normal subgroup of } G \text{ generated by}$$
$$\{-h - g + h + g \mid h \in H,\ g \in G\} \subset H$$
$$\text{and } \{-h + h \cdot \alpha \mid h \in H,\ \alpha \in \pi\} \subset H$$

$\Gamma_\phi H$ is normal and is also an invariant subgroup of G. Furthermore, $\Gamma_\phi H \subset H$ and the quotient $H/\Gamma_\phi H$ is an abelian group on which the operation of π, induced by ϕ, is trivial. We now define inductively $\Gamma_\phi^0 G = G$, $\Gamma_\phi^k G = \Gamma_\phi \Gamma_\phi^{k-1} G$, $k \geq 1$. We then have inclusions $\Gamma_\phi^{k+1} G \subset \Gamma_\phi^k G$, and therefore a tower of projections

$$0 = G/\Gamma_\phi^0 G \leftarrow \dots \leftarrow G/\Gamma^k G \leftarrow G/\Gamma_\phi^{k+1} G \leftarrow \dots$$

We say that the operation ϕ is <u>complete</u> when the canonical map $G \to \varprojlim G/\Gamma_\phi^k G$ is an isomorphism. We say ϕ is <u>nilpotent of order</u> $\leq k$, written $\mathrm{nil}(\phi) \leq k$, when $\Gamma_\phi^k G = 0$. If ϕ is complete but not nilpotent, we write $\mathrm{nil}(\phi) = \infty$. These concepts have their standard meaning for a group G if we take the operation of π to be trivial.

These group-theoretical notions allow us to make the

(1.5.11) <u>Definition</u>: Let $Y \to B$ be a fibration with fiber F, where F and Y are path-connected. We call the fibration $Y \to B$ <u>complete</u> when for $n \geq 1$ the operation ϕ_n of $\pi_1(Y)$ on $\pi_n(F)$ is such that $\mathrm{nil}(\phi_n) \leq \infty$. We call the fibration $Y \to B$ <u>nilpotent</u> when $\mathrm{nil}(\phi_n) < \infty$ for all $n \geq 1$.

(1.5.12) <u>Theorem</u>: <u>Let</u> $Y \to B$ <u>be a fibration of CW-spaces. If</u> $Y \to B$ <u>is nilpotent,</u> Y <u>is a Z-Postnikov space of order 1 over</u> B. <u>If</u> $Y \to B$ <u>is complete,</u> Y <u>is a Z-Postnikov space of order 2 over</u> B.

This theorem will be proved in connection with the general Postnikov decomposition, see (5.3.11). A consequence of (1.5.12) and the dual Whitehead theorem is the following result of Dror [26] .

(1.5.13) Corollary: Let $p : Y \to B$ and $p' : Y' \to B$ be fibrations which are either complete or nilpotent. Further let $h : Y \to Y'$ be a map over B and let $h_* : H_*(Y, Z) \cong (H_*(Y', Z)$ be an isomorphism. Then h induces an isomorphism $h_* : \pi_*(Y) \tilde{\cong} \pi_*(Y')$ of the homotopy groups as well.

Proof of (1.5.13): In virtue of (1.4.19) we can replace p and p' by fibrations between CW-spaces. Since h induces isomorphisms in homology, it follows from the universal coefficient theorem (0.5.1) that h also induces an isomorphism $h^* : H^*(Y', \pi) \to H^*(Y, \pi)$ for every $\mathbb{Z}$-module π. (1.5.13) then follows from the dual Whitehead theorem. □

Remark: Dror obtained this corollary in [26] for the case $B = *$, using other methods. Theorem (1.5.12) was proved for nilpotent fibrations by Hilton-Mislin-Roitberg in 2.14 p. 68 of [49], see also Bousfield-Kan [23] . Nilpotent spaces and theorem (1.5.12) are of great importance in the construction of localizations. This can be seen in the following result, which we will not prove.
The interested reader may compare the discussions of it in [124], [23] and especially [49] .

(1.5.14) Example: Let $R \subset \mathbb{Q}$ be a subring of the rational numbers and let Y be a nilpotent CW-space. Then there exists an R-Postnikov model $Y \to RY$ of Y.

We call a space Y nilpotent when the fibration $Y \to *$ is nilpotent. Let P be a set of primes and R the subring of the rationals consisting of fractions whose denominators are products of powers of primes outside P. Given a nilpotent space Y, we call an R-

Postnikov model $Y \to RY = Y_P$ of Y as in (1.5.14) a P-localization of Y. If $P = \phi$ so that $R = \mathbb{Q}$, we call $Y \to \mathbb{Q}Y = Y_o$ a <u>rationalization</u> of Y. The importance of localizations for the homotopy classification of maps can be seen in the following result: Given a nilpotent CW-space Y, for every $p \in \Pi$ (Π = set of primes) we have a diagram

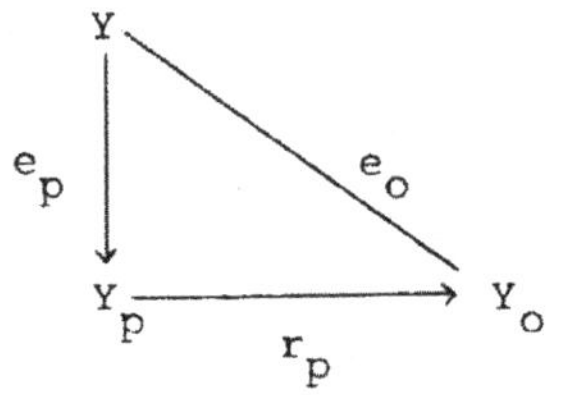

of localizations, where $e_o \simeq r_p e_p$ by uniqueness. Now let W be a finite connected CW-complex and Y a nilpotent CW-space of finite type (i.e. for all n the homotopy groups $\pi_n(Y)$ are finitely generated). Then the map

$$(1.5.15) \qquad e = (e_{p*} \mid p \in \Pi) : [W, Y] \to \times_{p \in \Pi} [W, Y_p]$$

is injective, and Im e with Im $e \approx [W, Y]$ is the set of all tuples $(\eta_p \mid p \in \Pi)$ with $r_{p*}(\eta_p) = r_{q*}(\eta_q)$ for $p, q \in \Pi$. This is the 'pullback theorem' proved by Hilton-Mislin-Roitberg in § 5 of [49] .

CHAPTER 2 : RELATIVE PRINCIPAL COFIBRATIONS AND FIBRATIONS

(2.1) Relative principal cofibrations

We here define relative principal cofibrations, and use them to obtain relative obstructions and differences by procedures quite similar to those of (1.2). It turns out that all the results of (1.2) can be put into a relative form. Relative principal cofibrations can also be used to solve the existence and classification problems of (1.1). The most important examples of relative principal cofibrations arise in (2.3) in the principal reduction of CW-decompositions.

Let A be a space together with maps

$$D \xrightarrow{\sigma} A \xrightarrow{\rho} D$$

such that $\rho\sigma = 1_D$. Following I.M. James in [59], we call the triple (A,σ,ρ), which we will simply denote as A, an ex-space (over D). We call A an ex-cofiber space when σ is a closed cofibration and accordingly D a subspace of A and ρ a retraction. We call A an ex-fiber space when ρ is a fibration and accordingly σ a section of this fibration. Given an ex-space A, the subspace $A_x = \rho^{-1}(x)$ with $x \in D$ is a pointed space with basepoint $\sigma(x)$. We can imagine such an ex-space A as a 'rope' fastened together out of the pointed subspaces A_x indexed by the elements $x \in D$. An ex-space is, in this sense, a generalization of a pointed space. The category $Top(D)$ of ex-spaces will be treated in detail in (2.5). It has essentially all the properties of the category Top^o of pointed spaces. For example, given an ex-space A over D, we can construct the cone C_DA, the suspension

$S_D A$, the path space $P_D A$ or the loop space $\Omega_D A$ by taking, for each $x \in D$, the cprresponding constructions CA_x, SA_x, PA_x of ΩA_x and 'fastening them together' in an appropriate way. Exact definitions of these relative constructions are given in this section and the following one. These constructions enable us to formulate the results on obstructions and differences in (1.2) and (1.3) for the relative case. In the following, we will in fact do this for relative principal cofibrations. In (2.2) we will then describe the dual properties of relative principal fibrations. Here we proceed in close analogy with (1.2) and so will not need to make the proofs quite so detailed as we did there. We leave it as an exercise to the reader to flesh them out.

Let X be a space under D, that is suppose we are given a map $d : D \to X$. We define the <u>product under</u> D with a space Z by means of

(2.1.1)
$$\begin{array}{ccc} Z \times X & \longrightarrow & Z \underset{D}{\ltimes} X \\ \uparrow {\scriptstyle 1 \times d} & \text{push} & \uparrow \\ Z \times D & \xrightarrow[\text{pr}]{} & D \end{array}$$

The product $I_D \ltimes X$ with the interval I is the <u>cylinder relative</u> D. Every homotopy $H : I \times X \to U$ relative D factors over $I_D \ltimes X$. If A is an ex-cofiber space over D, then $I_D \ltimes A$ is also an ex-cofiber space over D. The retraction is given by $(t, a) \to \rho(a)$ for $(t, a) \in I \times A$, and we have $(I_D \ltimes A)_x = I \ltimes A_x$ for $x \in D$. The <u>relative cone</u> $C_D A$ and the relative suspension $S_D A$ are obtained as quotients of $I_D \ltimes A$:

(2.1.2)
$$\begin{array}{ccc} I_D \ltimes A & \longrightarrow & C_D A \\ {\scriptstyle i_1}\big\uparrow & \text{push} & \big\uparrow \\ A & \xrightarrow[\rho]{} & D \end{array}$$

(2.1.3)
$$\begin{array}{ccc} I_D \ltimes A & \longrightarrow & S_D A \\ {\scriptstyle i_o \cup i_1}\big\uparrow & \text{push} & \big\uparrow \\ A \cup_D A & \xrightarrow[\rho \cup \rho]{} & D \end{array}$$

We have inclusions $i_t : A \subset I_D \ltimes A$ defined by $i_t(a) = (t, a)$. Of course, $S_D A$ is also a quotient of $C_D A$. $C_D A$ and $S_D A$ are ex-cofiber spaces over D as well, as can be seen from (0.1.6) and (0.1.7). The retractions are given by $(t, a) \mapsto \rho(a)$ just as for $I_D \ltimes A$, and again we have $(C_D A)_x = CA_x$ and $(S_D A)_x = SA_x$ for $x \in D$.

(2.1.4) <u>Lemma</u>: <u>The inclusion</u> $i_1 : D \subset C_D A$ <u>is a homotopy equivalence under and over</u> D. <u>The inclusion</u> $i_o : A \subset C_D A$ <u>is a closed cofibration with cofiber</u> $C_D A/_A = S_D A/_D = S(A/D)$.

<u>Proof</u>: By contracting the cone CA_x in the usual way for each $x \in D$, we thereby contract $C_D A$ onto D compatibly with ρ. Now $i_o : A \subset C_D A$ is induced by the inclusion $\{0, 1\} \times A \cup I \times D \subset I \times A$. which we know from (0.1.6) and (0.1.7) to be a closed cofibration. We then need only apply (0.0.1) (b). $\square$

Given an ex-cofiber space A and a map $f : A \to X$ in Top^o, we define $C_D f$ by means of the push-out

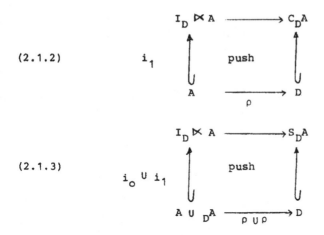

(2.1.5)

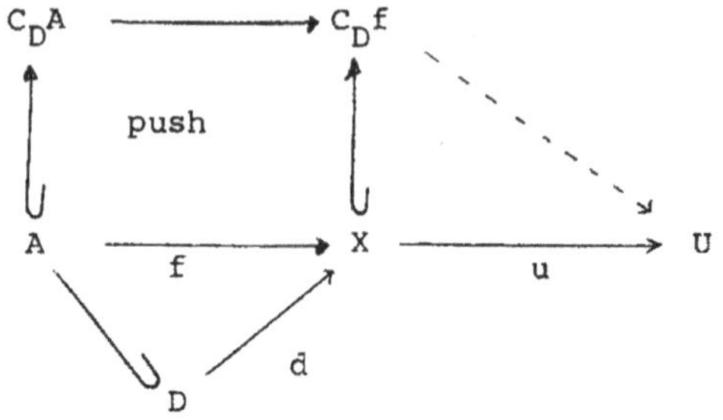

d is the restriction of f. We call C_Df a mapping cone relative D or relative d. By (2.1.4), $X \subset C_Df$ is a closed cofibration.

(2.1.6) <u>Definition</u>: We call a closed cofibration $X \subset Y$ a <u>principal cofibration relative</u> D or relative $d : D \to X$ when there exists an ex-cofiber space A over D and an extension $f : A \to X$ of d as above, such that $C_Df \simeq Y$ are homotopy equivalent under X. As before, we call f the attaching map. If $D = \{*\}$ consists of only one point, then a principal cofibration relative to $\{*\}$ is just a principal cofibration as in (1.2). The cofiber Y/X of a relative principal cofibration $X \subset Y$ is, by (2.1.4), again homotopy equivalent to a suspension.

We now investigate the extension and classification problems for relative principal cofibrations, proceeding analogously with the treatment of principal cofibrations in (1.2). Given an ex-cofiber space A over D and a map $w : D \to U$, we call $w\rho = 0 \in [A, U]^w$ the <u>zero element</u>. Due to (2.1.2), the cone C_DA has the following property. A map $v : A \to U$ extending w can be extended over C_DA exactly when $v \simeq w\rho$ relative D, that is when v represents the zero element. In terms of diagram (2.1.5), a map $u : X \to U$ such that $ud = w$ can be extended over C_Df exactly when

(2.1.7) $f^*(u) \in [A, U]^W$

is the zero element 0. We therefore call $f^*(u)$ the <u>relative primary obstruction</u> to extending u.

We now describe relative primary differences. The relative mapping cone $C_D f$ is a space under D. Taking its cylinder relative D, we see from the following lemma that the inclusion

$$I_D \dot{\ltimes} C_D f = \{0\} \times C_D f \cup I_D \ltimes X \cup \{1\} \times C_D f \subset I_D \ltimes C_D f$$

is a principal cofibration relative D. The attaching map is, generalizing (1.2.3), the map

(2.1.8) $w_f : S_D A \longrightarrow I_D \dot{\ltimes} C_D f$

under D. If $x \in D$, w_f maps the subspace SA_x just as w_f does in (1.2.3). Similarly to (1.2.4), we have a homotopy equivalence

(2.1.9) <u>Lemma</u>: $h : C_D w_f \simeq I_D \ltimes C_D f$ <u>under</u> $I_D \dot{\ltimes} C_D f$.

We assume now, as before, that we are given a map $w : D \to U$ and maps $u_o, u_1 : C_D f \to U$ under D, that is such that $w = u_o d = u_1 d$. A homotopy $H : u_{o|X} \simeq u_{1|X}$ relative D then gives us a map $u_o \cup H \cup u_1 : I_D \ltimes C_D f \to U$. We call

(2.1.10) $d(u_o, H, u_1) = w_f^*(u_o \cup H \cup u_1) \in [S_D A, U]^W$

the <u>relative primary difference</u> of (u_o, H, u_1). By (2.1.9) and (2.1.7), $d(u_o, H, u_1) = 0$ exactly when the homotopy H can be extended to a homotopy $u_o \simeq u_1$. Again we write $d(u_o, u_1) = d(u_o, upr, u_1)$, where $u_{o|X} = u_{1|X} = u$. Then (1.2.8) is generalized by

(2.1.11) Theorem: Let $u : X \to U$ be a map with $w = u \circ d$ and let $u_o \in [C_D f, U]^u$. Then $[C_D f, U]^u \xrightarrow{\approx} [S_D A, U]^w$ defined by $u_1 \longmapsto d(u_o, u_1)$ is a bijection.

The inverse of this bijection will be denoted, as in (1.2.8), by $\alpha \longmapsto u_o + \alpha$. This + is a group operation. The group multiplication in $[S_D A, U]^w$ is obtained as follows. We define a comultiplication relative D

$$\text{(2.1.12)} \qquad \mu : S_D A \longrightarrow S_D A \cup_D S_D A$$

by stipulating that (for every $x \in D$) the restriction of μ be the comultiplication $\mu : SA_x \to SA_x \vee SA_x$ of (0.2). In the group structure in $[S_D A, U]^w$ given by $+ = \mu^*$, we see that $w\rho = 0$ is the neutral element. If $A = S_D A'$ is a relative suspension, this group structure is abelian. All these statements can be proved in analogy with the non-relative case. This group multiplication + in $[S_D A, U]^w$ gives us the same addition formulas for relative differences as those in (1.2.6). We can describe the operation + defined immediately after (2.1.11), by means of a co-operation

$$\text{(2.1.13)} \qquad \mu : C_D f \longrightarrow C_D f \cup_D S_D A .$$

This map is defined on CA_x for every $x \in D$ to be $CA_x \to CA_x \vee SA_x$ as in (1.2.7).

The comultiplication μ also induces a group structure on section homotopy sets. Let $D \xrightarrow{w} \tilde{D} \xrightarrow{p} D$ be an ex-space (in the following, an ex-fiber space in fact) over D. The pulled-back fibration in

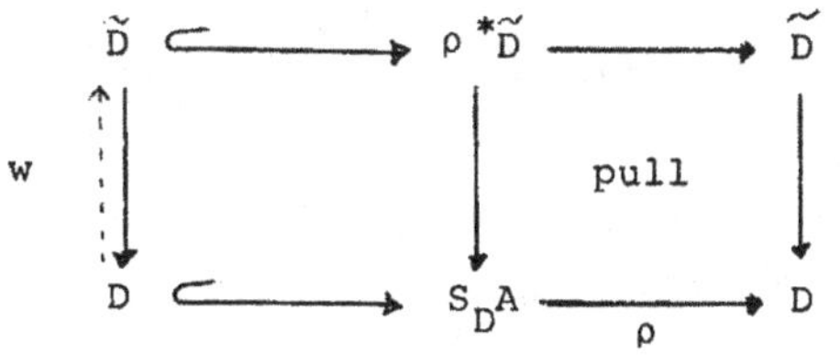

shows us that w is a section over D which is extended over $S_D A$
by the section $\rho^* w$. μ then induces on the section homotopy set
$\langle S_D A, \rho^* \widetilde{D} \rangle^w$ a group multiplication + :

(2.1.14)

$$\langle S_D A, \rho^* \widetilde{D} \rangle^w \times \langle S_D A, \rho^* \widetilde{D} \rangle^w \xrightarrow{\quad + \quad} \langle S_D A, \rho^* \widetilde{D} \rangle^w$$

$$\langle S_D A \cup_D S_D A, (\rho \cup \rho)^* \widetilde{D} \rangle^w$$

with $\rho^* w$ as neutral element. The bijection $\mathcal{H}$ is due to
excision, see (0.0.11). Since $(\rho \cup \rho)\mu = \rho$, we can define the
map μ^* which pulls back sections. As in (0.0.9), we have
$\langle S_D A, \rho^* \widetilde{D} \rangle^w = [S_D A, \widetilde{D}]_\rho^w = [S_D A, \widetilde{D}]_D^D$ which is an ex-homotopy
set over D, by (2.5).

The groups just described fit into the following <u>exact fiber</u>
<u>sequence relative</u> D. In the diagram

$$\begin{array}{ccc}
\widetilde{D} & \xrightarrow{\quad i \quad} & \widetilde{X} \\
{\scriptstyle w} \uparrow & \nearrow^{w} & \downarrow {\scriptstyle F} \\
D & \xrightarrow[\quad d \quad]{} & X
\end{array}$$

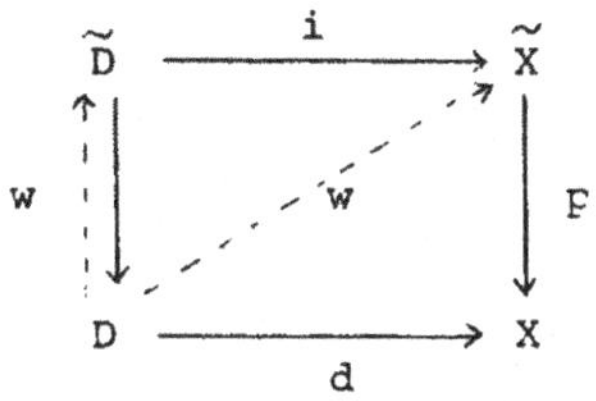

let p be a fibration and $\widetilde{D} = d^* \widetilde{X}$ the induced fibration with
section w. Then we have the long sequence

$$\cdots \xrightarrow{\delta} \langle S_D A, \rho^* \widetilde{D} \rangle^w \xrightarrow{i_*} [S_D A, \widetilde{X}]^w \xrightarrow{p_*} [S_D A, X]^d \xrightarrow{\delta} \cdots$$

$$\xrightarrow{\ \delta\ } \langle A,\ \rho^*\tilde{D}\rangle^{w} \longrightarrow [A,\ \tilde{X}]^{w} \xrightarrow{\ p_*\ } [A,\ X]^{d}$$

in which p_* is induced by p and i_* by i. The connecting homomorphism δ will be described in (2.5), where we will also discuss the following lemma.

(2.1.15) <u>Lemma: The above sequence is exact, from</u> S_DA <u>on it is an exact sequence of groups and homomorphisms and, from</u> S_D^2A <u>on, of abelian groups</u>. <u>If</u> $A = S_DA'$ <u>is a relative suspension, all the maps in the sequence are homomorphisms of groups.</u>

If $D = \{*\}$, the sequence is just the exact homotopy sequence of the fibration $\tilde{X} \to X$ with fiber F which we used in (1.2.11). In this case, $\langle A,\ \rho^*\tilde{D}\rangle^{w}$ is $[A,\ F]$.

We now describe relative obstructions and differences for the extension of sections, analogously with (1.2.11). Let $\tilde{C}_Df \to C_Df$ be a fibration. We then have fibrations induced as in the diagrams

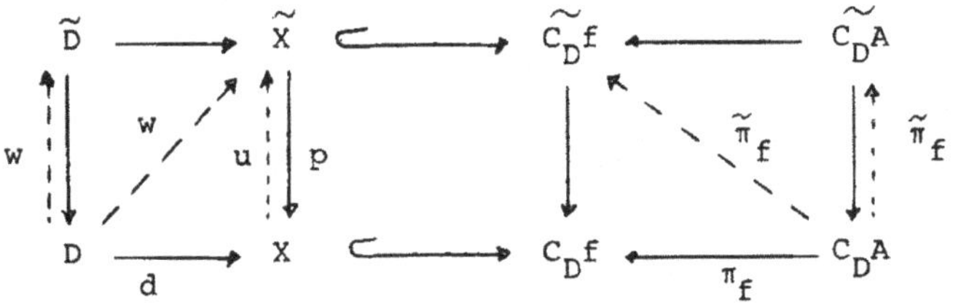

in which w is a given section. By (2.1.4) and (1.1.6), $\langle C_DA,\ \tilde{C}_DA\rangle^{w} \approx \langle D,\ \tilde{D}\rangle^{w}$ consists of exactly one element $\tilde{\pi}_f$. By restricting $\tilde{\pi}_f$ to $A \subset C_DA$, we obtain the element $\partial\tilde{\pi}_f = \tilde{\pi}_f|A \in [A,\ \tilde{X}]^{w}$ with $p_*\partial\tilde{\pi}_f = f$. Now assume that u is a section extension of w, so that $ud = w$. Then the exact sequence (2.1.15) splits by means of u_*,

$$\langle A, \rho^*\tilde{D}\rangle^w \xrightarrow{i_*} [A, \tilde{X}]^w \underset{p_*}{\overset{u_*}{\rightleftarrows}} [A, X]^d.$$

If A is a relative suspension, $[A, \tilde{X}]^w$ is a group and i_* is injective. Since $p_*(-\partial\tilde{\pi}_f + u_*(f)) = 0$, the element

(2.1.16) $$f^{\#}(u) = i_*^{-1}(-\partial\tilde{\pi}_f + u_*(f)) \in \langle A, \rho^*\tilde{D}\rangle^w$$

is well-defined. Generalizing (1.2.12), we obtain the

(2.1.17) Theorem: $f^{\#}(u) = 0$ exactly when the section u over X can be extended to a section u' over C_Df.

We again call $f^{\#}(u)$ a relative primary obstruction. As in (1.2.14), we see that such obstructions are natural. Now let $u_o, u_1 : C_Df \to \widetilde{C_Df}$ be section extensions of $w : D \to \tilde{D}$, and let $H : u_{o|X} \simeq u_{1|X}$ be a section homotopy relative D. As in (1.2.16), we then have a section $u_o \cup \bar{H} \cup u_1 : I_D \dot{\ltimes} C_Df \to \widetilde{I_D \dot{\ltimes} C_Df}$ and can define the relative primary difference to be

(2.1.18) $$d(u_o, H, u_1) = w_f^{\#}(u_o \cup \bar{H} \cup u_1) \in \langle S_DA, \rho^*\tilde{D}\rangle^w.$$

The section homotopy H can be extended to a section homotopy $u_o \simeq u_1$ exactly when $d(u_o, H, u_1) = 0$. A does not have to be a relative suspension in order for the difference to be defined. The additivity of (1.2.17) is preserved in this more general situation.

(2.1.19) Theorem: Let A be a suspension relative to D. Then in the abelian group $\langle S_DA, \rho^*\tilde{D}\rangle^w$ we have the equation $d(u_o, H, u_1) + d(u_1, H', u_2) = d(u_o, H + H', u_2)$.

In proving this theorem it should be remembered that $\partial\tilde{\pi}_f$ in (2.1.16) was not defined by means of a boundary operator as in (1.2.11).

Nevertheless, $\tilde{\pi}_f$ behaves in the proof just as it did in (1.2.17). Using the co-operation μ of (2.1.13), we can define an operation $+$, just as we did in (1.2.19), for which the following classification theorem holds.

(2.1.20) <u>Theorem</u>: <u>Let</u> A <u>be a suspension relative</u> D <u>and let</u> $u : X \longrightarrow \tilde{X}$ <u>be a section</u>. <u>Let</u> $w = d^*u : D \to \tilde{D}$ <u>be the induced section and</u> $u_o \in \langle C_D f, \widetilde{C_D f} \rangle^u$. <u>Then</u>

$$\langle C_D f, \widetilde{C_D f} \rangle^u \xrightarrow{\approx} \langle S_D A, \rho^* \tilde{D} \rangle^w$$

<u>is a bijection given by</u> $u_1 \to d(u_o, u_1)$ <u>with inverse</u> $\alpha \to u_o + \alpha$.

If A is a suspension relative to D, $+$ is actually a group operation. This is shown as in (1.2.22). The theorem can be proved as was the special case (1.2.21). It is a generalization of (2.1.11).

Just as in (1.2.26), we could go on to consider iterated relative principal cofibrations and define relative obstructions and differences of higher orders. The discussion at the end of (1.2) should be referred to in this connection.

(2.2) Relative principal fibrations

In this section we introduce relative principal fibrations and describe their basic properties. Relative obstructions and differences can then be defined. It will be seen that all the results of (1.3) on obstructions and differences for principal fibrations hold in the relative case when appropriately reformulated. Relative principal fibrations can again be used to solve the classification and existence problems of (1.1). They arise naturally in the Postnikov decomposition of a fibration, which we will discuss in (2.3). This section will be developed in strict duality to the preceding one. An item here numbered (2.2.i) will be the exact dual of the one there numbered (2.1.i).

Let X be a space over D, i.e. suppose we have a map $d : X \to D$. Given a space Y, we define the <u>mapping space over</u> D by means of

(2.2.1)
$$\begin{array}{ccc} X_D^Y & \longrightarrow & X^Y \\ \big\downarrow & \text{pull} & \big\downarrow d^1 \\ D & \xrightarrow[c]{} & D^Y \end{array}$$

c maps an element $x \in D$ into the constant map c_x with $c_x(y) = x$ for all $y \in Y$. We can take X_D^Y to be a subset of X^Y, since c is injective. With I the unit interval, we call X_D^I the <u>free path space over</u> D. Every homotopy $H : U \to X^I$ over D factors over X_D^I.

If $A = (A, \sigma, \rho)$ is an ex-fiber space over D, then A_D^I is also an ex-fiber space over D. The section $D \to A_D^I$ is given by $x \mapsto c_{\sigma_x}$.

Subspaces $A_x = \rho^{-1}(x)$ of A then have the property that $(A_D^I)_x = (A_x)^I$ for $x \in D$. The <u>relative path space</u> $P_D A$ and the <u>relative loop space</u> $\Omega_D A$ are subsets of A_D^I :

(2.2.2)
$$\begin{array}{ccc} P_D A & \longrightarrow & A_D^I \\ \big\downarrow & \text{pull} & \big\downarrow q_1 \\ D & \xrightarrow[\sigma]{} & A \end{array}$$

(2.2.3)
$$\begin{array}{ccc} \Omega_D A & \longrightarrow & A_D^I \\ \big\downarrow & \text{pull} & \big\downarrow (q_o, q_1) \\ D & \xrightarrow[(\sigma,\sigma)]{} & A \times_D A \end{array}$$

The projections $q_t : A_D^I \to A$ are given by $q_t(\tau) = \tau(t)$ for $t \in I$. Of course, $\Omega_D A$ is also a subset of $P_D A$. $\Omega_D A$ and $P_D A$ are also ex-fiber spaces over D, see the proof of (2.2.4). The sections are given by $x \mapsto c_{\sigma x}$ as for A_D^I. We also have $(P_D A)_x = PA_x$ and $(\Omega_D A)_x = A_x$ for $x \in D$. We thus obtain 'ropes' in the sense of (2.1).

(2.2.4) <u>Lemma</u>: <u>The projection</u> $q_1 : P_D A \to D$ <u>is a homotopy equivalence under and over</u> D. <u>The projection</u> $q_o : P_D A \to A$ <u>with</u> $q_o(\tau) = \tau(0)$ <u>is a fibration with fiber</u> $q_o^{-1}(*) = (\Omega_D A)_* = \Omega F$, <u>where</u> $* \in D \subset A$ <u>denotes the basepoint and</u> $F = \rho^{-1}(*)$ <u>the fiber of the ex-fiber space</u> A.

<u>Proof</u>: By contracting PA_x in the usual way for each $x \in D$, we find that q_1 is a homotopy equivalence under and over D. To see that q_o is a fibration, consider the cartesian squares

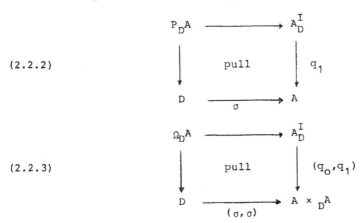

$$\begin{array}{ccccccc}
\Omega_D A & \longrightarrow & P_D A & \longrightarrow & A^I_D & \longrightarrow & A^I \\
\downarrow & & \downarrow{\scriptstyle q_o} & & \downarrow{\scriptstyle (q_o, q_1)} & & \downarrow{\scriptstyle \bar{\rho}} \\
D & \longrightarrow & A & \xrightarrow[(1,\ \sigma\rho)]{} & A \times_D A & \xrightarrow[\bar{c}]{} & A \times A \times D^I .
\end{array}$$

We define $\bar{\rho}(\tau) = (\tau(0), \tau(1), \rho\tau)$ and $\bar{c}(a, b) = (a, b, c_{\rho a})$, where $(a, b) \in A \times_D A \subset A \times A$ and $\rho(a) = \rho(b)$. Thus it suffices to show that $\bar{\rho}$ is a fibration when ρ is. To see that $\bar{\rho}$ has the homotopy lifting property, we start from the commutative diagram

$$\begin{array}{ccc}
A^I & \xleftarrow{\ h\ } & U \times \{0\} \\
\downarrow & \overset{G}{\nwarrow} & \cap\!\downarrow \\
A \times A \times D^I & \xleftarrow[(H_o,\ H_1,\ H)]{} & U \times I
\end{array}$$

in which we want to construct an extension G. Adjunction gives us the diagram

$$\begin{array}{ccc}
A & \xleftarrow{\ H_o \cup H_1 \cup \bar{h}\ } & U \times I \times \{0,1\} \cup U \times \{0\} \times I = U' \\
{\scriptstyle \rho}\downarrow & & \cap\!\downarrow{\scriptstyle j} \\
D & \xleftarrow[\bar{H}]{} & U \times I \times I
\end{array}$$

We have the usual homeomorphism g in

$$\begin{array}{ccc}
U' \times I & \overset{g}{\approx} & U \times I \times\times I \\
\cup\!\uparrow & \nearrow{\scriptstyle j} & \\
U' \times \{0\} & &
\end{array}$$

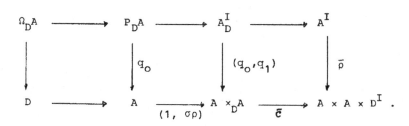

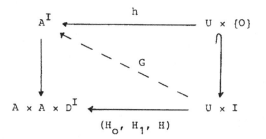

extending the inclusion j. Since ρ is a fibration, $\bar{H}g$ has a lifting G'. The adjoint of $G'g^{-1}$ is the required lifting G. □

Given an ex-fiber space A over D and a map $f : X \to A$ in Top^o, we define $P_D f$ by the pullback

(2.2.5)

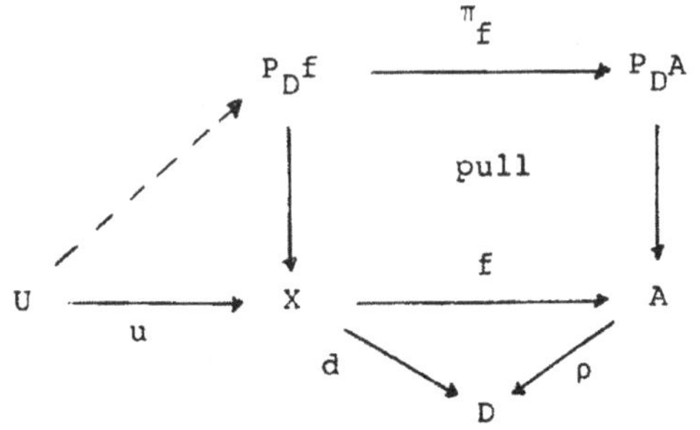

where $d = \rho \circ f$. We call $P_D f$ the <u>mapping path space relative</u> D or relative d. By (2.2.4), $P_D f \to X$ is a fibration.

(2.2.6) <u>Definition</u>: We call a fibration $Y \to X$ a <u>principal fibration relative</u> D or relative $d : X \to D$, if there exist an ex-fiber space A over D and a lifting $f : X \to A$ of d as above, such that $P_D f \simeq Y$ are homotopy equivalent over X. As before, we call f a classifying map. If $D = *$ consists of only one point, then a principal fibration relative $*$ is just a principal fibration as in (1.3.1). The fiber of a relative principal fibration is again homotopy equivalent to a loop space, by (2.2.4).

We now investigate the lifting and classification problems for relative principal fibrations, proceeding analogously with the treatment of principal fibrations in (1.3). Given an ex-fiber space A over D and a map $w : U \to D$, we call $\sigma w = 0 \in [U, A]_w$ the <u>zero element.</u> The path space $P_D A$ then has the following characteristic property due to (2.2.2). A map $v : U \to A$ lifting w can be lifted to $P_D A$ exactly when $v \simeq \sigma w$ over D, that is when v represents the zero element. In terms of diagram (2.2.5), a map $u : U \to X$ with $du = w$ can be lifted to $P_D f$ exactly when

(2.2.7) $$f_*(u) \in [U, A]_w$$

is the zero element. We therefore call $f_*(u)$ the <u>relative primary obstruction to lifting</u> u.

We now describe relative primary differences. If X is a space over D, so is the mapping path space $P_D f$. The free path space $(P_D f)^I_D$ over D is the total space of a principal fibration p relative D where p is defined by the diagram

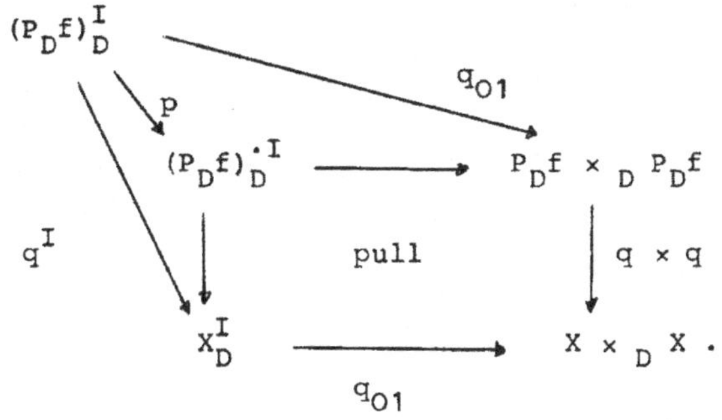

The classifying map for p is, generalizing (1.3.3), the map

(2.2.8) $$w_f : (P_D f)^{\cdot I}_D \longrightarrow \Omega_D A$$

over D. The map w_f is defined similarly to the w_f of (1.3.3). If $x \in D$ and $y = (\tau, (x_0, \tau_0), (x_1, \tau_1)) \in (P_D f)^{\cdot I}_D \subset X^I \times P_D f \times P_D f$ are such that $d \circ \tau = c_x$, then $w_f(y) = -\tau_0 + f\tau + \tau_0$ is a loop in $\Omega A_x \subset \Omega_D A$. Just as in (1.3.4), we obtain a

(2.2.9) <u>Lemma</u>: p <u>is a fibration, and there is a homotopy equivalence</u> $h : (P_D f)^I_D \simeq P_D w_f$ <u>over</u> $(P_D f)^{\cdot I}_D$, <u>that is</u> w_f <u>is a classifying map for</u> p.

As before, we now let $u_o, u_1 : U \longrightarrow P_Df$ be liftings of $w : U \to D$, that is $dqu_o = dqu_1 = w$. A homotopy $H : qu_o \simeq qu_1 : U \to X$ over D then gives us a map $\bar{H} \underline{\times} (u_o, u_1) : U \to (P_Df)_D^{\cdot I}$. We call

(2.1.10) $\quad d(u_o, H, u_1) = w_f{*}(\bar{H} \underline{\times} (u_o, u_1)) \in [U, \Omega_DA]_w$

the <u>relative primary difference</u> of (u_o, H, u_1). By (2.2.9) and (2.2.7), $d(u_o, H, u_1) = 0$ exactly when the homotopy H can be lifted to a homotopy $u_o \simeq u_1$. We generalize (1.3.8) in the

(2.2.11) <u>Theorem</u>: <u>Let</u> $u : U \to X$ <u>be a map with</u> $w = d \circ u$ <u>and let</u> $u_o \in [U, P_Df]_u$. <u>Then</u> $[U, P_Df]_u \xrightarrow{\approx} [U, \Omega_DA]_w$ <u>defined by</u> $u_1 \longmapsto d(u_o, u_1)$ <u>is a bijection.</u>

The inverse of this bijection will be denoted, as in (1.3.8), by $\alpha \longmapsto u_o + \alpha$. This $+$ is a group operation. The group multiplication $+ = \mu_*$ in $[U, \Omega_DA]_w$ is obtained as follows. We define a multiplication relative D

(2.2.12) $$\mu : \Omega_DA \times_D \Omega_DA \to \Omega_DA$$

by stipulating that for every $x \in D$ the restriction of μ be the multiplication $\mu : \Omega A_x \times \Omega A_x \to \Omega A_x$ as in (0.2). $\sigma w = 0$ is then the neutral element. The group operation of (2.2.11) can also be characterized by means of an operation

(2.2.13) $$\mu : P_Df \times_D \Omega_DA \to P_Df .$$

This operation generalizes the μ of (1.3.7), being defined as there on $PA_x \times \Omega A_x$ to be addition of paths, for every $x \in D$.

More generally, we find that the multiplication μ induces a group structure on retraction homotopy sets. Let $D \xrightarrow{i} \bar{D} \xrightarrow{w} D$ be an ex-space (in the following, an ex-cofiber space in fact) over D.

The induced cofibration $\sigma_*\bar{D}$

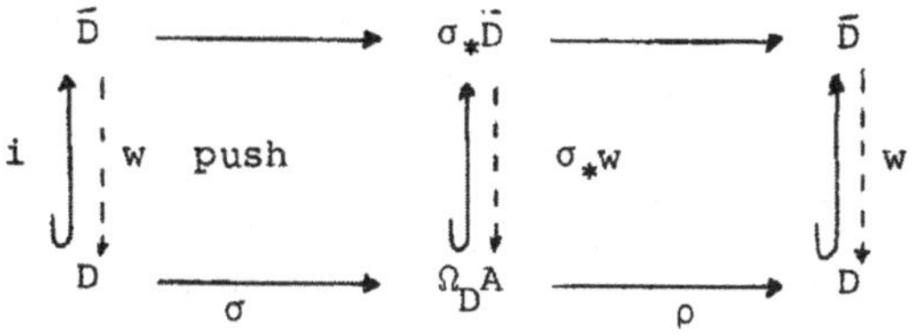

gives us a retraction over D which is lifted to $\Omega_D A$ by $\sigma_* w$. μ then induces on the retraction homotopy set $\langle \sigma_*\bar{D}, \Omega_D A \rangle_w$ a group multiplication

$$\begin{array}{ccc} \langle \sigma_*\bar{D}, \Omega_D A\rangle_w \times \langle \sigma_*\bar{D}, \Omega_D A\rangle_w & \xrightarrow{+} & \langle \sigma_*\bar{D}, \Omega_D A\rangle_w \\ \mathfrak{X} \;\Vert & \nearrow \mu_* & \\ & \langle (\sigma, \sigma)_* \bar{D}, \Omega_D A \times_D \Omega_D A\rangle_w & \end{array} \qquad (2.2.14)$$

with $\sigma_* w = 0$ as neutral element. The bijection $\mathfrak{X}$ is due to excision, see (0.0.12). Since $\mu(\sigma, \sigma) = \sigma$, the map μ_* is defined. As in (0.0.9), we have $\langle \sigma_*\bar{D}, \Omega_D A\rangle_w = [\bar{D}, \Omega_D A]^{\sigma}_w = [\bar{D}, \Omega_D A]^D_D$ which is an ex-homotopy set over D, by (2.5).

The groups just described fit into the following long <u>exact cofiber sequence over</u> D. In the diagram

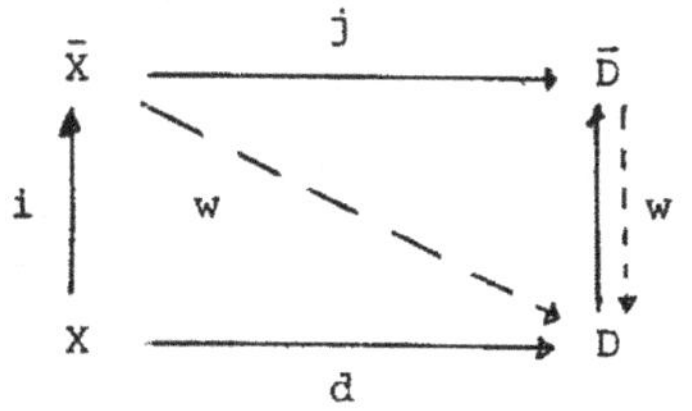

let i be a closed cofibration and $\bar{D} = d_*\bar{X}$ the induced cofibration, and let w be a retraction. Then given an ex-fiber space A over D, we have a long sequence

$$\cdots \longrightarrow \langle \sigma_*\bar{D}, \Omega_D A\rangle_w \xrightarrow{j^*} [\bar{X}, \Omega_D A]_w \xrightarrow{i^*} [X, \Omega_D A]_d \xrightarrow{\delta}$$

$$\longrightarrow \langle \sigma_*\bar{D}, A\rangle_w \xrightarrow{j^*} [\bar{X}, A]_w \xrightarrow{i^*} [X, A]_d$$

in which i^* is induced by i and j^* by j. The connecting homomorphism δ will be described in (2.5), where we will also discuss the following lemma.

(2.2.15) Lemma: The above sequence is exact, from $\Omega_D A$ on it is an exact sequence of groups and homomorphisms and, from $\Omega_D^2 A$ on, of abelian groups. If $A = \Omega_D A'$ is a relative loop space, all the maps in the sequence are homomorphisms of groups.

If $D = \{*\}$, the sequence is just the exact homotopy sequence of a cofibration $X \subset \bar{X}$ with cofiber F which we used in (1.3.12). In this case, $\langle \sigma_*\bar{D}, A\rangle_w = [F, A]$. We will see furthermore in (5.2.4) that the long exact sequence of a pair, in cohomology with local coefficients, is a special case of the sequence in (2.2.15).

We will now describe relative obstructions and differences for lifting retractions, analogously with (1.3.12). Let $P_D f \subset \overline{P_D f}$ be a closed cofibration. We then have cofibrations induced as in the diagrams

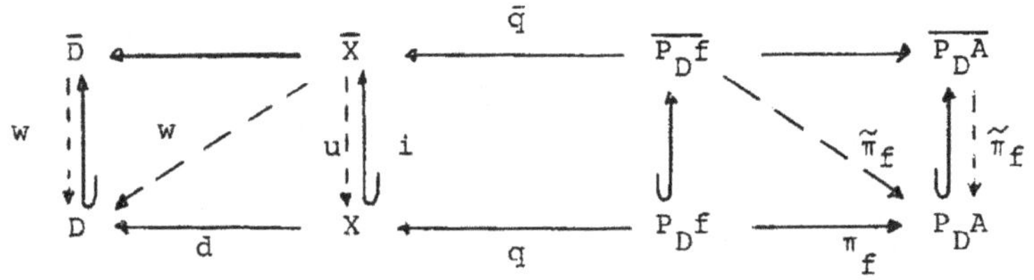

in which we let w be a retraction. By (2.2.4) and (1.1.7), $\langle \overline{P_D A}, P_D A\rangle_w \approx \langle \bar{D}, D\rangle_w$ consists of exactly one element $\tilde{\pi}_f$. From $\tilde{\pi}_f$ we obtain the element $\delta\tilde{\pi}_f \in [\bar{X}, A]_w$ with $i^*(\delta\tilde{\pi}_f) = f$, that is $\delta\tilde{\pi}_f$ is given on the pushout $\bar{X} = (X \cup P_D f)/\sim$ by $f \cup q_o\tilde{\pi}_f$. Now assume u is a retraction lifting of w, so that $du = w$. Then the exact sequence (2.2.15) splits by means of u^*,

$$\langle \sigma_* \bar{D}, A\rangle_w \xrightarrow{j^*} [\bar{X}, A]_w \underset{u^*}{\overset{i^*}{\rightleftarrows}} [X, A]_d .$$

If A is a relative loop space, $[\bar{X}, A]_w$ is a group and j^* is injective. Since $i^*(-\delta\tilde{\pi}_f + u^*(f)) = 0$, the element

(2.2.16) $$f_\#(u) = (j^*)^{-1}(-\delta\tilde{\pi}_f + u^*(f)) \in \langle \sigma_* \bar{D}, A\rangle_w$$

is well-defined. Generalizing (1.3.13), we obtain as the strict dual of (2.1.17) the

(2.2.17) <u>Theorem:</u> $f_\#(u) = 0$ <u>exactly when the retraction</u> $u : \bar{X} \to X$ <u>can be lifted to a retraction</u> $u' : \overline{P_D f} \to P_D f$.

Although we have defined $\delta\tilde{\pi}_f$ here otherwise than in (1.3.12), it functions in the proof of the theorem as did $\delta\tilde{\pi}_f$ in the proof of (1.3.13). The relative primary obstruction $f_\#(u)$ is, in analogy with (1.3.15), natural with respect to maps.

Now let $u_0', u_1' : P_D f \to P_D f$ be retraction liftings of $u_0, u_1 : \bar{X} \to X$, themselves retraction lfitings of $w : \bar{D} \to D$. As in (1.3.17), a retraction homotopy $H : u_0 \simeq u_1$ over w leads to a retraction $\bar{H} \underline{\times} (u_0', u_1') : \overline{(P_D f)_D^{\cdot I}} \to (P_D f)_D^{\cdot I}$ which we use to define the relative primary difference

(2.2.18) $$d(u_0', H, u_1') = w_{f\#}(\bar{H} \underline{\times} (u_0', u_1')) \in \langle \sigma_* D, \Omega_D A\rangle_w .$$

The retraction homotopy H can be lifted to a retraction homotopy $u_0' \simeq u_1'$ exactly when $d(u_0', H, u_1') = 0$. A does not have to be a relative loop space in order for this difference to be defined.

The additivity of (1.3.14) is preserved in the more general situation.

(2.2.19) Theorem: Let A be a relative loop space over D. Then in the abelian group $\langle \sigma_* \bar{D}, \Omega_D A \rangle_w$ we have the equation
$d(u_o, H, u_1) + d(u_1, H', u_2) = d(u_o, H + H', u_2)$.

This theorem can be proved as was (1.3.19).

(2.2.20) Theorem: Let A be a relative loop space over D and let $u : \bar{X} \to X$ be a retraction and $w = d_* u : \bar{D} \to D$ the induced retraction. Taking an element $u_o \in \langle \overline{P_D f}, P_D f \rangle_u$, we obtain a bijection

$$\langle \overline{P_D f}, P_D f \rangle_u \xrightarrow{\approx} \langle \sigma_* \bar{D}, \Omega_D A \rangle_w$$

defined by $u_1 \longmapsto d(u_o, u_1)$.

This theorem is proved exactly as was the special case (1.3.23), by first defining a group operation + from the operation μ of (2.2.13). Then $u_o \longmapsto u_o + \alpha$ is the inverse of the bijection in (2.2.20).

Just as in (1.3.28), we could go on to consider iterated relative principal fibrations and define relative obstructions and differences of higher orders. Examples of iterated relative principal fibrations arise naturally in the Postnikov decomposition of a fibration, which we will treat in the next section.

(2.3) Postnikov decompositions, CW-decompositions, and their principal reductions

In (2.1) and (2.2) we investigated relative principal cofibrations and fibrations. In this section we show how they arise naturally in CW-decompositions and Postnikov decompositions. In particular, the "principal reduction" of such decompositions provides examples of relative principal fibrations and cofibrations. As a result of this principal reduction, we obtain new decompositions of a cofibration or fibration which in their turn can be used to solve the existence and classification problems of (1.1). We will not yet be able to prove fully the reduction theorems (2.3.4) and (2.3.5) in this section, but we will prove particular versions for the unstable case. We will also defer the existence proof of the general Postnikov decomposition. We bring in these theorems in advance of their proofs in order to have certain important examples at our disposal for the general constructions in this and the following chapter.

Although as we saw in (1.4.5) every closed cofibration has a CW-decomposition, it is not in general possible to find a Z-Postnikov model for a given fibration $Y \to B$ that is built up only using principal fibrations, see (1.5.2). However, there is a fundamental decomposition theorem for fibrations:

(2.3.1) Theorem (Postnikov decomposition): *Let $Y \to B$ be a fibration with fiber F and let Y, F and B be path-connected CW-spaces. Then there exist fibrations q_n and maps h_n making the diagram*

$$\begin{array}{ccccccccc} Y & & & & & & & & \\ \downarrow & \overset{h_1}{\searrow} & & \overset{h_{n-1}}{\searrow} & & & & \overset{h_n}{\searrow} & \\ B & \xleftarrow{q_1} & Y_1 & \longleftarrow & Y_2 & \cdots \longleftarrow & Y_{n-1} & \xleftarrow{q_n} & Y_n \end{array}$$

<u>commute, and such that for</u> $n \geqslant 1$

(i) q_n <u>is a fibration with</u> $K(\pi_n(F), n)$ <u>as fiber,</u>

(ii) h_n <u>is</u> $(n+1)$-<u>connected</u>.

A system of fibrations $\{q_n\}$ with maps h_n as in the theorem is called a <u>Postnikov decomposition</u> of $Y \to B$, or just of Y is $B = *$. We will prove this theorem in (5.3) and exhibit there further properties of the Postnikov decomposition. We will show that the Y_n are uniquely determined up to homotopy equivalence by conditions (i) and (ii). We will also prove the additional statements contained in the next two notes.

(2.3.2) <u>Note</u>: If the operation φ_n of (1.5.9) is trivial, $n \geqslant 2$, then the fibration $Y_n \to Y_{n-1}$ in the Postnikov decomposition is a principal fibration with classifying map

$$k_n : Y_{n-1} \longrightarrow K(\pi_n(F), n+1).$$

These k_n are the classical k-invariants of a Postnikov decomposition. If all $Y_n \to Y_{n-1}$ are principal fibrations, we speak of an orientable Postnikov decomposition.

(2.3.3) <u>Note</u>: In the general case the fibration $Y_n \to Y_{n-1}$, $n \geqslant 2$, is a principal fibration relative to the map

$$Y_{n-1} \to Y_1 \xrightarrow{d} K = K(\pi_1(Y), 1).$$

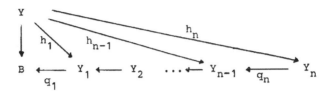

Here d is the map, determined up to homotopy, which induces the isomorphism $h_{1*} : \pi_1(Y) \cong \pi_1(Y_1)$ of fundamental groups. The classifying map k_n of $Y_n \to Y_{n-1}$ maps into an ex-fiber space $L(\varphi_n, n+1)$ over K determined solely by the operation φ_n, see (5.2.7).

The Postnikov decomposition thus essentially consists of relative principal fibrations $Y_n \to Y_{n-1}$, $n \geq 2$. (The fibrations $Y_1 \to B$ plays a special role.) This fact and the following theorems together point up the fundamental importance of principal fibrations and cofibrations relative to a space D. They will enable us to apply the relative theory of (2.1) and (2.2) in many situations.

(2.3.4) <u>Theorem</u> (relative principal reduction of a Postnikov decomposition): <u>Let</u> $Y \to B$ <u>be a fibration with fiber</u> F, <u>and let</u> $\{Y_n\}$ <u>be a Postnikov decomposition of it. Suppose</u> $Y_r = B$ <u>for some</u> $r \geq 1$ (i.e. <u>the fiber</u> F <u>is</u> r-<u>connected</u>) <u>and that</u> $d : B \to D$ <u>for some</u> **D** <u>is an</u> $(r+1)$-<u>connected map. Then the composed fibration</u> $Y_{n+r} \to Y_n$ <u>is a principal fibration relative</u> $Y_n \to B \xrightarrow{d} D$.

<u>Remark:</u> Such a principal reduction of a Postnikov decomposition relative to a space D is to be found in a paper of McClendon's see [82]. We will give a new presentation here.

The following theorem on CW-decompositions is dual to that on principal reduction of Postnikov decompositions.

(2.3.5) <u>Theorem</u> (relative principal reduction of a CW-decomposition): <u>Let</u> A <u>be path-connected and let</u> (X, A) <u>be a relative CW-complex with skeletons</u> (X_n, A). <u>Suppose</u> $X_{r-1} = A$ <u>for some</u> $r \geq 1$ <u>and that</u> $d : D \to A$ <u>is an</u> $(r-1)$-<u>connected map. Then, for</u> $n \geq 0$,

<u>the inclusion</u> $X_n \subset X_{n+r}$ <u>is a principal cofibration relative</u> $D \xrightarrow{d} A \subset X_n$.

One of the main goals of this book is to prove the preceding theorems on principal reduction. The proofs in the appendices to (3.4) and (6.4) are essentially based on properties of the functional loop operation of (6.4) and the functional suspension of (3.4).

(2.3.6) <u>Corollary</u>: <u>Let</u> $p : Y \to B$ <u>be a fibration of path-connected CW-spaces with fiber</u> F, <u>and let</u> $\pi_i(F) \neq 0$ <u>only in dimensions</u> i <u>with</u> $r < i \leq 2r$. <u>Then</u> $p : Y \to B$ <u>is a principal fibration relative</u> B

<u>Example</u>: Suppose $\{Y^n\}$ is a Postnikov decomposition of Y. Then the fibration $Y^{2r} \to Y^r$ is a principal fibration relative Y^r.

(2.3.7) <u>Consequence</u>: Let $p : Y \to B$ be a fibration of path-connected CW-spaces with fiber F, and let $\pi_i(F) \neq 0$ only in dimensions i with $r < i \leq 2r$. Given $u : U \to B$ and $u_o \in [U, Y]_B$, we obtain a group structure on the homotopy set $[U, Y]_B$ with u_o as neutral element. This follows from (2.3.6) and (2.2.11). There is of course a corresponding statement for retraction homotopy sets that arises by using (2.2.20) instead of (2.2.11). This group structure on $[U, Y]_B$ has also been described by J.C. Becker [21] and L.L. Larmore [65]. J.F. McClendon [82] has pointed out that this structure can be obtained only by means of principal reduction.

Dual to (2.3.6) is the

(2.3.8) <u>Corollary</u>: <u>Let</u> A <u>be path-connected and let</u> (X, A) <u>be a relative CW-complex with cells only in dimensions</u> i <u>with</u> $r \leq i \leq 2r - 1$. <u>Then the inclusion</u> $A \subset X$ <u>is a principal co-</u>

<u>fibration relative to</u> A.

<u>Example</u>: Let X be a path-connected CW-complex with skeletons X^n. Then the inclusion $X^{r-1} \subset X^{2r-1}$ is a principal cofibration relative to X^{r-1}.

(2.3.9) <u>Consequence</u>: Let A be path-connected and let (X, A) be a relative CW-complex with cells only in dimensions i with $r \leq i \leq 2r - 1$. Given $u : A \to U$ and $u_o \in [X, U]^u$, we obtain a group structure on the homotopy set $[X, U]^u$ with u_o as neutral element. This follows from (2.3.7) and (2.1.11). Similarly, (2.1.20) tells us that if we have a fibration $\tilde{X} \to X$ and take a section $u_o : A \to \tilde{A}$ and an $u_o \in \langle X, \tilde{X} \rangle^u$, we obtain a group structure on $\langle X, \tilde{X} \rangle^u$ with u_o as neutral element.

If in (2.3.4) and (2.3.5) we set $D = *$, we obtain the following special cases of principal reduction in which we actually end up with <u>principal</u> cofibrations and fibrations.

(2.3.10) <u>Corollary</u> (principal reduction of a Postnikov decomposition): <u>Let</u> $Y \to B$ <u>be a fibration between path-connected CW-spaces with fiber</u> F, <u>and let</u> F <u>and</u> B <u>be r-connected. Let</u> $\{Y_n\}$ <u>be the Postnikov decomposition of</u> $Y \to B$. <u>Then the composed fibration</u> $Y_{n+r} \to Y_n$ <u>is a principal fibration</u>.

(2.3.11) <u>Consequence</u>: Let $F \subset Y \to B$ be a fibration of CW-spaces, and let F and B be r-connected. The Postnikov decomposition of $Y \to B$ then gives us a tower

$$B = Y_r \leftarrow Y_{2r} \leftarrow Y_{3r} \leftarrow \cdots \leftarrow Y_{nr} \leftarrow \cdots$$

of principal fibrations. This fact, in light of condition (1.1.1) (B), can be used in solving existence and classification problems.

(2.3.12) Corollary (Principal reduction of a CW-decomposition): Let A be $(r-1)$-connected, $r \geq 1$. Let (X, A) be a relative CW-complex with skeletons (X_n, A) and suppose $X_{r-1} = A$. Then the inclusion $X_n \subset X_{n+r}$ for $n > 0$ is a principal cofibration.

(2.3.13) Consequence: Let (X, A) be a relative CW-complex with $X_{r-1} = A$ and A $(r-1)$-connected. Then

$$A = X_{r-1} \subset X_{2r-1} \subset X_{3r-1} \subset \dots \subset X_{nr-1} \subset \dots$$

is a filtration of principal cofibrations that, in light of condition (1.1.1)(A), can be used in solving existence and classification problems.

Remark: The results (2.3.11) and (2.3.13) immediately imply results of T. Ganea and P.J. Hilton on the category and co-category of a space, cf. [33], [47].

Principal reduction is especially simple in the unstable case.

(2.3.14) Definition: Let (X, A) be a strictly pointed relative CW-complex. We call X unstable if all attaching maps $f_e : S^n \to X_n$ vanish when suspended, that is, when $Sf_e \simeq 0$.

(2.3.15) Example: Let X be a simply connected CW-space. Then there exists an unstable CW-complex X and a map $h : \bar{X} \to X$ that is a rational homotopy equivalence, that is h induces an isomorphism $h_* : \pi_*(\bar{X}) \otimes \mathbb{Q} \cong \pi_*(X) \otimes \mathbb{Q}$. We constructed such an unstable model $\bar{X}$ of X in [19]. If X and X' are unstable CW-complexes, then so is $X \times X'$. In particular, the product $T = S_1 \times \dots \times S_n$ of spheres S_i is an unstable CW-complex, see [19]. The dual definition is as follows.

(2.3.16) Definition: Let $p : Y \to B$ be a fibration with fiber F and possessing an orientable Postnikov decomposition $\{Y_n\}$. We call the fibration p Ω-unstable if all the k-invariants $k_n : Y_{n-1} \longrightarrow K(\pi_n(F), n+1)$ vanish when looped, that is, when $\Omega k_n \simeq 0$.

(2.3.17) Example: Let Y be a simply-connected rational CW-space, that is $\pi_n(Y)$ is a $\mathbb{Q}$-vector space for all n. (For example, the rationalization X_0 of a simply-connected CW-space X is a rational space, see (1.5.14)). Then the Postnikov decomposition of Y is orientable and Ω-unstable.

(2.3.18) Theorem: (principal reduction of an unstable CW-decomposition): Let A be $(r-1)$-connected, $r \geq 1$. Let (X, A) be a relative unstable CW-complex with skeletons X_n and let $X_{r-1} = A$. Then the inclusion $X_n \subset X_{n+r}$ for $n \geq 0$ is a principal cofibration with attaching map $f : \bigvee_e S^{|e|-1} \to X_n$, where e runs through all cells in $X_{n+r} - X_n$ and $|e|$ is the dimension of the cell e. The composition $S^{|e|-1} \subset \bigvee_e S^{|e|-1} \xrightarrow{f} X_n \subset X_{|e|-1}$ is homotopic to the attaching map of the cell e.

Remark: The attaching map f in the theorem is not uniquely defined. In 3.5 of [14] we characterize the indeterminancy of f.

(2.3.19) Theorem (principal reduction of an Ω-unstable Postnikov decomposition): Let $Y \to B$ be an Ω-unstable fibration between CW-spaces with fiber F, and let Y be r-connected. Then $Y_{n+r} \to Y_n$ in the Postnikov decomposition of $Y \to B$ is a principal fibration with $f : Y_n \to \prod_{i=n+1}^{n+r} K(\pi_i(F), i+1)$ as classifying map.

<u>The composition</u> $Y_{j-1} \to Y_n \to \bigtimes_{i=n+1}^{n+r} K(\pi_i(F), i+1) \to K(\pi_i(F), i+1)$ <u>is homotopic to the</u> k-<u>invariant</u> k_j <u>for</u> $n < j \leqslant n + r$.

This is an example of a modified Postnikov tower as defined by M. Mahowald in [72].

We now prove the principal reduction theorems for the unstable case. The proof already contains ideas which will be used in the proofs of the general reduction theorems. The argumentation need only be refined in the appropriate manner.

<u>Proof of</u> (2.3.18): The theorem results from inductive application of the following two lemmas:

(2.3.20) <u>Lemma</u>: Let $i : X \subset Y$ be a principal cofibration with classifying map $f : A \to X$ and let $g : B \to Y$ be homotopic to a map $i\,g' : B \to X \subset Y$. Then $X \subset Cg$ is a principal cofibration with classifying map $(f, g') : A \vee B \to X$.

(2.3.21) <u>Lemma</u>: Let (X, A) be a relative CW-complex with skeletons X_j. Suppose X_n is $(r-1)$-connected $(r \geqslant 2)$ and let $j \leqslant \text{Min}\,(2n, n + r + 1)$. Then given a map f in

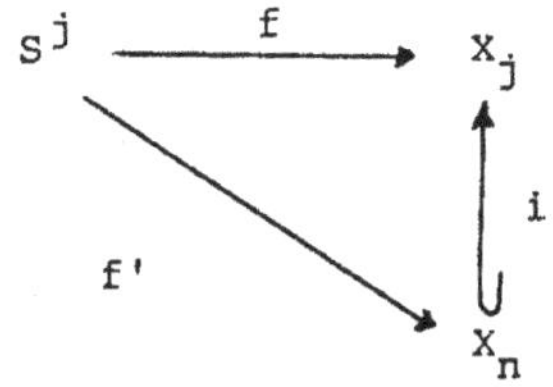

with $Sf \simeq 0$, there exists a map f' with $if' \simeq f$.

<u>Proof</u>: Consider the commutative diagram

$$\begin{array}{ccccc}
\pi_j(X_n) & \xrightarrow{i_*} & \pi_j(X_j) & \xrightarrow{j_*} & \pi_j(X_j, X_n) \\
 & & & & \cong\big\downarrow\beta \\
 & & \big\downarrow S' & & \pi_j(X_j/X_n) \\
 & & & & \cong\big\downarrow S \\
 & & \pi_{j+1}(SX_j) & \xrightarrow{\beta'} & \pi_{j+1}(SX_j/SX_n)
\end{array}$$

in which the upper row is exact. The maps β, β' are induced by quotient maps, S and S' are suspensions. By the Blakes-Massey theorem (3.4.9), β is an isomorphism for $j \leq n + r - 1$, and the Freudenthal suspension theorems tell us that S is an isomorphism for $j \leq 2n$. $\square$

<u>Proof</u> of (2.3.19): This theorem results from inductive application of the following two lemmas.

(2.3.22) <u>Lemma</u>: Let $p : Y \to X$ be a principal fibration with classifying map $f : X \to A$ and let $g : Y \to B$ be homotopic to a map $g'p : Y \to X \to B$. Then $P_g \to X$ is a principal fibration with classifying map $(f, g') : X \to A \times B$.

To prove this lemma, we use the adjoint map $\bar{H}$ defined from the homotopy $H : g \simeq g'p$ as in (1.3.14).

(2.3.23) <u>Lemma</u>: Let $Y \to X$ be a fibration of CW-spaces. Suppose that the fiber F is n-connected and X is r-connected, and let $j \leq \mathrm{Min}(2n + 1, n + r + 1)$. Then given a map f

$$\begin{array}{ccc}
Y & \xrightarrow{f} & K(\pi, j) \\
p\big\downarrow & \nearrow_{f'} & \\
X & &
\end{array}$$

with $\Omega f \simeq 0$, there exists a map f' with $f'p \simeq f$.

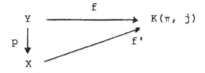

Proof: Let $K = K(\pi, j)$. Consider the commutative diagram

$$\begin{array}{ccccc}
[F, K] & \xleftarrow{i^*} & [Y, K] & \xleftarrow{j^*} & [C_i, K] \\
\Omega \downarrow & & \Omega' \downarrow & \overset{p^*}{\nwarrow} & \cong \uparrow q^* \\
[\Omega F, \Omega K] & \xleftarrow{(\Omega i)^*} & [\Omega Y, \Omega K] & & [X, K]
\end{array}$$

in which the upper row is exact. $i : F \subset Y$ is the inclusion and $j : Y \subset Y \cup_i CF = C_i$ is the inclusion into the mapping cone. Let $q : Y \cup CF \to X$ be the extension of the fibration $p : Y \to X$, so that $q|CF = *$. By a theorem of T. Ganea (6.4.9), q is then $(n + r + 2)$ connected. Therefore, by (1.4.12), q induces a bijection q^* in the diagram for $n + r + 2 \geqslant j + 1 = m_K$. By a theorem of W. Barcus, the evaluation map $R : S\Omega F \to F$, $R(t, \sigma) = \sigma(t)$, is then $(2n + 1)$-connected, so R induces an injection Ω for $2n + 1 \geqslant j = m_K - 1$, see (6.1.4). The lemma follows from exactness. $\square$

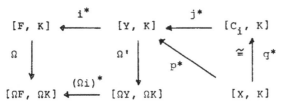

(2.4) The exact classification sequences of a principal cofibration

In this section, we condense into exact sequences various properties of obstructions and differences in principal cofibrations. These long exact sequences generalize the cofibration or Puppe sequence. We will present them in four different guises: for principal cofibrations and for relative principal cofibrations, where in each of these cases homotopy sets of maps and of sections are considered. These sequences are then used to classify homotopy sets, whereby the calculation of the operators Σf, f^* and $f^\#$ is the crucial task. Later on, in chapters (3) and (4) we will discuss properties of these operators. At the end of this section we will classify, as an example, the homotopy set $[T_n, S^2]$ of maps from an n-dimensional torus T_n into the 2-sphere, using the homotopy groups $\pi_*(S^2)$.

We now describe the four classification sequences of a principal cofibration. Given an inclusion $Y \subset X$, we define as in (2.1.1) the ex-space $\Sigma_Y X = S^1{}_Y \ltimes X$, $X \overset{\sigma}{\subset} \Sigma_Y X \overset{\rho}{\to} X$ where $\sigma(x) = (*, x)$ and $\rho(t, x) = x$. This space is a quotient of the relative cylinder $I_Y \ltimes X$. Thus, given a map H on $\Sigma_Y X$ such that $u = H|X$, we obtain a self-homotopy $H : u \simeq u$ relative to Y.

Let $* \in Y \subset X$ and $f : A \to X$ be given. A map $u_o : C_f \to U$ from the mapping cone then leads to the sequence

$$(A) \quad [\Sigma_Y X, U]^u \xrightarrow{\Sigma f} [SA, U] \xrightarrow{u_o^+} [C_f, U]^v \xrightarrow{j} [X, U]^v \xrightarrow{f*} [A, U]$$

where $u = u_o|X$ and $v = u|Y$. $f*$ is induced from f, and j is the restriction. Furthermore $u_o^+(\alpha) = u_o + \alpha$, and $(\Sigma f)H = d(u_o, H, u_o)$ is the difference, see (1.2.7) and (1.2.5).

If in (A) we take $u_o = 0 : C_f \longrightarrow * \in U$ to be the trivial map and set $Y = *$, we obtain the Puppe sequence [100]

$$(A_o) \qquad [SX, U] \xrightarrow{(Sf)^*} [SA, U] \xrightarrow{0^+} [C_f, U] \xrightarrow{j} [X, U] \xrightarrow{f^*} [A, U]$$

To see this, one need only look at (0.0.6) and the definition of w_f in (1.2.3). On replacing f by Sf in (A_o), we obtain by induction the long Puppe sequence. The Puppe sequence for the inclusion $A \vee B \subset A \times B$ leads easily to the exact sequence of (0.3).

We now describe three different generalizations of the sequence (A) and thus of the Puppe sequence (A_o).

Let A be a co-H-group and let $\tilde{C}_f \to C_f$ be a fibration with fiber F Given a section $u_o : C_f \to \tilde{C}_f$, we obtain the sequence

$$(B) \quad \langle \Sigma_Y X, \rho^*\tilde{X} \rangle^u \xrightarrow{\Sigma f} [SA, F] \xrightarrow{u_o^+} \langle C_f, \tilde{C}_f \rangle^v \xrightarrow{j} \langle X, \tilde{X} \rangle^v \xrightarrow{f^\#} [A, F]$$

where $u = u_o|X$ and $v = u|Y$ and j is the restriction, $f^\#$, $u_o^+(\alpha) = u_o + \alpha$ and $(\Sigma f)H = d(u_o, H, u_o)$ are defined as in (1.2.12), (1.2.18) and (1.2.16) respectively. (Sequence (B) can be regarded as a special case of M.H. Eggar's sequence 2.4. in [30].)

Now we generalize (A) and (B) for the relative case. Let $D \subset A \overset{\rho}{\to} D$ be an ex-cofiber space over D and let $f : A \to X$ be an extension of $d : D \to Y$. Then given a map $u_o : C_D f \to U$ from the relative mapping cone $C_D f$, we obtain the sequence

$$(C) \quad [\Sigma_Y X, U]^u \xrightarrow{\Sigma f} [S_D A, U]^w \xrightarrow{u_o^+} [C_D f, U]^v \xrightarrow{j} [X, U]^v \xrightarrow{f^*} [A, U]^w$$

where again $u = u_o|X$ and $v = u|Y$, and additionally $w = ud$. f^* is induced from f and j is the restriction. Furthermore, $u_o^+(\alpha) = u_o + \alpha$ and $(\Sigma f)H = d(u_o, H, u_o)$ are defined as in (2.1.11) and (2.1.10). Now suppose $A = S_D A'$ is in fact a relative suspension over D, let $\widetilde{C_D f} \to C_D f$ be a fibration whose restriction is $\tilde{X} \to X$ and let $\tilde{D} = d^*\tilde{X}$. Given a section $u_o : C_D f \to \widetilde{C_D f}$, we obtain the sequence

$$(D)\quad \langle \Sigma_Y X, \rho^*\tilde{X}\rangle^u \xrightarrow{\Sigma f} \langle S_D A, \rho^*\tilde{D}\rangle^w \xrightarrow{u_o^+} \langle C_D f, \widetilde{C_D f}\rangle^v \xrightarrow{j} \langle X, \tilde{X}\rangle^v$$
$$\xrightarrow{f^{\#}} \langle A, \rho^*\tilde{D}\rangle^v$$

where $u = u_o|X$, $v = u|Y$ and we set $w = d^*u$. $f^{\#}$, $u_o^+(\alpha) = u_o + \alpha$ and $(\Sigma f)H) = d(u_o, H, u_o)$ are defined as in (2.1.16), (2.1.20) and (2.1.18). The sequences (B) and (C) are special cases of (D), in which we set $D = *$ or $C_D f = (C_D f) \times U$ respectively.

(2.4.1) <u>Theorem</u>: <u>The classification sequences</u> (A), (B), (C), (D) <u>are exact, that is</u>

i) $\mathrm{Im}\,(j) = f^{*-1}(0)$ <u>or</u> $\mathrm{Im}\,(j) = f^{\#-1}(0)$

ii) $j^{-1}(u) = \mathrm{Im}\,(u_o^+) = \{u_o + \alpha \mid \alpha\}$

iii) $\mathrm{Im}\,(\Sigma f) = (u_o^+)^{-1}(u_o) = \{\alpha | u_o + \alpha = u_o\}$

<u>In each sequence,</u> Σf <u>is a homomorphism of groups, where the group structure in</u> $[\Sigma_Y X, U]^u$ <u>and</u> $\langle \Sigma_Y X, \rho^*\tilde{X}\rangle^u$ <u>is defined to be addition of self-homotopies of</u> u. <u>The operation</u> + <u>in the definition of</u> u_o^+ <u>is also a group operation</u>.

We call (A), (B), (C), (D) classification sequences because we can use them to calculate homotopy sets. The homotopy set in the middle of a sequence, for instance $[C_f, U]^v$ in (A), can be decomposed into subsets using the maps Σf and f^*. That is, to every $u \in \mathrm{Ker}\,(f^*)$ we have the subset $j^{-1}(u) \subset [C_f, U]^v$ which is an orbit of the operation of $[SA, U]$, by ii). The isotropy subgroup of this operation at $u_o \in j^{-1}(u)$ is precisely $\mathrm{Im}\,(\Sigma f)$, by iii). Therefore

u_o^+ defines a bijection

(2.4.2) $$[SA, U]/\mathrm{Im}\,(\Sigma f) \xrightarrow{\approx} j^{-1}(u)$$

(We define the quotient G/H, where H is a subgroup of the group G, to be the set $\{H\cdot g \mid g \in G\}$.) Notice that the homomorphism Σf in (2.4.2) is defined when we are given a map $u_o : C_f \to U$ that represents the class $u_o \in j^{-1}(u)$. Thus to every $u \in \mathrm{Ker}\,(f^*)$ we choose a map u_o and herewith obtain a set $[SA, U]/\mathrm{Im}\,(\Sigma f)$. The union of these sets is $[C_f, U]^V$. Similar considerations hold for the sequences (B), (C), (D).

Proof of (2.4.1): We prove this theorem only for sequence (A), since corresponding arguments suffice for (B), (C), and (D). To start with, i) follows from the obstruction property of f^* enunciated in (1.2.2). ii) follows from the classification (1.2.8), see (1.2.9). iii) is obtained as follows. Let $u_o + \alpha = u_o$, that is suppose we have a homotopy $H : (u_o, \alpha)\,\mu \simeq u_o$ relative Y. Since $(u_o, \alpha)\,\mu|_X = u$, the restriction of H to X is a self-homotopy $H' : u \simeq u$ relative Y for which

$$\begin{aligned} 0 &= d(u_o + \alpha, H', u_o) && \text{see (1.2.5)} \\ &= d(u_o + \alpha, u_o) + d(u_o, H', u_o) && \text{see (1.2.6)} \\ &= -d(u_o, u_o + \alpha) + (\Sigma f)H' && \text{see (1.2.6) and the def. of } \Sigma f, \\ &= -\alpha + (\Sigma f)H' && \text{see (1.2.8)} \end{aligned}$$

Thus $\alpha \in \mathrm{Im}\,(\Sigma f)$. On the other hand, if $\alpha \in \mathrm{Im}\,(\Sigma f)$ then an H' exists such that $0 = -\alpha + (\Sigma f)H'$. The equations above then imply that $0 = d(u_o + \alpha, H', u_o)$. Therefore H' can be extended to a homotopy $u_o + \alpha \simeq u_o$, and so $u_o^+(\alpha) = u_o$. Finally, Σf is a homomorphism of groups by (1.2.6), and $+$ is a group operation in virtue of (1.2.7). ▭

The next theorem considerably simplifies the task of computing Σf.

(2.4.3) Theorem: *If* Σf *maps into an abelian group (which is always the case in sequences* (B) *and* (D)), *then* Σf *depends only on* $u = u_o|X$. *In this case we write* $\Sigma f = \Sigma(u, \Sigma f)$. *If* $Y \subset X$ *is a cofibration and* u *is homotopic to* u', *then* $\operatorname{Im} \Sigma(u, f) = \operatorname{Im} \Sigma(u', f)$.

Proof: We prove the theorem only for sequence (A), since the obvious modifications suffice to prove it for the other three sequences. Let u_o and u_o' be two extensions of $u : X \to U$ over C_f. Then, using (1.2.6), we have the equalities

$$\begin{aligned}(\Sigma f)H &= d(u_o, H, u_o)\\ &= d(u_o, u_o') + d(u_o', H, u_o') + d(u_o', u_o)\\ &= -d(u_o', u_o) + d(u_o', H, u_o') + d(u_o', u_o)\\ &= d(u_o', H, u_o') .\end{aligned}$$

Thus Σf depends only on u and not on the extension u_o of u. Of course u must be extendable over C_f in order for Σf to be defined.

Now let $Y \subset X$ be a cofibration and $G : u \simeq u'$ a homotopy. Then $X \subset \Sigma_Y X$ is a cofibration as well. Given a map $F_o : \Sigma_Y X \to U$ extending u, there exists a homotopy $F : F_o \simeq F_1$ extending G. In the commutative diagram, $G^{\#}$ with $G^{\#}F_o = F_1$ is an isomorphism of groups.

$$\begin{array}{ccccc} [\Sigma_Y X,\ U]^u & \xrightarrow{p^*} & [\Sigma_* X,\ U]^u & \searrow^{\Sigma_*(u,\ f)} & \\ \downarrow G^{\#} & & \downarrow \underline{G}^{\#} & & [SA,\ U] \\ [\Sigma_Y X,\ U]^{u'} & \xrightarrow{p^*} & [\Sigma_* X,\ U]^{u'} & \nearrow_{\Sigma_*(u',\ f)} & \end{array} \quad .$$

If $p : \Sigma_* X \to \Sigma_Y X$ denotes the identification map, then $\Sigma(u,\ f) = \Sigma_*(u,\ f) \circ p^*$. Defining $\underline{G}^{\#}$ as $\underline{G}^{\#}(H) = -G + H + G$, we have $\underline{G}^{\#} p^* = p^* G^{\#}$. The theorem then follows from the equations

$$\begin{aligned} \Sigma_*(u,\ f)(H) &= d(u_o,\ H,\ u_o) \\ &= -d(u_o,\ G,\ u_o') + d(u_o,\ H,\ u_o) + d(u_o,\ G,\ u_o') \\ &= d(u_o',\ -G + H + G,\ u_o') \\ &= \Sigma_*(u',\ f) \circ \underline{G}^{\#}(H) \end{aligned}$$

where u_o and u_o' are extensions of u and u' respectively over C_f. □

Theorems (2.4.3) and (2.4.2), applied in particular to the classification sequence (B), yield the formula

$$(2.4.4) \qquad \langle C_f,\ \tilde{C}_f \rangle^V \approx \coprod_{u\ \in\ \mathrm{Ker}\ (f^{\#})} [SA,\ F]/\mathrm{Im}\ \Sigma(u,\ f)$$

where the bijection arises from the operation $+$ of $[SA,\ F]$. There are similar formulas for the other classification sequences. Thus, we solve an existence problem by calculating $\mathrm{Ker}\ (f^{\#})$, and we wsolve a classification problem by calculating $\Sigma(u,\ f)$.

We can obtain exact extensions of the classification sequences just as in the case of the Puppe sequence (A_o). To see this, we first define, in accordance with (2.1.1), the ex-space

$$\Sigma_Y^n\ X \ = \ S_Y^n \ltimes X, \quad X \overset{\sigma}{\subset} \Sigma_Y^n\ X \overset{\rho}{\to} X$$

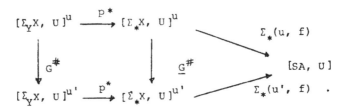

where $\sigma(x) = (*, x)$ and $\rho(t, x) = x$. Then $\Sigma_Y^0 X = X \cup_Y X$ and there is a canonical homeomorphism

(2.4.5) $\Sigma_Y^n X = S_X^n (\Sigma_Y^0 X), \quad n \geq 0$

of ex-spaces over X, where $S_X^n = S_X \circ S_X^{n-1}$ is the n-fold suspension relative X, see (2.1.3). Then $\Sigma_Y X = \Sigma_Y^1 X$, and addition of self-homotopies is the same as the comultiplication (2.1.12) on $S_X (\Sigma_Y^0 X)$. We will now extend sequences (A) and (D); the reader may extend (B) and (C) for himself. The map f in sequenc A gives us the map

(2.4.6) $$Wf : S^n A \xleftarrow[w_1]{\simeq} S^n \ltimes A \cup CA \xrightarrow[w_2]{} S^n \ltimes X \cup C_f \xrightarrow[w_3]{} \Sigma_Y^n X \cup C_f$$

where w_1 identifies CA to a point, w_2 is the restriction of $1 \ltimes \pi f$, and w_3 is the restriction of $S^n \ltimes C_f \to \Sigma_Y^n X$. Just as in (1.2.4), we obtain herewith a homotopy equivalence

(2.4.7) $C_{Wf} \simeq \Sigma_Y^n C_f$ under $\Sigma_Y^n X \cup C_f$.

Choosing a map $u_0 : C_f \to U$, we arrive at the sequence $(n \geq 1)$

($\underline{A}$) $$\xrightarrow{\Sigma^{n+1} f} [S^{n+1} A, U] \xrightarrow{u_0^+} [\Sigma_Y^n C_f, U]^{u_0} \xrightarrow{j} [\Sigma_Y^n X, U]^u \xrightarrow{\Sigma^n f} [S^n A, U]$$

where j is the restriction and $u_0^+(\alpha) = \rho^*(u_0) + \alpha$ is defined using (2.4.7). We define $(\Sigma^n f)H = (Wf)^*(H \cup u_0)$. By (1.2.5), $\Sigma^1 f = \Sigma f$ and so ($\underline{A}$) extends sequence (A).

We can extend sequence (D) in a very similar manner. The map f in it leads to

(2.4.8) $$Wf : S_D^n A \xleftarrow[w_1]{\simeq} S_D^n \ltimes A \cup C_D A \xrightarrow[w_2]{} S_D^n \ltimes X \cup C_D f \xrightarrow[w_3]{} \Sigma_Y^n X \cup C_D f$$

where the homotopy equivalence w_1 under and over D is defined by the projection $\rho : C_D A \to D$. w_2 is the restriction of $1_D \propto \pi_f$ with π_f as in (2.1.5), and w_3 is the restriction of the identification map (notice that $D \subset Y$). As a generalization of (2.4.7), we then have the homotopy equivalence

(2.4.9) $$C_D Wf \simeq \Sigma_Y^n C_D f \quad \text{under} \quad \Sigma_Y^n X \cup C_D f.$$

Choosing a section $u_o : C_D f \to \widetilde{C_D f}$, we arrive at the sequence $(n \geq 1)$

$$(\underline{D}) \xrightarrow{\Sigma^{n+1} f} \langle S_D^{n+1} A, \rho^* \widetilde{D} \rangle \xrightarrow{w} \xrightarrow{u_o^+} \langle \Sigma_Y^n C_D f, \rho^* \widetilde{C_D f} \rangle \xrightarrow{u_o \; j} \langle \Sigma_Y^n X, \rho^* \widetilde{X} \rangle^u \xrightarrow{\Sigma^n f}$$

where j is the restriction and $u_o^+(\alpha) = \rho^*(u_o) + \alpha$ is defined using (2.4.9). We define $(\Sigma^n f)H = (Wf)^{\#}(H \cup u_o)$. By (2.1.18), $\Sigma^1 f = \Sigma f$ and so $(\underline{D})$ extends sequence (D).

(2.4.10) Theorem: Sequence $(\underline{D})$ (and therefore $(\underline{A})$, $(\underline{B})$, $(\underline{C})$) is an exact sequence of group homomorphisms and groups, where the groups are abelian for $n \geq 2$.

Proof: We will prove the theorem for sequence $(\underline{A})$. The same arguments can be used for $(\underline{D})$, where the only difference is there is more pencil-pushing to do.

First we replace the notations in sequence (A) according to the scheme

$$\begin{array}{ccccc} f, & A, & X, & Y, & u_o \\ Wf, & S^n A, & \Sigma_Y^n X \cup C_f, & C_f, & (u_o \rho) h \end{array}$$

where h is the homotopy equivalence of (2.4.7). This leads to the sequence $(\underline{A})$. A consideration of the definitions shows that all we have to prove is that the diagram

$$\begin{array}{ccc}
[\Sigma_{C_f}(\Sigma_Y^n X \cup C_f),\, U]^{\bar{u}} & \xrightarrow{\ \Sigma(Wf)\ } & [SS^n A,\, U] \\
\mathfrak{v} \ \wr & \nearrow_{\Sigma^{n+1} f} & \\
[\Sigma_Y^{n+1} X,\, U]^u & &
\end{array}$$

commutes, where $\bar{u}$ is the restriction of $u_o \rho h$, that is of $u_o\rho$, to $\Sigma_Y^n X \cup C_f$. The bijection $\mathfrak{v}$ is thus given by (0.0.6). Clearly $\Sigma^n f$ and j are homomorphisms of groups. The map u_o^+ in (A) is also a homomorphism, since the diagram

$$\begin{array}{ccccc}
\Sigma_Y^n C_f \simeq C_{Wf} & \xrightarrow{\ \mu\ } & C_{Wf} \vee S^{n+1}A \simeq \Sigma_Y^n C_f \vee S^{n+1}A & & \\
\downarrow 1\times\mu & & & \searrow \rho\vee\varepsilon & \\
\Sigma_Y^n (C_f \vee SA) & = & \Sigma_Y^n C_f \vee S^{n+1} \ltimes SA & \xrightarrow{\ \rho\vee\pi\ } & C_f \vee S^{n+1}A
\end{array} \tag{2.4.11}$$

is homotopy commutative. Here $\varepsilon = (-1)^{n+1}$ and π is the identification, and μ denotes the appropriate comultiplication. □

(2.4) Example: Classification of maps from an n-dimensional torus T_n into the 2-sphere S^2.

We will see that the homotopy set $[T_n, S^2]$ can be characterized fairly easily by use of the exact sequence (A). We will need additionally only the following result of [12]. Let $w_n : S^{n-1} \to T_n^{n-1}$ be the attaching map of the n-cell of the torus T_n. The induced map $w_n^* : [T_n^{n-1}, S^2] \to \pi_{n-1}(S^2)$ then has the property that

(2.4.12) Lemma: $w_n^* = 0$ for $n \geq 5$.

The k-skeleton T_n^k of the torus T_n is the set of points

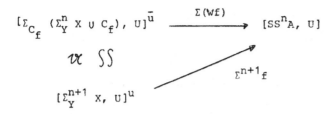

$t \in T_n = S^1 \times \ldots \times S^1$ in which at least $(n - k)$ coordinates t_i are equal to the basepoint of S^1. To every subset $a \subset \{1,\ldots,n\}$ of cardinality $\# a$ there is a cell of dimension $\# a$. Let $f_k : \bigvee_{\# a = k} S^{k-1} \to T_n^{q-1}$ be the wedge of the attaching maps of the k-cells in T_n. We write $\pi_k = \pi_k(S^2)$. Because $T_n^k = C_{f_k}$ for every f_k, we have by (2.4.4) the bijection $(Y = *)$

$$[T_n^k, S^2] \xleftarrow{\chi} \bigsqcup_{u \in \operatorname{Ker}(f_k^*)} \pi_k^{\binom{n}{a}} / \operatorname{Im} \Sigma(u, f_k).$$

(2.4.12) tells us that $f_k^* = 0$ for $k \geqslant 5$. It also shows that $\Sigma(u, f_k) = 0$ for $k \geqslant 4$, since Wf_k is given by the attaching maps of $(k+1)$-cells in $S^1 \ltimes T_n$ (see (2.4.7)) and since all attaching maps of k-cells in T_n or $S^1 \ltimes T_n$ factor over w_k in (2.4.12). Thus we obtain the bijection

$$(2.4.13) \qquad [T_n, S^2] \approx \operatorname{Ker}(f_4^*) \times \pi_4^{\binom{n}{4}} \times \ldots \times \pi_n^{\binom{n}{n}} .$$

where $\operatorname{Ker} f_4^* \subset [T_n^3, S^2]$. Using (4.4.6) we can determine $\operatorname{Ker}(f_4^*)$.

(2.5) The exact classification sequences of a principal fibration

In this section we show how the properties of obstructions and differences for principal fibrations are reflected in certain long exact sequences which are generalizations of the fiber sequence of a map. These have four different versions: for principal fibrations and relative principal fibrations, and in each of these cases for either homotopy sets of maps or of retractions. We will use these sequences to classify homotopy sets. The crucial step is always the computation of the operators $\underline{\Omega}f$, f_*, $f_{\#}$ in the sequences. It should be noted that the initial segment of the first of these sequences has also been studied by I.H. James, E. Thomas [55], [56] and J.W. Rutter [106].

We will again develop this section in strict duality to the preceding one. An item here numbered (2.5.i), $i \geqslant 1$ will be the exact dual of the one there numbered (2.4.i). We start by listing the four classification sequences of a principal fibration. Given a space $X \overset{\eta}{\rightarrow} Y$ over Y, we define, in accordance with (2.2.1), the ex-space $\underline{\Omega}_Y X = X_Y^{S^1}$, $X \overset{\sigma}{\rightarrow} \underline{\Omega}_Y X \overset{\rho}{\rightarrow} X$ where $\sigma(x) = c_x$ and $\rho(\tau) = \tau(*)$. This is a space over the free path space X_Y^I. Therefore, given a map H into $\underline{\Omega}_Y X$ such that $u = \rho H$, we obtain a self-homotopy $H : u \simeq u$ over Y. If $Y = *$ then $\underline{\Omega}_* X = X^{S^1}$ is the free loop space.

Now suppose we are given a map $f : X \rightarrow A$. Then a map $u_o : U \rightarrow P_f$ into the mapping path space P_f leads to the sequence

$$(A') \quad [U, \underline{\Omega}_Y X]_u \xrightarrow{\underline{\Omega}f} [U,\Omega A] \xrightarrow{u_o^+} [U,P_f]_v \xrightarrow{q_*} [U,X]_v \xrightarrow{f_*} [U,A]$$

in which $u = qu_o$ and $v = \eta u$ and $U \xrightarrow{u_o} P_f \xrightarrow{q} X \xrightarrow{\eta} Y$. f_*

and q_* are induced by f and q, and $u_o^+(\alpha) = u_o + \alpha$ and $(\underline{\Omega}f)H = d(u_o, H, u_o)$ are defined as in (1.3.7) and (1.3.5). If in (A') we take $u_o = 0 : U \to * \in P_f$ to be the trivial map and set $Y = *$, we obtain as a special case the fiber sequence

$$(A_o') \quad [U, \Omega\underline{X}] \xrightarrow{(\Omega f)} [U,\Omega A] \xrightarrow{0^+} [U,P_f] \xrightarrow{q_*} [U,X] \xrightarrow{f_*} [U,A]$$

which was studied in detail by Nomura in [90], see also [24]. To arrive at (A_o'), we use (0.0.7) and the definition of w_f in (1.3.3). If we replace f by Ωf in (A_o') and proceed inductively, we obtain the long fiber sequence of the map f. A special case of this is the sequence (0.2.1), see (0.1.11).

We discuss three different generalizations of the sequence (A') and accordingly of the fiber sequence (A_o'). Let A be an H-group and $P_f \subset \overline{P_f}$ a closed cofibration with cofiber F. Given a retraction $u_o : \overline{P_f} \to P_f$, we can construct the sequence

$$(B') \quad \langle\sigma_*\overline{X}, \underline{\Omega}_Y X\rangle_u \xrightarrow{\underline{\Omega} f} [F,\Omega A] \xrightarrow{u_o} \langle\overline{P}_f,P_f\rangle_v \xrightarrow{q_*} \langle\overline{X},X\rangle_v \xrightarrow{f_\#} [F,A]$$

$X \subset \overline{X}$ is the cofibration induced by q, $u = q_* u_o$ and $v = \eta_* u$ are induced retractions, and q_* is induced by q. $f_\#$, $u_o^+(\alpha) = u_o + \alpha$ and $(\underline{\Omega}f)H = d(u_o, H, u_o)$ are defined as in (1.3.12), (1.3.20) and (1.3.18).

We now generalize (A') and (B') for the relative case. Let A together with $D \xrightarrow{\sigma} A \xrightarrow{\rho} D$ be an ex-fiber space over D, and let $f : X \to A$ be a lifting of $X \xrightarrow{\eta} Y \xrightarrow{d} D$. Given a map $u_o : U \to P_D f$ we can construct the sequence

$$(C') \quad [U, \underline{\Omega}_Y X]_u \xrightarrow{\underline{\Omega} f} [U,\Omega_D A]_w \xrightarrow{u_o^+} [U,P_D f]_v \xrightarrow{q_*} [U,X]_v \xrightarrow{f_*} [U,A]_w$$

where $w : U \xrightarrow{u_o} P_D f \xrightarrow{q} X \xrightarrow{\eta} Y \xrightarrow{d} D$, $u = qu_o$ and $v = \eta u$. $u_o^+(\alpha) = u_o + \alpha$ and $(\underline{\Omega}f)H = d(u_o, H, u_o)$ are defined as in (2.2.11) and (2.2.10).

Now suppose that $A = \Omega_D A'$ is in fact a relative path space over D, and let $P_D f \subset \overline{P_D f}$ be a closed cofibration inducing $X \subset \bar{X}$ and $D \subset \bar{D}$. Given a retraction $u_o : \overline{P_D f} \to P_D f$, we can construct the sequence

$$(D') \quad \langle \sigma_* \bar{X}, \underline{\Omega}_Y X \rangle_u \xrightarrow{\underline{\Omega}f} \langle \sigma_* \bar{D}, \Omega_D A \rangle_w \xrightarrow{u_o^+} \langle \overline{P_D f}, P_D f \rangle_v \xrightarrow{q_*}$$
$$\langle \bar{X}, X \rangle \xrightarrow{f_\#} \langle \sigma_* \bar{D}, A \rangle_w$$

in which, again $u = q_* u_o$, $v = \eta_* u$ and $w = d_* v$ are induced retractions. $f_\#$, $u_o^+(\alpha) = u_o + \alpha$ and $(\underline{\Omega}f)H = d(u_o, H, u_o)$ are defined as in (2.2.16), (2.2.10) and (2.2.18). The sequence (B') and essentially (C') as well are special cases of (D'), as is seen by setting $D = *$ and $\overline{P_D f} = P_D f \vee U$ respectively.

(2.5.1) <u>Theorem:</u> <u>The classification sequences are exact, that is</u>

i) $\operatorname{Im}(q_*) = f_*^{-1}(0)$ <u>or</u> $\operatorname{Im}(q_*) = f_\#^{-1}(0)$

ii) $q_*^{-1}(u) = \operatorname{Im}(u_o^+) = \{u_o + \alpha \mid \alpha\}$

iii) $\operatorname{Im}(\underline{\Omega}f) = (u_o^+)^{-1}(u_o) = \{\alpha \mid u_o + \alpha = u_o\}$.

<u>In every sequence</u> $\underline{\Omega}f$ <u>is a homomorphism of groups, where the group structures on</u> $[U, \underline{\Omega}_Y X]_u$ <u>and</u> $\langle \sigma_* \bar{X}, \underline{\Omega}_Y X \rangle_u$ <u>are given by addition of self-homotopies of</u> u. <u>The operation</u> $+$ <u>in the definition of</u> u_o^+ <u>is a group operation.</u>

The remarks following Theorem (2.4.1) can be adapted to apply here. In particular, u_o^+ determines in the sequence (A) a bijection

$$(2.5.2) \qquad [U, \Omega A]/\operatorname{Im} \underline{\Omega}f \approx q_*^{-1}(u)$$

There are similar bijections arising from the sequences (B'), (C'), (D'). The proof of (2.5.1) is dual to that of (2.4.1) - the arguments can be transferred without difficulty. This is also true for the

(2.5.3) Theorem: *If Ωf maps into an abelian group, which is automatically the case in sequences (B'), (D'), then Ωf depends only on $u = q_*(u_o)$. We then write $\Omega f = \underline{\Omega}(u, f)$. If u is homotopic to u' over Y, then $\operatorname{Im} \underline{\Omega}(u, f) = \operatorname{Im} \underline{\Omega}(u', f)$.*

For example, we can apply (2.5.3) and (2.5.2) to extract from classification sequence (A') the formula

$$(2.5.4) \quad [U, P_f]_V \approx \bigsqcup_{u \in \operatorname{Ker} f_*} [U, \Omega A]/\operatorname{Im} \underline{\Omega}(u, f)$$

The bijection is created by the operation + of $[U, \Omega A]$. Similar formulas arise from the other classification sequences. Formula (2.5.4) was proved in 3.5 of [56]. I.M.James - E. Thomas call the elements of $\operatorname{Im} \underline{\Omega}(u, f)$ "f-*correlated to* u".

Just as the fiber sequence (A_o') can be extended to a long exact sequence, so also can the other classification sequences. To see this, we first define, in accordance with (2.2.1), the ex-space

$$\underline{\Omega}_Y^n X = X_Y^{S^n}, \quad X \xrightarrow{\sigma} \underline{\Omega}_Y^n X \xrightarrow{\rho} X$$

with $\sigma(x) = c_x$ and $\rho(\tau) = \tau(*)$. τ and c_x are maps $S^n \to X$, and $c_x(t) = x$. We have $\underline{\Omega}_Y^o = X \times_Y X$, and there is a canonical homeomorphism of ex-spaces over X

$$(2.5.5) \qquad \underline{\Omega}_Y^n X = \Omega_X^n (\underline{\Omega}_Y^o X), \quad n \geq 0.$$

Here $\Omega_X^n = \Omega_X \circ \Omega_X^{n-1}$ is the n-fold relative loop space relative X, see (2.2.3). Thus $\underline{\Omega}_Y^1 X = \underline{\Omega}_Y X$, and addition of self-homotopies can also be characterized as multiplication in $\Omega_X(\underline{\Omega}_Y^o X)$, see (2.2.12). We will extend only sequence (A'). The sequences (B'), (C'), (D') can be extended in a similar fashion, whereby the extended D' becomes the dual of sequence ($\underline{D}$) in (2.4). Starting with the map f in the sequence A, we define as the dual to (2.4.6) the map

$$\text{(2.5.6)} \quad Wf : \underline{\Omega}_Y^n X \times_X P_f \xrightarrow[w_3]{} \underline{\Omega}_*^n X \times_X P_f \xrightarrow[w_2]{} \underline{\Omega}_*^n A \times_A PA \xleftarrow[w_1]{\simeq} \Omega^n A,$$

where w_3 is induced by $Y \to *$ and w_2 is induced by $\pi_f : P_f \to PA$. w_1 is the inclusion of the fiber. Since PA is contractible, w_1 is a homotopy equivalence. As in (1.3.4) we have a homotopy equivalence

$$\text{(2.5.7)} \quad P_{Wf} \simeq \underline{\Omega}_Y^n P_f \quad \text{over} \quad \underline{\Omega}_Y^n X \times_X P_f .$$

Given a map $u_o : U \to P_f$, we can construct the sequence $(n \geq 1)$

$$\text{(A')} \quad \xrightarrow{\underline{\Omega}^{n+1} f} [U, \Omega^{n+1} A] \xrightarrow{u_o^+} [U, \underline{\Omega}_Y^n P_f]_{u_o} \xrightarrow{q_*} [U, \underline{\Omega}_Y^n X]_u \xrightarrow{\underline{\Omega}^n f} [U, \Omega^n A]$$

in which q_* is induced by $q : P_f \to X$, and $u_o^+(\alpha) = (\sigma_* u_o) + \alpha$ is defined using (2.5.7). $(\underline{\Omega}^n f)H = (Wf)_*(\bar{H} \underline{\times} u_o)$ is defined analogously with (1.3.5). Thus $\underline{\Omega}^1 f = \underline{\Omega} f$ and ($\underline{A}$') extends (A').

A proof dual to that of (2.4.10) can be given for the

(2.5.8) Theorem: The sequence ($\underline{D}$') (and therefore ($\underline{A}$'), ($\underline{B}$'), ($\underline{C}$')) is an exact sequence of group homomorphisms and groups, and of abelian groups for $n \geq 2$.

(2.6) The fiber and cofiber sequences in the category of ex-spaces

The fiber and cofiber sequences in the category Top^o of pointed spaces can be generalized to sequences in the category Top(D) of ex-spaces. We will show that they are essentially special cases of the classification sequences (D) and (D'). J. F. McClendon, I.M. James and others have made clear the importance of the category Top (D) for obstruction theory. The term 'ex-space' ('ex' from 'extract') is taken over from I.M. James.

Definition: Top(D) denotes the category of ex-spaces over D. The objects A, B are ex-spaces over D, and the morphisms or ex-maps are maps $f : A \to B$ over and under D. We write $\pi(A, B) = [A, B]_D^D$ for the homotopy set of ex-maps from A to B.

As we pointed out at the beginning of (2.1), the category Top(D) is closely related to the category $Top(*) = Top^o$ of pointed spaces. The cone, the path space, the suspension nand the loop space can be constructed in Top(D). In particular, adjunction again gives us an isomorphism

$$(2.6.1) \qquad \pi(S_D A, B) \cong \pi(A, \Omega_D B)$$

of groups. The group structures were described in (2.1.12) and (2.2.12).

Let X be an ex-space over D. If we take an ex-map $f : A \to X$ and define the mapping cone $C_D f$ as in (2.1.5), we again get an ex-space over D. We then obtain the Puppe (or cofiber) sequence

$$(D_o) \qquad A \xrightarrow{f} X \xrightarrow{i} C_D f \xrightarrow{q} S_D A \xrightarrow{S_D f} S_D X \to \dots$$

of ex-maps. i is the inclusion and q the projection defined by $C_DA \to S_DA$. S_Df is the relative suspension of f. By repeated application of the functor S_D to the sequence (D_o), we obtain a long sequence. Notice that C_Df in (2.1.5) in general is not an ex-space over D.

Similarly, if we take an ex-map $f : X \to A$ and define the mapping path space P_Df as in (2.2.5), we again get an ex-space over D. We thus arrive at the <u>fiber sequence</u>

$$(D_o') \qquad A \xleftarrow{f} X \xleftarrow{q} P_Df \xleftarrow{i} \Omega_DX \xleftarrow{\Omega_Df} \Omega_DA \longleftarrow \cdots$$

of ex-maps. q is the projection and i is the inclusion defined by $\Omega_DA \subset P_DA$. Ω_Df is the relative loop map of f. Applying the functor Ω_D to the sequence (D_o'), we obtain inductively a long sequence.

We can generalize the sequences (A_o) of (2.4) and (A_o') of (2.5) to arrive at the

(2.6.2) <u>Theorem</u>: <u>Let U be an ex-space. When we apply the contravariant functor $\pi(\ldots, U)$ and the covariant functor $\pi(U, \ldots)$ to the cofiber and fiber sequences (D_o) and (D_o') respectively, we obtain long exact sequences of sets which from the fourth term onwards are sequences of groups and group homomorphisms</u>.

<u>Proof</u>: We here prove the theorem only for the cases when U is an ex-cofiber or an ex-fiber space. For the general case see the remark immediately following the proof. Let U be an ex-fiber space. Then the sequence $\pi(D_o, U)$ is essentially a special case of the sequence (D) of (2.4). We can first replace $\widetilde{C_Df} \to C_Df$ there by

the fibration induced from $U \to D$. We then let u_o be the section pulled back from $\sigma : D \to U$, and $Y = D$. Then (D) of (2.4) is exactly the sequence $\pi(D_o, U)$. (The condition $A = S_D A'$ for the sequence (D) of (2.4) was used only to define $f^{\#}$. In this special case however, $f^{\#} = f^*$ and so we can dispense with the condition.) Exactness at the first term of $\pi(D_o, U)$ was already a consequence of the obstruction property (2.1.7). Now suppose U is an ex-cofiber space. Then in (D') of (2.5) we replace $P_D f \subset \overline{P_D f}$ by the cofibration induced from $D \subset U$, and we take u_o to be the retraction induced from $D \to U$ and set $Y = D$. We obtain herewith the sequence $\pi(U, D_o')$. Again we can dispense with the condition $A = \Omega_D A'$. It should be noticed that Lemmas (2.1.15) and (2.2.15) are not involved in this proof. They were used only to define $f^{\#}$, $w_f^{\#}$ and $f_{\#}$, $w_{f\#}$, whereas in the present situation $\# = *$ and so $f^{\#} = f^*$, $f_{\#} = f_*$. □

<u>Remark</u>: There is another way to show the exactness of the fiber and cofiber sequences in Top(D). Using the methods of Puppe (cf. [100]) and Nomura (for $D = *$ see 14.6 of [24]) we can show that the sequences

$$A \xrightarrow{f} X \xrightarrow{i_1} C_D f \xrightarrow{i_2} C_D i_1 \longrightarrow C_D i_2 \longrightarrow \cdots$$

$$A \longleftarrow X \xleftarrow{q_1} P_D f \xleftarrow{q_2} P_D q_1 \longleftarrow P_D q_2 \longleftarrow \cdots$$

are homotopy equivalent in Top(D) to the sequences D_o and D_o' respectively. The exactness of $\pi(D_o, U)$ and $\pi(U, D_o)$ at each term then follows from the obstruction properties (2.1.7) and (2.2.7) This was how I.M. James (7.2. [59]) and J.F. McClendon (3.3 [79]) proved Theorem (2.6.2), see also [29], [68], [126].

We can now use the exact fiber and cofiber sequences in Top(D) to prove Lemmas (2.1.15) and (2.2.15) respectively. (See in this connection the remark following the proof of (2.6.2)).

(2.6.3) Proof of (2.1.15) and (2.2.15): We first prove (2.2.15). Given a closed cofibration $i : X \subset \bar{X}$, we can construct the ex-spaces $X \vee D$ and $\bar{X} \vee D$ over D, where σ is the inclusion of D, and $\rho = (d, 1)$ and $\rho = (w, 1)$ respectively. Then $f = i \vee 1 : X \vee D \subset \bar{X} \vee D$ is an ex-map. Taking the ex-fiber space A of (2.2.15), we see that the exact cofiber sequence $\pi(D_o, A)$ is isomorphic to the exact sequence of (2.2.15). It is clear that

$$\pi(X \vee D, A) = [X,A]_d \,, \pi(\bar{X} \vee D, A) = [X, A]_w \,.$$

Furthermore, there is a bijection

$$\langle \sigma_* \bar{D}, A \rangle_w = \pi(\bar{D}, A) \xrightarrow[\approx]{r^*} \pi(C_D f, A)$$

since $C_D f$ is the same as the double mapping cylinder $Z(i, d)$. The canonical map $r : Z(i, d) \to \bar{D}$ into the push-out $\bar{D}$ of (i, d) is a homotopy equivalence under D, since $X \subset \bar{X}$ is a cofibration. By (1.1.8), r induces the above bijection. Applying (2.6.2), we arrive at the claim of (2.2.15). The proof of (2.1.15) is dual. We start with a fibration $p : \tilde{X} \to X$ and consider the ex-map $f = p \times 1 : \tilde{X} \times D \to X \times D$ where $\sigma = (w, 1_D)$ and $\sigma = (d, 1_D)$, and ρ is the projection onto D. □

We now want to draw attention to further relationships between the classf ication sequences and the fiber and cofiber sequences in Top (D). The reader may provide the connecting details for himself.

(2.6.4) Let $f : A \to X$ and $Y \subset X$ be as in the sequences (A) and (B) respectively of (2.4). Then $\tilde{f} = i_1 \cup 1 : A \vee C_f \subset Z_f \cup_Y C_f$

is an ex-map in $Top(C_f)$ between ex-spaces over C_f, where σ is the inclusion of C_f and $\rho = (*, 1)$ or $\rho = \pi \cup 1$, $\pi : Z_f \to Z_f/A = C_f$ the identification map. $\tilde{f}$ then gives us the long cofiber sequence in Top (C_f). If we apply to it the functors $\pi(\ldots, C_f \times U)$ and $\pi(\ldots, \tilde{C}_f)$, we obtain exact sequences equivalent to the classification sequences (A) and (B) respectively of (2.4). Here $C_f \times U$ and $\tilde{C}_f$ are ex-spaces over C_f, for which ρ is the projection and σ is the section u_o. If A is an ex-cofiber space over D and $d : D \to Y \subset X$, then we have in Top (C_Df) similarly the ex-map $\tilde{f} : A \cup_D C_Df \subset Z_Df \cup_Y C_Df$. From the exact cofiber sequence for f we can extract sequences isomorphic to (C) and (D) of (2.4). The dual statements also hold. Let $f : X \to A$ and $X \to Y$ be as in (A') and (B') of (2.5). Then we have in Top (P_f) an ex-map $\bar{f} = q_1 \times 1 : W_f \times_Y P_f \to A \times P_f$. The fiber sequence for $\bar{f}$ gives us sequences isomorphic to (A') and (B') of (2.5). We obtain similarly the sequences (C') and (D') of (2.5).

Remark: We could have used the above method at the start to derive the classification sequences, but in view of the overall plan of this book it did not seem natural to put the exact fiber and cofiber sequences in Top(D) so into the foreground. It is our intent to use differences in order to describe the operators Σf, $\underline{\Omega} f$ in the classification sequences. Differences are obstructions to the existence of homotopies and so are of natural importance in homotopy classification. Also, in the following chapters, it will be very useful to us to have introduced so explicitly the operators Σf, $\underline{\Omega} f$, $f^{\#}$ and $f_{\#}$ in the classification sequences.

CHAPTER 3: ITERATED PRINCIPAL COFIBRATIONS

Chapter 6 and this one are dual and are developed along parallel lines. The reader will find it easier to understand the discussion if he refers back and forth between the two chapters.

(3.1) The partial suspension

We will now describe various properties of a generalized Freudenthal suspension called the partial suspension. An important application is the result (3.1.11) that the partial suspension of a Whitehead product map is also a Whitehead product map. In an appendix we define in analogy with stable homotopy theory the partially stable homotopy groups of a space. We will show that the algebra of partially stable maps is actually a Pontrjagin algebra. The latter is dual to the Massey-Peterson algebra described in the appendix to (6.1). In the sections to follow it will be seen that the partial suspension plays a central role in the homotopy classification of maps and sections. In (3.4) a relative version of the partial suspension will be defined. This more general object has properties similar to and generalizing those of the partial suspension.

Let $\rho : B \vee Y \to Y$ be the retraction onto Y with $\rho(B) = *$. We say that $\xi \in [A, B \vee Y]$ is trivial on Y when $\rho_* \xi = 0$ and we write $\pi_n^A (B \vee Y)_2 = [S^n A, B \vee Y]_2 = \text{Kern}\ \rho_*$ for the set of elements trivial on Y. Let A be a co-H-group. In the diagram

$$\begin{array}{ccc}
 & \pi_1^A(CB \vee Y, B \vee Y) & \xrightarrow[\cong]{\partial} \pi_o^A (B \vee Y)_2 \\
 & \downarrow \pi_{o*} & \\
\pi_1^A (SB \vee Y)_2 & \xrightarrow[\cong]{j} \pi_1^A (SB \vee Y, Y) &
\end{array}$$

the isomorphisms j and ∂ come from the exact homotopy sequences of the indicated pairs, and $\pi_o : CB \vee Y \longrightarrow (CB \vee Y)/B = SB \vee Y$ is the identification map. We call the homomorphism

$$(3.1.1.)\qquad E : \pi_o^A(B \vee Y)_2 \longrightarrow \pi_1^A(SB \vee Y)_2 , \quad E = j^{-1}\pi_{o*}\partial^{-1}$$

the partial suspension relative to Y. A representative of $E\xi$ is obtained as follows. Let $\xi : A \to B \vee Y$ be a map that is trivial on Y. Then $\rho\xi$ can be extended to $\bar{\xi} : CA \to Y$. If $\sigma : Y \subset SB \vee Y$ denotes the inclusion, then

$$H_\xi : I \ltimes A \xrightarrow{1\ltimes\xi} I \ltimes B \vee I \times Y \xrightarrow{\pi \vee pr} SB \vee Y$$

is a self-homotopy of $\sigma\rho\xi$. Then

$$(3.1.2)\qquad E\xi = d(\sigma\bar{\xi}, H_\xi, \sigma\bar{\xi}) \in [SA, SB \vee Y]_2$$

is the associated difference, cf. (1.2.5).

The partial suspension E generalizes the Freudenthal suspension S. More precisely, we find that the diagram

$$(3.1.3)\qquad \begin{array}{ccc} \pi_o^A(B \vee Y)_2 & \xrightarrow{E} & \pi_1^A(SB \vee Y)_2 \\ \uparrow i_{1*} & & \uparrow i_{1*} \\ \pi_o^A(B) & \xrightarrow{S} & \pi_1^A(SB) \end{array}$$

commutes. The Freudenthal suspension theorems generalize to

(3.1.4) Theorem: If B is $(b-1)$-connected, then

$$E : \pi_{n-1}(B \vee Y)_2 \to \pi_n(SB \vee Y)_2 \text{ is an } \begin{cases} \text{epimorphism for } n \leqslant 2b \\ \text{isomorphism for } n < 2b. \end{cases}$$

We will prove this theorem in a yet more general form in (3.4.7). On repeated application of the partial suspension E we obtain the homomorphism

$$E^n : \pi_0^A(B \vee Y)_2 \longrightarrow \pi_n^A(S^n B \vee Y)_2$$

Since diagram (3.1.1) is natural for maps $u : Y' \to Y$, $v : B' \to B$ and $w : A \to A'$, the partial suspension is natural for these maps too. That is, we have equalities

(3.1.5)
$$(S^n v \vee u)_* \circ E^n = E^n \circ (v \vee u)_*$$
$$(S^n w)^* \circ E^n = E^n \circ w^*$$

E is natural in another way too. To formulate this, we let A and B be co-H-groups. Given a map $k : X \to Y$ and an element $\xi \in \pi_0^A(D \vee Y)_2$, we obtain homomorphisms

$$\begin{array}{ccc} \pi_0^B(A \vee X)_2 & \xrightarrow{(\xi,k)_*} & \pi_0^B(D \vee Y)_2 \\ \downarrow E & & \downarrow E \\ \pi_1^B(SA \vee X)_2 & \xrightarrow{(E\xi,\, k)_*} & \pi_1^B(SD \vee Y)_2 \end{array}$$

(3.1.6) Theorem: The diagram is commutative, that is, if $\eta \in \pi_0^B(A \vee X)_2$ then we have $E((\xi, k) \circ \eta) = (E\xi, k) \circ E\eta$.

Proof: Let $\bar{\xi} : (CA, A) \to (CD \vee Y, D \vee Y)$ be such that $\partial\bar{\xi} = \xi$. Then we have $j(E\xi) = \pi_{0*}\bar{\xi}$ by definition of $E\xi$, and so it follows that the diagram of pairs

$$\begin{array}{ccc}
(CA \vee X,\ A \vee X) & \xrightarrow{(\bar{\xi},\ k)} & (CD \vee Y,\ D \vee Y) \\
\downarrow{\scriptstyle \pi_o} & & \downarrow{\scriptstyle \pi_o} \\
(SA \vee X, X) & \xrightarrow[(E\xi, k)]{} & (SD \vee Y,\ Y)
\end{array}$$

is homotopy commutative. By applying the functor $\pi_1^B(\ldots)$ to this diagram, we obtain essentially the diagram immediately preceeding (3.1.6). □

We now generalize the partial suspension to obtain a partial smash product. Let A be a co-H-group as before, with comultiplication $\mu: A \to A \vee A$. If D is any space then $D \ltimes A$ and $D \vee A$ are also co-H-groups with $1_D \ltimes \mu$ and $1_D \wedge \mu$ as comultiplications. From the cofiber sequence (2.1.11) we have a short exact sequence

$$0 \longrightarrow [D \wedge A, U] \xrightarrow{\pi^*} [D \times A,\ U] \xrightarrow{\sigma^*} [A,\ U] \longrightarrow 0$$

of group homomorphisms, in which σ^* is induced from the inclusion $\sigma : A \subset D \ltimes A$ and π^* is induced from the identification π : $D \ltimes A \to D \wedge A$. Suppose $\xi : A \to B \vee Y$ is trivial on Y. Since π^* is a monomorphism, the diagram

$$\begin{array}{ccc}
D \ltimes A & \xrightarrow{1 \ltimes \xi} & D \ltimes B \vee D \ltimes Y \\
{\scriptstyle \pi}\downarrow & & \downarrow{\scriptstyle \pi \vee \rho} \\
D \wedge A & \xrightarrow{D \hat{\wedge} \xi} & D \wedge B \vee Y
\end{array}$$

has, up to homotopy, exactly one commutative extension $D \wedge \xi$. We thus have a homomorphism

(3.1.7) $$D \hat{\wedge} \ :\ [A,\ B \vee Y]_2 \longrightarrow [D \wedge A,\ D \wedge B \vee Y]_2$$

with $D \wedge \xi = (\pi^*)^{-1}(\pi \vee \rho) \circ (1 \ltimes \xi))$, which we call the <u>partial smash product</u> with D. It is easy to see that

(3.1.8) $D' \mathbin{\hat{\wedge}} (D \mathbin{\hat{\wedge}} \xi) = (D' \wedge D) \mathbin{\hat{\wedge}} \xi$

(3.1.9) <u>Theorem:</u> $E^n \xi = S^n \mathbin{\hat{\wedge}} \xi$

<u>Proof:</u> By (3.1.8) it suffices to show that $E\xi = S^1 \wedge \xi$. This follows from (3.1.2) and the fact that the diagram

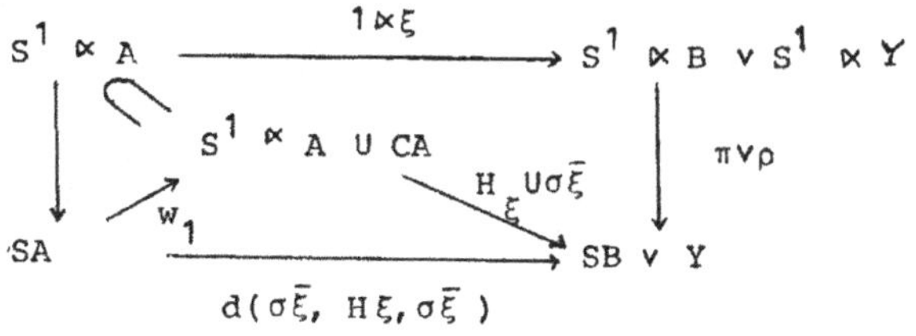

is homotopy commutative. □

We can obtain a representative of $E^n\xi$ by basically the same procedure as in (3.1.2). Since ξ is trivial on Y, we can extend the retraction $\rho : B \vee Y \to Y$ to obtain $\bar{\xi} : C_\xi \to Y$. This $\bar{\xi}$ gives us the map $\pi \cup \bar{\xi}$ in the composition

$$S^n A \xrightarrow{W\xi} \Sigma^n_Y (B \vee Y) \cup C_\xi = S^n \ltimes B \cup {}_B C_\xi \xrightarrow{\pi \cup \bar{\xi}} S^n B \vee Y$$

where π is the identification and $W\xi$ is defined as in (2.4.6). It follows easily from (3.1.9) that

(3.1.10) <u>Theorem:</u> $E^n \xi = (W\xi)^*(\pi \cup \bar{\xi})$

When $n = 1$ this is precisely the claim of (3.1.2)

The Whitehead product map $w_{A,B} : SA \wedge B \to SA \vee SB$ and the co-operation $\mu^A_B = \mu_B : SA \wedge B \to SA \wedge B \vee SB$ in (0.3) and (0.4) are both trivial on SB.

(3.1.11) <u>Theorem:</u> $E\, w_{A,B} = w_{SA,B}$, $E\mu^A_B = \mu^{SA}_B$.

This theorem is very useful in calculations. Denoting as in (0.3) the relative Whitehead Product map $\tilde{w}_{A,B} = \partial^{-1} w_{A,B}$, we see from the definition of partial suspension:

(3.1.12) <u>Corollary</u>: <u>The induced map</u> $(\pi_o \vee 1_{SB})_*$: $\pi_2^{A \wedge B}(CSA \vee SB,\ SA \vee SB) \longrightarrow \pi_2^{A \wedge B}(SSA \vee SB,\ SB)$, <u>formed from the quotient map</u> $\pi_o : CSA \to SSA$, <u>is such that</u> $j\, w_{SA,\,B} = (\pi_o \vee 1_{SB})_* \tilde{w}_{SA,\,B}$.

This corollary is equivalent to the equation $Ew_{A,B} = w_{SA,B}$.

<u>Proof of (3.1.11)</u>: We will show that

$$\begin{array}{ccc}
D \wedge SA \wedge B & \xrightarrow{\;D \,\hat{\wedge}\, w_{A,B}\;} & D \wedge SA \vee SB \\
\Big\downarrow \tau & (1) & \Big\downarrow \tau \vee 1 \\
SD \wedge A \wedge B & \xrightarrow[\;w_{D \wedge A,B}\;]{} & SD \wedge A \vee SB
\end{array}$$

is homotopy commutative. τ is the map exchanging S^1 and D. The claim $E\,w_{A,B} = w_{SA,B}$ then follows from (3.1.9), in consideration of the fact that $\tau = -1$ when $D = S^1$. We have first of all the commutative diagram

$$\begin{array}{ccc}
D \ltimes S(A \times B) & \xrightarrow{\;1 \ltimes (-P_A - P_B + P_A + P_B)\;} & D \ltimes (SA \vee SB) \\
\Big\uparrow \tau & & \\
S(D \times A \times B) & (2) & \Big\downarrow (\tau\pi) \vee pr \\
\Big\downarrow v & & \\
S((D \wedge A) \times B) & \xrightarrow{\;-P_{D\wedge A} - P_B + P_{D\wedge A} + P_B\;} & SD \wedge A \vee SB
\end{array}$$

in which v is the quotient map $D \times A \to D \wedge A$, and the maps P_A, P_B, $P_{D \wedge A}$ are given by projection and inclusion. Since $\bar{v}$: $S(D \times A \times B) \to S(D \wedge A \wedge B)$ induces a monomorphism $\bar{v}^*$, the commutativity of (2) implies the homotopy commutativity of (1), cf. the definition of $w_{A,B}$ in (0.4). Similarly, the equation $(\tau \vee 1)(D \hat{\wedge} \mu_B^A) = \mu_B^{D \wedge A}\tau$ can be used to show that $E\mu_A^B = \mu_B^{SA}$. □

(3.1) Appendix: The algebra of stable and partially stable maps and the Pontrjagin algebra

Let $\sigma_*^Y(A)$ be the graded abelian group in which

(3.1.13) $$\sigma_q^Y(A) = \varinjlim \pi_{q+k}^Y(S^kA), \quad q \in \mathbb{Z},$$

where the direct limit is taken over the suspension homomorphisms $S : \pi_{q+k}^Y(S^kA) \longrightarrow \pi_{q+k+1}^Y(S^{k+1}A)$. An element of $\sigma_q^Y(A)$ is called a stable map of degree q, In particular, $\pi_q^S(A) = \sigma_q^{S^0}(A)$ is the q-th stable homotopy group of A. In the same manner we define a a graded abelian group $\sigma_*^Y(A; X)$ in which

(3.1.14) $$\sigma_q^Y(A; X) = \varinjlim \pi_{q+k}^Y(S^kA \vee X)_2$$

is the direct limit taken over the partial suspensions E. By (3.1.3) we have an inclusion $\sigma_q^Y(A) \subset \sigma_q^Y(A; X)$, and if we let X be a point * then $\sigma_q^Y(A) = \sigma_q^Y(A; *)$. We call an element of $\sigma_q^Y(A; X)$ a partially stable map of degree q. By (3.1.6) we see that partially stable maps can be composed, and so there is a bilinear map

(3.1.15) $$\circ : \sigma_p^A(D; X) \times \sigma_q^B(A; X) \to \sigma_{p+q}^B(D, X)$$

where $(\xi, \eta) \longmapsto \xi \circ \eta$. If η and ξ are represented by η' : $S^{N+q}B \to S^N A \vee X$ and ξ' : $S^{M+p}A \to S^M D \vee X$ respectively, then $\xi \circ \eta$ is represented by

$$S^{M+N+p+q}B \xrightarrow{E^{M+p}\eta'} S^{M+N+p}A \vee X \xrightarrow{(E^N\xi', 1)} S^{M+N}D \vee X.$$

In particular, the pairing $\circ$ is an associative multiplication on $\sigma_*^A(A; X)$. Thus $\sigma_*^A(A; X)$ is a graded algebra with $\sigma_*^A(A)$ and $\sigma_0^A(A; X)$ as subalgebras.

For $A = S^0$ we will show that the algebra is canonically isomorphic to the Pontrjagin algebra of ΩX. To define the latter we recall the bilinear pairing

$$(3.1.16) \qquad \wedge : \pi_p^S(A) \times \pi_q^S(B) \longrightarrow \pi_{p+q}^S(A \wedge B)$$

which takes the pair (ξ, η) whose components are represented by $\xi' : S^{M+p} \to S^M A$ and $\eta' : S^{N+q} \to S^N B$ to the class $\xi \wedge \eta$ represented by

$$S^{M+N+p+q} \xrightarrow{S^{M+p}\eta'} S^{M+N+p}B \xrightarrow{S^N\xi' \wedge 1_B} S^{M+N}A \wedge B$$

Therefore the loop space multiplication μ in ΩX yields an associative multiplication

$$(3.1.17) \qquad \circ : \pi_*^S(\Omega X)^+ \otimes \pi_*^S(\Omega X)^+ \xrightarrow{\wedge} \pi_*^S(\Omega X \times \Omega X)^+ \xrightarrow{\mu_*} \pi_*^S(\Omega X)^+$$

since if $A^+ = A + \{*\}$ then $A^+ \wedge A^+ = (A \times A)^+$. $\pi_*^S(\Omega X)^+$ with this multiplication is called the Pontrjagin algebra of ΩX. There is a canonical isomorphism of algebras

(3.1.18) Theorem: $\sigma_*^{S^0}(S^0, X) = \pi_*^S(\Omega X)^+$

To prove this theorem we will need several lemmas. Denoting by E^∞ the canonical map into the direct limit, we have that (see 3.1.4).

(3.1.19) Lemma: $E^{\infty} : \pi_{N+q}(S^N A \vee X)_2 \longrightarrow \sigma\mathcal{C}_q^{S^o}(A;X)$ is an isomorphism when $q + 2 \leq N$.

We need further the following result of Ganea (lemma 5.1 in [34]). Let $R_Y : S\Omega Y \to Y$ be the evaluation map, that is $R_Y(t, \sigma) = \sigma(t)$. For $B \vee Y$ we construct the Whitehead Product $[R_B, R_Y] : S(\Omega B \wedge \Omega Y) \to B \vee Y$, where R_B, R_Y are also the evaluation maps followed by the inclusions i_q, i_2 of B and Y.

(3.1.20) Lemma: Let A be a co-H-group. Then $\Theta : \pi_o^A(B) \times \pi_o^A(Y) \times \pi_o^A(S\Omega B \wedge \Omega Y) \to \pi_o^A(B \vee Y)$ with $\Theta = i_{1*} + i_{i*} + [R_B, R_Y]_*$ is a bijection.

Proof: Denoting by $i : B \vee Y \subset B \times Y$ the inclusion, we see that $P_i = PB \times \Omega Y \cup \Omega B \times PY$, the fiber of i, is homotopy equivalent to $S\Omega B \wedge \Omega Y$, therefore $S\Omega B \wedge \Omega Y \simeq P_i \to B \vee Y$ is equal to $[R_B, R_Y]$, see (0.3.2). The lemma then follows from the exact fiber sequence A_o' of (2.5). $\square$

The multiplication $\mu : \Omega X \times \Omega X \to \Omega X$ gives us the Hopf-construction $H\mu \in [S\Omega X \wedge \Omega X, S\,\Omega\, X]$ defined by

(3.1.21) $\sigma(H\mu) = -(Sp_1) - (Sp_2) + S\,\mu \in [S(\Omega X \times \Omega X), S\Omega X]$

$\sigma = (Ss)^*$ is the homomorphism induced by the quotient map $s : \Omega X \times \Omega X \to \Omega X \wedge \Omega X$ as in the exact sequence (0.3), and p_1 and p_2 are the projections onto the first and second component.

(3.1.22) Lemma: Let $i : S^n \subset S^n \vee X$ and $R_X : S\Omega X \to X \subset S^n \vee X$. If $n \geq 1$ then the equation

$$[[i, R_X], R_X] = [i, R_X] \circ S^{n-1}(H\mu)$$

holds in $[S^n\Omega X \wedge \Omega X, S^n \vee X]$.

Proof : By (3.1.11) and (3.1.6), it suffices to consider the case $n = 1$. Since $\sigma = s^*$ is injective, we need only show that

(1) $s^* [[i, R_X], R_X] = \underline{\mu}^*[i, R_X], \underline{\mu} = \sigma(H\mu) = (H\mu)s$

in $[S(\Omega X \times \Omega X), S^1 \vee X]$. The operation of the fundamental group $i \in \pi_1(S^1 \vee X)$ is such that

(2) $[i, R_X] = (-R_X)^i + R_X \in [S\Omega X, S^1 \vee X]$.

Moreover

(3) $R_X \circ (S\mu) = R_X \circ (sp_1) + R_X \circ (Sp_2) \in [S(\Omega X \times \Omega X), X]$

as is easily seen. We now set $p_i = Sp_i$ $(i = 1, 2)$. It follows from (1) and the definition of the Whitehead product (0.4) that the equations

(4) $-p_i^*[i, R_X] - p_2^*R_X + .p_1^*[i, R_X] + p_2^*R_X = \bar{\mu}^*[i, R_X]$

(5) $p_2^*[i, R_X] - p_2^*R_X + p_1^*[i, R_X] + p_2^*R_X = (p_2+p_1+\bar{\mu})^*[i, R_X] =$

$= (S\mu)^*[i, R_X]$

(6) $p_2^* (-R_X)^i + p_1^*(-R_X)^i + p_1^*R_X + p_2^*R_X = -(S\mu)^*(R_X)^i + (S\mu)^*R_X$

are equivalent. (6) follows from (3), and (4) follows from (2) and (3.1.21). ☐

(3.1.23) Definition: We use the Hopf construction $H\mu$ of (3.1.20) to define the semidirect sum algebra $\pi_*^S(\Omega X) \tilde{\oplus} \pi_*^S(S^0)$. As a graded abelian group this algebra is $\pi_*^S(\Omega X) \oplus \pi_*^S(S^0)$. The multiplication is defined by

$$(d, a)(e, b) = (db + (-1)^* ea + S^{-1}(H\mu)_* S(d \wedge e), ab)$$

where $* = (\deg e)(\deg a)$. If $a, b \in \pi_*^S(S^0)$ then ab, db and ea

are defined by composition as in (3.1.15). Furthermore $d \wedge e$ is the operation of (3.1.16), and S is the suspension. (For graded abelian groups $v = \{v_q\}$, $w = \{w_q\}$ we set $v \oplus w = \{v_q \oplus w_q\}$)

For $(\Omega X)^+ = \Omega X + \{*\}$ there are two canonical pointed maps $\varepsilon : (\Omega X)^+ \to S^0$ and $i : (\Omega X)^+ \to \Omega X$, with respectively $\varepsilon(\Omega X) = S^0 - \{*\}$ and $i|\Omega X = \text{id}$.

(3.1.24) <u>Lemma</u>: $(i_*, \varepsilon_*) : \pi_*^S(\Omega X)^+ \to \pi_*^S(\Omega X) \mathbin{\tilde{\oplus}} \pi_*^S(S^0)$ <u>is an algebra isomorphism.</u>

<u>Proof:</u> Since $(\, , \varepsilon) : S^N A^+ = S^N \ltimes A \simeq S^N A \vee S^N \subset S^N A \times S^N$ is defined via a homotopy equivalence, it is easy to see that (i_*, ε_*) is an isomorphism of groups in the lemma. We leave it to the reader himself to show that (i_*, ε_*) is also an algebra homomorphism. We only point out that $a \wedge e = (-1)^* ea$. □

To complete the proof of theorem (3.1.18) we now show that

(3.1.25) <u>Theorem</u>: <u>There is a canonical isomorphism</u>

$$\alpha : \sigma_*^{S^0}(S^0, X) \cong \pi_*^S(\Omega X) \mathbin{\tilde{\oplus}} \pi_*^S(S^0)$$

<u>of algebras.</u>

<u>Proof</u>: Let $\xi \in \sigma_p^{S^0}(S^0;X)$. We construct $\alpha\xi = (\xi_2, \xi_1)$ as follows. If $\xi' : S^{M+p} \to S^M \vee X$ represents ξ, then the elemma of Ganea (3.1.20) gives us maps $\xi_1' : S^{M+p} \to S^M$ and $\xi_2' : S^{M+p} \to S^M \Omega X$ such that

(1) $\quad \xi' = i\xi_1' + [i, R_X]\xi_2'$.

ξ' determines ξ_1' and ξ_2' up to homotopy. Suppose that ξ_1' represents ξ_1 and that ξ_2' represents ξ_2. It then follows from

(3.1.11) and (3.1.6) that α is a well-defined homomorphism between abelian groups. Now let $\eta \in \sigma_q^{S^o}(S^o; X)$ be represented by $\eta' : S^{N+q} \to S^N \vee X$, and suppose η_1' and η_2' are the maps in (1) arising from $\alpha\eta = (\eta_2, \eta_1)$. Then $\xi \circ \eta$ is represented by

$$(E^N \xi', 1_X)(E^{M+p} \eta') : S^{M+N+p+q} \to S^{M+N+p} \vee X \to S^{M+N} \vee X$$

(2) $= (iS^N \xi_1' + [i, R_X](S^N \xi_2'), 1_X) \circ (i'S^{M+p} \eta_1' + [i', R_X](S^{M+p} \eta_2'))$

(3) $= i(S^N \xi_1')(S^{M+p} \eta_1') + [i, R_X](S^N \xi_2')(S^{M+p} \eta_1') +$

$$[i\, S^N \xi_1' + [i, R_X](S^N \xi_2'), R_X](S^{M+p} \eta_2').$$

i' is the inclusion of S^{M+N+p}, and i the inclusion of S^{M+N}. (2) follows from (3.1.11), (3.1.6) and equation (1). Examining the terms of (3) we find that

(4) $[iS^N \xi_1', R_X](S^{M+p} \eta_2') = [i, R_X](S^N \xi_1' \wedge 1_{\Omega X})(S^{M+p} \eta_2')$,

(5) $[[i, R_X](S^N \xi_2'), R_X](S^{M+p} \eta_2') = [[i, R_X], R_X](S^N \xi_2' \wedge 1_{\Omega X})(S^{M+p} \eta_2')$

(6) $= [i, R_X](S^{N-1} H\mu)(S^N \xi_2' \wedge 1_{\Omega X})(S^{M+p} \eta_2')$.

(4) and (5) follow from the corresponding naturality of the Whitehead product, and (6) follows from lemma (3.1.22). In accordance with the definition in (1) we can conflate equations (2) through (6) to obtain

$$\alpha(\xi \circ \eta) = (\xi_2 \circ \eta_1 + \xi_1 \wedge \eta_2 + S^{-1}(H\mu)_* S(\xi_2 \wedge \eta_2)\ ,\ \xi_1 \circ \eta_1).$$

Since $\xi_1 \wedge \eta_2 = (-1)^* \eta_2\, \xi_1$, it follows from definition (3.1.23) that $\alpha(\xi \circ \eta) = (\alpha\xi) \circ (\alpha\eta)$. $\square$

As a consequence of the last result we have ring isomorphisms

(3.1.26) <u>Corollary:</u> $\pi_o^S (\Omega X)^+ = \sigma_o^{S^o}(S^o, X) = \mathbb{Z}[\pi]$.

$\mathbb{Z}[\pi]$ is the group ring of $\pi = \pi_1(X)$. These isomorphisms arise from $\pi_o^S(\Omega X)^+ = H_o(\Omega X, \mathbb{Z}) = \mathbb{Z}[\pi]$, or can be derived as follows. If $n \geq 2$, $\pi = \pi_1(S^n \vee X)$ operates on $i \in \pi_n(S^n \vee X)$, where i is the inclusion of S^n as before. An element $\alpha \in \mathbb{Z}[\pi]$ has the form $\alpha = n_1\alpha_1 + \ldots + n_k\alpha_k$ with $\alpha_i \in \pi$ and $n_i \in \mathbb{Z}$.

(3.1.27) <u>Lemma:</u> <u>If</u> $n \geq 2$ <u>then there is an isomorphism</u> σ : $\mathbb{Z}[\pi] \longrightarrow \pi_n(S^n \vee X)_2$ <u>defined by</u> $\sigma(n_1\alpha_1 + \ldots + n_k\alpha_k) = n_1\, i^{\alpha_1} + \ldots + n_k i^{\alpha_k}$.

This also implies the preceeding lemma in view of (3.1.11) since σ is also a homomorphism between the ring structures.

(3.2) <u>The classification of maps and sections by means of iterated principal cofibrations and the associated spectral sequence.</u>

In this section we use the exact classificationssequences $(A, \underline{A})$ and $(B, \underline{B})$ of (2.4) to construct a spectral sequence for an iterated principal cofibration $Y \subset X$, from which we can in principle derive the solution of the classification problem of (1.1). The first differential of this spectral sequence consists of certain primary homotopy operations, which are defined in terms of partial suspensions and the attaching maps. The reader will see that this spectral sequence, which apparently has gone unnoticed, can be set up for the relative case in essentially the same way by using the classification sequences $(C, \underline{C})$ and $(D, \underline{D})$ of (2.4). In chapter 4 we will apply the spectral sequence to cases where (X, Y) is a relative CW-complex. In these examples it will become clear that the spectral sequence provides a context in which to understand many results of classical obstruction theory.

Let A be a co-H-group. If a map $\xi : A \to B \vee X$ is trivial on X, then we can take partial suspensions $E^n\xi : S^nA \to S^nB \vee X$. Given a space U, or a fibration $p : \tilde{X} \to X$ with fiber $i : F \subset \tilde{X}$, we have the following products induced by $E^n\xi$, $n \geq 0$.

$$[\ ,\ ]^n_\xi : [S^nB,U] \times [X,U] \to [S^nA,U], \quad [\beta,u]^n_\xi = (E^n\xi)^*(\beta,u)$$

(3.2.1)

$$\langle\ ,\ \rangle^n_\xi : [S^nB,F] \times \langle X,\tilde{X}\rangle \to [S^nA,F], \quad \langle\beta,u\rangle^n_\xi = i_*^{-1}[i_*\beta,u]^n_\xi .$$

Here $i_* : \pi^A_n(F) \to \pi^A_n(\tilde{X})$ is injective if a section $u \in \langle X,\tilde{X}\rangle$ exists. Since $p_*[i_*\beta,u]^n_\xi = [p_*i_*\beta,u]^n_\xi = [0,u]^n_\xi = 0$ and $\operatorname{Ker} p_* = \operatorname{Im} i_*$, the element $\langle\beta,u\rangle^n_\xi$ is well-defined. For fixed u, $[\ ,u]^n_\xi$ and $\langle\ ,u\rangle^n_\xi$ are homomorphisms of groups when $n \geq 1$. This follows from (3,1.2) and (1.2.6). We call $[\ ,\ ]^n_\xi$ the product induced by ξ, and $\langle\ ,\ \rangle^n_\xi$ the <u>twisted</u> product induced by ξ. Some properties of these products will be presented in detail in (3.3).

Suppose now that we have a double mapping cone

$$\begin{array}{ccccc} & & C_f & & \\ & & \cup & & \\ A & \xrightarrow{\ f\ } & C_g & = & X \\ & & \cup & & \\ B & \xrightarrow{\ g\ } & X_o & \supset & Y \end{array}$$

where A is a co-H-group. The element ∇f is defined as follows. If we take the inclusions $\sigma_1 : SB \subset SB \vee Cg$ and $\sigma_2 : Cg \subset SB \vee Cg$, then $\sigma_2 + \sigma_1 : Cg \to SB \vee Cg$ is just the cooperation μ of SB defined in (1.2.7), except for the order in which the spaces appear in the wedge. The difference element

(3.2.2) $\nabla f = -f^*(\sigma_2) + f^*(\sigma_2 + \sigma_1) \in \pi_o^A(SB \vee C_g)_2$

is then trivial on C_g. From ∇f we can construct the partial suspensions $E^n \nabla f \in \pi_n^A(S^{n+1}B \vee C_g)_2$, and so the operations (3.2.1) are defined when $\xi = \nabla f$.

Let $u : X \to U$ be a map from $X = C_g$ that can be extended over C_f, or let $u : X \to \tilde{X}$ be a section of a fibration $\tilde{C}_f \to C_f$ with fiber F that can be extended as a section over C_f. Then for $n \geqslant 1$ we have homomorphisms

(3.2.3)
$$\nabla^n(u, f) : [S^{n+1}B, U] \xrightarrow{\gamma} [\Sigma_Y^n X, U]^u \xrightarrow{\Sigma^n f} [S^n A, U]$$
$$\tilde{\nabla}^n(u,f) : [S^{n+1}B, F] \xrightarrow{\gamma} \langle \Sigma_Y^n X, \rho^* \tilde{X} \rangle^u \xrightarrow{\Sigma^n f} [S^n A, F]$$

As in (2.4.3), $\Sigma^n f = \Sigma^n(u,f)$ denotes a homomorphism from the classification sequence of C_f, and $\gamma = (-1)^{n+1} u^+$ is a homomorphism from the classification sequence of C_g, see (2.4.7) (A), (B). The operations defined in (3.2.1) have the property that

(3.2.4) Theorem:
$$\nabla^n(u,f)(\beta) = [\beta, u]_{\nabla f}^n ,$$
$$\tilde{\nabla}^n(u,f)(\beta) = \langle \beta, u \rangle_{\nabla f}^n .$$

If $Y = X_o$ then γ in (3.2.3) is an isomorphism - this follows from (1.2.8) and (1.2.21). Thus we can use γ to replace the relevant groups in the classification sequences (A, $\underline{A}$) and (B,$\underline{B}$). The homomorphism $\Sigma^n f = \Sigma^n(u,f)$ then becomes $\nabla^n(u,f)$ and $\tilde{\nabla}^n(u,f)$ respectively. This results in two new classification sequences, which we will not write out explicitly however. We derive from them a classification theorem corresponding to (2.4.4).

(3.2.5) Corollary: Let $v = u|X_o$. Then there are bijections

$$[C_f, U]^V \approx \bigcup_\beta [SA,U]/\text{Im } [\ , u + \beta]^1_{\nabla f} \ ,$$

$$\langle C_f, \widetilde{C}_f \rangle^V \approx \bigcup_\beta [SA,F]/\text{Im } \langle\ , u + \beta\rangle^1_{\nabla f} \ .$$

The first union is taken over all $\beta \in [SB, U]$ with $f^*(u + \beta) = 0$.

The second union over all $\beta \in [SB, F]$ with $f^\#(u + \beta) = 0$.

The bijections are defined by $\alpha \mapsto u_\beta + \alpha$, where u_β is an extension of $u + \beta$ over C_f. In (3.5.11) we describe a condition on double mapping cones under which the homomorphism $\langle\ , u + \beta\rangle^1_{\nabla f}$ is independent of β. We will generalize (3.2.5) for iterated mapping cones in the form of the spectral sequence defined a few pages on.

Proof of (3.2.4): Let $i : C_g \subset C_f$ be the inclusion. Consider the diagram

$$
\begin{array}{ccc}
\Sigma^n_Y C_g \cup C_f & \xrightarrow{\ \bar\mu \cup 1\ } & S^{n+1}B \vee C_f \\
{\scriptstyle Wf}\uparrow & \quad\textcircled{1}\quad \nearrow^{E^n\xi}\quad \textcircled{2}\ \ \Big\uparrow{\scriptstyle j' = 1 \vee i} & \\
S^n A & \xrightarrow[\ E^n\nabla f\ ]{} & S^{n+1}B \vee C_g
\end{array}
$$

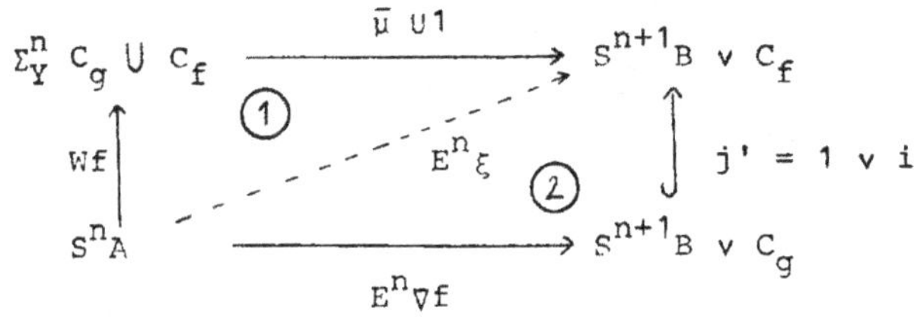

in which the map $\bar\mu$ on $\Sigma^n_Y C_g$ is the composite in (2.4.11). The composite $(\bar\mu \cup 1) \circ Wf$ then induces $\nabla^n(u,f)$ and $\widetilde{\nabla}^n(u,f)$ respectively, by definition of u^+ and $\Sigma^n f$. The map j' is the inclusion. To prove the theorem we need only show that the above diagram is homotopy commutative. To this end we construct the map

$$\xi = j\mu f : A \xrightarrow{\ f\ } C_g \xrightarrow{\ \mu\ } SB \vee C_g \subset SB \vee C_f$$

with $\mu = \sigma_2 + \sigma_1$, $j = 1 \vee i$. The map ξ is trivial on C_f, and so the partial suspension E^n in the diagram is indeed defined.

We now show that the subdiagrams ① and ② are homotopy commutative. First of all

$$j_*(\nabla f) = j_*(-\sigma_2 f + \mu f) = j_*\mu f = \xi,$$

but by naturality (3.1.5) it follows that

$$E^n\xi = E^n(j_*\nabla f) = j'_*(E^n\nabla f),$$

and this equivalent to subdiagram ② being homotopy commutative. Since ξ is trivial on C_f, in fact by means of a canonical homotopy, there exists a map $\bar{\xi}$ which fits into the commutative diagram

$$\begin{array}{ccccccc}
C_f & & \xrightarrow{\quad\lambda\quad} & & C_\xi & & \\
\cup\uparrow & & & & \cup\uparrow & \searrow^{\bar{\xi}} & \\
C_g & \xrightarrow{\mu} & SB \vee C_g & \xrightarrow{j} & SB \vee C_f & \xrightarrow{\rho} & C_f
\end{array}$$

and is such that $\bar{\xi}\lambda$ is homotopic relative X_o to the identity on C_f, where λ is induced by the identity on CA, see (1.2.13). Therefore the diagram

$$\begin{array}{ccccc}
S^nA & \xrightarrow{W\xi} & \Sigma^n_{C_f}(SB \vee C_f) \cup C_\xi & = & S^n \ltimes SB \cup_{SB} C_\xi \\
\downarrow{\scriptstyle Wf} & \nearrow{\scriptstyle (1\times j\mu)\cup\lambda} & & & \downarrow{\scriptstyle \pi\cup\bar{\xi}} \\
\Sigma^n_Y C_g \cup C_f & & \xrightarrow{\qquad \mu\,\cup\,1 \qquad} & & S^{n+1}B \vee C_f
\end{array}$$

is homotopy commutative too. This follows from the definition of $\widetilde{\mu}$ in (2.4.11). $(1 \times j\mu)$ is induced from the inclusion $Y \subset C_f$ and from $j\mu$. By (3.1.10) it follows that subdiagram ① is homotopy commutative. Clearly the first equation in (3.2.4) is a consequence of the homotopy commutativity of ① and ② . The second equation arises from the following equalities. If $\widetilde{\rho} : \rho^*\widetilde{C}_f \longrightarrow \widetilde{C}_f$ denotes the map over $\rho : \Sigma^n_Y C_f \to C_f$, $j : F \subset \rho^*\widetilde{C}_f$ the inclusion of the fiber, and $i = \widetilde{\rho}j$, then

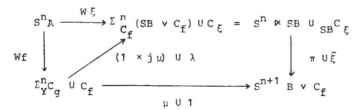

$$\tilde{\nabla}^n(u,f)(\beta) = (Wf)^{\#}((\rho^*(u)+\bar{\beta})\cup u_o), \qquad \bar{\beta} = (-1)^{n+1}\beta,$$

(1) $$= i^{-1}_{*}\tilde{\rho}_*(-\partial\tilde{\pi}_{Wf}+((\rho^*(u)+\bar{\beta})\cup u_o)_*Wf)$$

(2) $$= i^{-1}_{*}\tilde{\rho}_*((\rho^*(u)+\bar{\beta})\cup u_o)_*Wf$$

(3) $$= i^{-1}_{*}(Wf)^*(\bar{\mu}\cup 1)^*(i_*\beta,u_o)$$

(4) $$= i^{-1}_{*}(E^n\nabla f)^*j'^*(i_*\beta,u_o)$$

(5) $$= i^{-1}_{*}(E^n\nabla f)^*(i_*\beta,u) = \langle\beta,u\rangle^n_{\nabla f}$$

(1) holds by definition of $(Wf)^{\#}$, since $j^{-1}_{*} = i^{-1}_{*}\tilde{\rho}_*$. From $\rho_*\pi_{Wf} = 0$ it follows that $\tilde{\rho}_*\tilde{\pi}_{Wf} = 0$, and so (2) holds. (3) follows from (1.2.19) and (2.4.11). (4) derives from the homotopy commutativity of ① and ② , and (5) from the definition in (3.2.1). ▭

We will now introduce a spectral sequence by means of an exact couple. Exact couples are discussed more fully in [48].

An abelian group E together with an endomorphism $d : E \to E$ such that $d^2 = dd = 0$ is called a <u>differential group</u>. Let $Z = \operatorname{Ker} d$ be the subgroup of cycles of E, and $B = \operatorname{Im} d$ be the subgroup of boundaries of E. B is clearly a subgroup of Z, and the quotient $H(E) = Z/B$ is called the homology group of (E, d). A <u>spectral sequence</u> is a sequence of differential groups $\{E_n, d_n;\ n \geq 0\}$ such that $E_{n+1} = H(E_n, d_n)$, $n \geq 0$. One method of obtaining spectral sequences is by considering <u>exact couples</u>, first introduced by Massey. Let E be an abelian group and D a group. Suppose there are group homomorphisms making the triangle

(3.2.6)
$$\begin{array}{ccc} D & \xrightarrow{\ \alpha\ } & D \\ & {}_{\gamma}\nwarrow \quad \swarrow_{\beta} & \\ & E & \end{array} \qquad \begin{array}{l} \operatorname{Ker}\alpha = \operatorname{Im}\gamma \\ \operatorname{Ker}\beta = \operatorname{Im}\alpha \\ \operatorname{Ker}\gamma = \operatorname{Im}\beta \end{array}$$

exact. Then $(D, E, \alpha, \beta, \gamma)$ is called an exact couple. D is often required in the definition to be abelian, but in the sequel we will want to treat cases where D is not. Notice that the homomorphism $d = \beta\gamma : E \to E$ is such that $d^2 = \beta\gamma\beta\gamma = 0$, since by exactness $\gamma\beta = 0$. Let $E_1 = H(E, d)$. It is easy to verify that $Z(E) = \gamma^{-1}(\alpha D)$ and $B(E) = \beta(\text{Ker } \alpha)$, and so

$$E_1 = \gamma^{-1}(\alpha D)/\ \beta(\text{Ker } \alpha)$$

We now set $D_1 = \alpha D$ and define $\alpha_1 : D_1 \to D_1$ to be the restriction of α to D_1, and we define homomorphisms β_1, γ_1 in

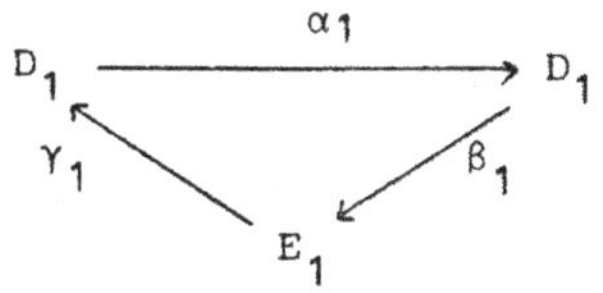

by $\beta_1 : \alpha x \mapsto \beta x + \beta(\text{Ker } \alpha)$ $(x \in D)$ and $\gamma_1 : y + \beta(\text{Ker } \alpha) \mapsto \gamma(y)$ $(y \in E)$. A diagram chase show that β_1 and γ_1 are well-defined and that the triangle is exact. The exact couple

$(D_1, E_1, \alpha_1, \beta_1, \gamma_1)$ is called the <u>derived couple</u> of $(D, E, \alpha, \beta, \gamma)$.

The successive derived couples give us groups $E_2, E_3, \ldots$ such that $E_{n+1} = H(E_n, d_n)$, and so we have a spectral sequence. A direct description of the groups E_r is possible without carrying out the successive derivations. For any integer $r \geqslant 0$ we have the r-th derived couple

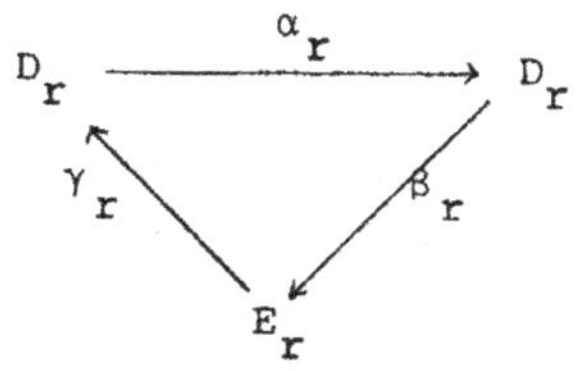

Let $D_r = \alpha^r D$, where α^r is the composite of α with itself r times, and let

$$E_r = \gamma^{-1}(\alpha^r D)/\beta(\mathrm{Ker}\ \alpha^r). \tag{3.2.8}$$

Define α_r to be the restriction of α to D_r, $\beta_r : \alpha^r x \longmapsto \beta x + \beta(\mathrm{Ker}\ \alpha^r)$ $(x \in D)$ and $\gamma_r : y + \beta(\mathrm{Ker}\ \alpha^r) \longmapsto \gamma(y)$, $(y \in \gamma^{-1}\alpha^r D)$. Now the differential

$$d_r = \beta_r\gamma_r : E_r \longrightarrow E_r \tag{3.2.9}$$

is seen to be induced by $\beta\,\alpha^{-r}\gamma$.

We will associate a spectral sequence to an iterated principal cofibration $Y = X_o \subset X_1 \subset \ldots \subset X = \varinjlim X_i$ with attaching maps $f_i : A_i \to X_{i-1}$, under the additional assumption that A_i is a co-H-group for $i \geqslant 2$. Let u be a map $X \to F$, or let u be a section $X \to \tilde{X}$ of a fibration $\tilde{X} \to X$ with fiber F. Define $u_i = u|X_i$ and $v = u_o$ to be the indicated restrictions of u. Denoting by

$$j_n : [X_n, F]^v \longrightarrow [X_{n-1}, F]^v \quad \text{and} \quad j_n : \langle X_n, \tilde{X}_n\rangle^v \longrightarrow \langle X_{n-1}, \tilde{X}_{n-1}\rangle^v$$

the restriction maps, we can define the abelian group.

$$J_n(X,Y;u) = j_n^{-1}(u_{n-1}) \cong [SA_n, F]/\mathrm{Im}\ \Sigma(u_{n-1}, f_n), \tag{3.2.10}$$

where A_n is a co-H-group. The group operation gives us a group structure on $J_n(X, Y;u)$ in which u_n is the zero element, see (2.4.2). By (2.4.3) this group depends only on the homotopy class $u \in [X, F]$, that is on $u \in \langle X, \tilde{X}\rangle$. There now follows the definition of a spectral sequence which can be used to compute the abelian group $J_n(X, Y;u)$. Such a computation for every n amounts to a solution of the classification problem of (1.1).

First we define bigraded groups $E_1^{**} = \{E_1^{p,q}\}$ and $D^{**} = \{D^{p,q}\}$ with $p,q \in \mathbb{Z}$, where the E_1^{**} are abelian. Let $E_1^{p,q} = [S^{q+1}A_p, F]$

for $q \geq 0$ and $p \geq 1$, and let

$$D^{p,q} = [\Sigma_Y^q X_p, F]^{u_p} \quad \text{or} \quad = \langle \Sigma_Y^q X_p, \rho * \tilde{X}_p \rangle^{u_p}$$

for $q \geq 1$ and $p \geq 1$. If $q = 0$ and $p \geq 1$ let $D^{p,0} = \bigoplus_{n=1}^{p} J_n(u)$ with $J_n(u) = J_n(X,Y;u)$. Finally we define

$$E_1^{p,q} = D^{p,q} = 0 \qquad \text{for } q < 0 \text{ or } p \leq 0.$$

(If A_1 is not a co-H-space, then we set $E^{1,0} = D^{1,0} = J_1(u) = 0$. This ensures that E^{**} remains abelian. We want to leave room for this special case because in the skeletal filtration of a CW-complex, A_1 consists only of 0-spheres and so is not a co-H-group.)

The classification sequences $(A, \underline{A})$ and $(B, \underline{B})$ of (2.4) give us, for every pair $q, p \in \mathbb{Z}$, a long exact sequence of groups and group homomorphisms

$$\cdots \to E_1^{p,q} \xrightarrow{\gamma^{p,q}} D^{p,q} \xrightarrow{\alpha^{p,q}} D^{p-1,q} \xrightarrow{\beta^{p-1,q}} E_1^{p,q-1} \xrightarrow{\gamma}$$

If $q \geq 1$, $p \geq 1$, then

$$\gamma^{p,q} = (-1)^{q+1} u_p^+ \qquad \text{as in (3.2.3)}$$

$$\alpha^{p,q} = j \qquad \text{the restriction}$$

$$\beta^{p-1,q} = \Sigma^q f_p \qquad \text{as in (3.2.3)}$$

If $q \neq 0$ and $p \geq 1$, then

$$\gamma^{p,0} = i_p \circ \gamma_p \quad : \quad E_1^{p,0} \twoheadrightarrow J_p(u) \subset \bigoplus_{i \leq p} J_i(u)$$

$$\alpha^{p,0} = pr \quad : \quad \bigoplus_{i \leq p} J_i(u) \twoheadrightarrow \bigoplus_{i \leq p-1} J_i(u)$$

γ_p is the surjective homomorphism that defines the group structure of $J_p(u)$, as discussed above. i_p is the inclusion and pr the projection. In all other cases, the homomorphisms in the sequence are zero.

These sequences can be presented simultaneously as an exact couple

(3.2.11)

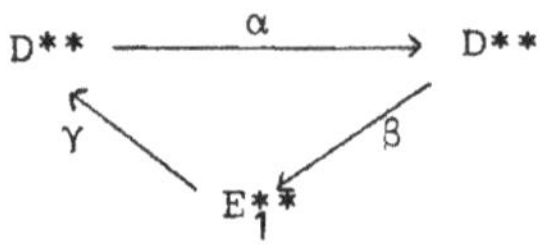

in which α, β, γ have bidegree $(-1,0)$, $(+1,-1)$, $(0,0)$ respectively.

<u>Note:</u> By definition, the first differential $d_1 = \beta\gamma$, $d_1^{p,q}$: $E_1^{p,q} \to E_1^{p+1,q-1}$ for $p,q \geqslant 1$ is just the homomorphism $\nabla^q(u_p, f_{p+1})$, or or $\tilde{\nabla}^q(u_p, f_{p+1})$ respectively, from (3.2.3). By proving theorem ((3.2.4), we proved that this differential on E_1^{**} is induced by the q-th partial suspensions of the element $\nabla f_{p+1} \in [A_{p+1}, SA_p \vee X_p]_2$. Accordingly, the homology $E_2^{**} = H(E_1^{**}, d_1)$ can be computed with the aid of the primary operations (3.2.1). In analogy with the construction of the cochain complex (E_1^{**}, d_1), we can use the elements ∇f_{p+1} to obtain a chain complex, described in the appendix to this section.

In the manner of (3.2.8), the exact couple (3.2.11) gives us a spectral sequence $\{E_r^{**}, d_r; r \geqslant 1\}$ in which $d_r : E_r^{**} \to E_r^{**}$ is induced by $\beta\alpha^{-(r-1)}\gamma$, as we saw in (3.2.9), and so has bidegree $(r,-1)$. The reader will note that our spectral sequence begins with the E_1-term, a slight alteration of (3.2.9).

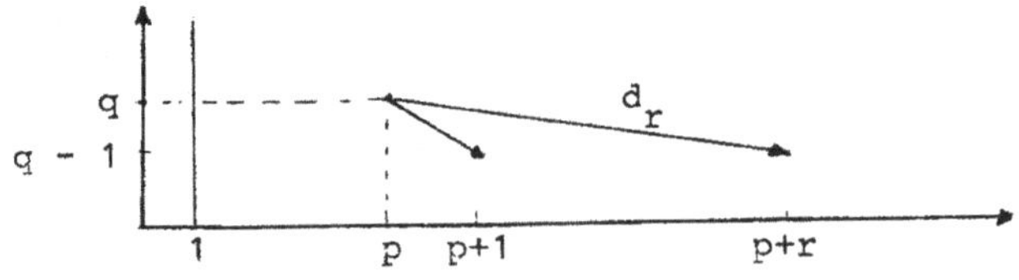

(3.2.12) <u>Theorem:</u> <u>The spectral sequence depends only on the homotopy class</u> $u \in [X, F]$, <u>that is</u> $u \in \langle X, \tilde{X} \rangle$. <u>We write</u> $E_r^{**} =$

$E_r^{**}(X, Y;u)$ and $d_r = d_r(u)$.

Proof: As in the proof of (2.4.3), one shows that a homotopy $u \simeq u'$ induces an isomorphism between the exact sequence $(B, \underline{B})$ for u and u' thus also between the exact couples (3.2.11) . This isomorphism is the identity on E_1^{**}. □

Because of the bidegree of the differential, we see that $E_p^{p,0} = E_{p+1}^{p,0} = \ldots =: E_\infty^{p,0}$.

(3.2.13) Theorem: $J_p(X, Y; u) = E_\infty^{p,0}(X, Y;u)$ for $p \geq 1$.

By computing the differentials $d_1, d_2, \ldots, d_{p-1}$ on $E_1, E_2, \ldots, E_{p-1}$ we can work out the group $J_p(u)$ of (3.2.10). Examples of this computation will be given in (4.3).

Proof: We have

$$E_p^{p,0} = \gamma^{-1}(\alpha^{p-1}(D^{2p-1,0}))/\beta(\mathrm{Ker}\ \alpha^{p-1}) = E_1^{p,0}/\mathrm{Im}\ \beta = J_p(u). \quad \square$$

If (X, Y) is a relative strictly pointed CW-complex, than the skeletal filtration $Y = X_0 \subset \ldots \subset X_n \subset \ldots \subset X$ is an iterated principal cofibration. The above spectral sequence is thus defined in this important case, which we will treat more fully in (4.4). We will prove that in this case the E_2-term of the spectral sequence is given by cohomology groups with local coefficients, see also the appendix to this section.

In (1.2.29) we defined higher order differences. These are connected with the spectral sequence (3.2.12) in the following way.

(3.2.14) Theorem: Let $u, v : X \to \tilde{X}$ be sections restricting to $u^r, v^r : X_r \to \tilde{X}_r$. Suppose $H : u^p \simeq v^p$ is a section homotopy. Then the r-th order difference

$$D_r(u,\ H,\ v) \subset [SA_{p+r},\ F] \ = \ E_1^{p+r,0}$$

is either empty or else a coset of $\operatorname{Im} d_{r-1}(u)$, so that

$$D_r(u,\ H,\ v) \in E_r^{p+r,0}(X,\ Y;\ u)\ .$$

Proof: If $D_r(u, H, v)$ is not empty, H can be extended to $G : u^{p+r-1} \simeq v^{p+r-1}$. Therefore $D_r(u, H, v) = D_r(u, u) + d(u^{p+r}, G, v^{p+r})$, where we have $D_r(u, u) = \operatorname{Im} d_{r-1}(u)$. ☐

(3.2) Appendix: Iterated principal cofibrations and the associated chain complex

As before, suppose $Y = X_o \subset X_1 \subset \ldots \subset X = \varinjlim X_i$ is an iterated principal cofibration with attaching maps $f_i : A_i \to X_{i-1}$ and that A_i is a co-H-group when $i \geq 2$. Then the first differential d_1 of the spectral sequence discussed is induced by the element $\nabla f_{p+1} \in [A_{p+1}, SA_p \vee X_p]_2$, see the note following (3.2.11). The property $d_1 d_1 = 0$ of this differential also follows from

(3.2.15) Theorem: The composition

$$(E\ \nabla f_p, 1) o (\nabla f_{p+1}) : A_{p+1} \to SA_p \vee X_p \longrightarrow S^2 A_{p-1} \vee X_p$$

is null-homotopic.

Proof: Let i be the inclusion of X_p. Then

$$(\nabla f_{p+1})^* (E \nabla f_p, 1) = (-if_{p+1} + \mu f_{p+1})^* (E \nabla f_p, 1) =$$

$$= -if_{p+1} + f^*_{p+1}\ \mu^*(E \nabla f_p, 1).$$ Moreover

$\mu^*(E\nabla f_p, 1) = i + E(\nabla f_p) = i + d(i, H, i) = i$, see (3.1.2) and (1.2.8). H is the self-homotopy $I \times X_{p-1} \longrightarrow I \times SA_{p-1} \longrightarrow S^2 A_{p-1}$ ☐

<u>Definition</u>: We define $df_{p+1} \in \sigma_{-1}^{A_{p+1}}(A_p; X)$ to tbe the image of ∇f_{p+1} under the map

$$[A_{p+1}, SA_p \vee X_p]_2 \xrightarrow{E^\infty} \sigma_{-1}^{A_{p+1}}(A_p; X_p) \xrightarrow{j} \sigma_{-1}^{A_{p+1}}(A_p; X)$$

where j is induced by the inclusion $X_p \subset X$.

We can then say about the pairing (3.1.15) that

(3.2.16) <u>Corollary</u>: <u>If</u> $Y \subset X$ <u>is an iterated principal cofibration with attaching maps</u> f_p, <u>then the sequence of elements</u>

$$df_p \in \sigma_{-1}^{A_{p+1}}(A_p; X) \quad \underline{\text{satisfies}} \quad (df_p)\circ(df_{p+1}) = 0 \ .$$

In the stable range, that is when $j\,E^\infty$ in the preceeding definition is a bijection, the differential d_1 is induced by the sequence of elements df_p. We consider two important cases.

(A) Every A_p is a wedge of spheres $A_p = \bigvee_{e \in Z_p} S^{(\dim e)-1}$

where $\dim e \geqslant p$ for all $e \in Z_p$.

(B) Condition (A) holds and also $\dim e = p$ for all $e \in Z_p$.

Thus in each of these cases X is a CW-complex and Z_p is the set of cells attached to X_{p-1}. In (B), Z_p is the set of p-cells of X, and X_p is the p-skeleton of X. When condition (A) or (B) is satisfied, the group $\sigma_p^{A_p}(A_{p-1}; X)$ can be decomposed as in the

(3.2.17) <u>Lemma</u>: $\sigma_q^{Y \vee Y'}(A; X) = \sigma_q^{Y}(A; X) \times \sigma_q^{Y'}(A; X)$ <u>and</u> $\sigma_q^{Y}(A \vee A'; X) = \sigma_q^{Y}(A; X) \times \sigma_q^{Y}(A'; X)$ <u>when</u> Y <u>is a finite-dimensional CW-complex.</u>

This is easily seen to be a consequence of (3.1.19) and (3.1.20). Since

(3.2.17)

$\sigma_q^{S^i}(S^o; X) = \sigma_{i+q-j}^{S^o}(S^o; X)$, when (A) or (B) holds we obtain elements

$$df_{p+1} \in \bigtimes_{e \in Z_{p+1}} \bigoplus_{d \in Z_p} \sigma_{|e|-|d|-1}^{S^o}(S^o; X)$$

as in (3.2.16), where $|e| = \dim e$. In case (B) we always have $|e| - |d| - 1 = 0$. By (3.1.26), $\sigma_o^{S^o}(S^o; X) = \mathbb{Z}[\pi]$ is the group ring of $\pi = \pi_1(X)$. Thus in case (B) the elements df_p yield homomorphisms of free $\mathbb{Z}[\pi]$-modules

$$(B')\quad \ldots \to \bigoplus_{e\in Z_{p+1}} \mathbb{Z}[\pi] \xrightarrow{df_{p+1}} \bigoplus_{e\in Z_p} \mathbb{Z}[\pi] \xrightarrow{df_p} \bigoplus_{e\in Z_{p-1}} \mathbb{Z}[\pi] \to \ldots$$

via $\mathbb{Z}[\pi]$-linear extension, $p \geqslant 2$.

Now let $\sigma_* X = \sigma_*^{S^o}(S^o; X) = \pi_*^S(\Omega X)^+$ be the graded ring in (3.1.18). Similarly to (B'), in the more general case (A) the elements df_p can be $\sigma_* X$-linearly extended to homomorphisms

$$(A')\quad \ldots \to \bigoplus_{e \in Z_{p+1}} \sigma_* X \xrightarrow{df_{p+1}} \bigoplus_{e \in Z_p} \sigma_* X \xrightarrow{df_p} \bigoplus_{e \in Z_{p-1}} \sigma_* X \to \ldots$$

of free $\sigma_* X$-modules $p \geqslant 2$. As a consequence of (3.2.16) we then have

(3.2.17) Corollary: <u>(B') is a chain complex over $\mathbb{Z}[\pi]$, and (A') is a chain complex over $\sigma_* X$. That is, in each case the composition $(df_p) \circ (df_{p+1}) = 0$ is zero.</u>

We will show in (4.1.10) that the chain complex (B') is in fact the cellular chain complex of the homology of X with local coefficients. We will calculate as an example the chain complex (A') for the complex projective space $X = \mathbb{C}P_\infty$, for which $\Omega(\mathbb{C}P_\infty) \simeq S^1$. The results are stated in (3.6.17).

Remark: The $\sigma_* X$ -chain complex (A') is the starting point for a general theory of partially stable higher order homotopy operations. It can be drawn up on parallel lines to the theories of Maunder [77] and McClendon [79] for cohomology operations. The reader may refer to the dual discussion in the appendix to (6.2).

(3.3) The first differential of the spectral sequence, Whitehead products and relative Whitehead products.

In the prec eding section we exhibited a spectral sequence whose first differential (on the E_1-term) is defined in terms of the operators $\nabla^n(u, f)$ and $\tilde{\nabla}^n(u, f)$. We now want to show that these operators are additive in f, and we give a composition formula for them. We will then express the operators in terms of Whitehead products and relative Whitehead products. Finally, we calculate $\nabla^n(u, f)$ and $\tilde{\nabla}^n(u, f)$ for a map $f : SA \to SB$ with the aid of the Hilton-Hopf invariants of f and the Hilton-Milnor theorem. In an appendix we classify fibrations over a suspension, and use these results to obtain a twisting map for relative Whitehead products.

Let A be a co-H-group and B a co-H-space. From inclusions $\sigma_1, \sigma_2 : B \subset B \vee B$ we form $\sigma_2 + \sigma_1 : B \to B \vee B$ which is, apart from the order of addition, just the comultiplication on B. Given a map $f : A \to B$, we define the difference element

$$(3.3.1) \qquad \nabla f = -f_*(\sigma_2) + f_*(\sigma_2 + \sigma_1) \in \pi_o^A(B \vee B)_2$$

similarly to (3.2.2). This element is trivial on $\sigma_2(B)$. As in (3.2.1) we obtain for each partial suspension $E^n \nabla f$ the operations

$$\nabla^n(u, f) = [\ ,\ u]^n_{\nabla f} \ :\ [S^nB, U] \longrightarrow [S^nA, U]$$

$$\tilde{\nabla}^n(u, f) = \langle\ ,\ u\rangle^n_{\nabla f} \ :\ [S^nB, F] \longrightarrow [S^nA, F]$$

in which $u \in [B, U]$ or $u \in \langle B, \tilde{B}\rangle$ denotes a section of a fibration $\tilde{B} \to B$ with fiber F. When $n \geq 1$ these operations are homomorphisms of abelian groups. Let $\tilde{C}_f \to C_f$ be a fibration with fiber F. A partial extension of the classification theorem (3.2.5) is the

(3.3.2) <u>Classification theorem</u>: <u>There are bijections</u>

$$[C_f, U] \approx \bigcup_u [SA, U]/\mathrm{Im}\ \nabla(u, f),$$

$$\langle C_f, \tilde{C}_f\rangle \approx \bigcup_u [SA, F]/\mathrm{Im}\ \tilde{\nabla}(u, f),$$

<u>where the first union is taken over all</u> $u \in [B, U]$ <u>with</u> $f^*(u) = 0$, <u>and the second over all</u> $u \in \langle B, \tilde{B}\rangle$ <u>with</u> $f^{\#}(u) = 0$.

The bijections are defined by $\alpha \mapsto u_o + \alpha$, where u_o is any extension over C_f of u. If $B = SB'$ is a suspension, this result expresses the special case of (3.2.5) in which we set $X_o = *$. The reader may adapt the proof of (3.2.5) to apply to a co-H-space B instead of SB', and so arrive at a proof of (3.3.2).

(3.3.3) <u>Note</u>: Suppose that B' is a co-H-space and that $B = SB'$. If a section $u_o : C_f \to \tilde{C}_f$ exists, then the second bijection of (3.3.2) can be expressed as

$$\langle C_f, \tilde{C}_f\rangle \approx \bigcup_\beta [SA, F]/\mathrm{Im}\ \tilde{\nabla}(u + \beta, f),$$

in which $u = u_o | B$. The union is to be taken over all $\beta \in [B, F]$ with $\langle \beta, u\rangle^o_{\nabla f} = 0$.

<u>Proof</u>: By (1.2.21), the map $u \mapsto u + \beta : [B, F] \approx \langle B, \tilde{B}\rangle$ is a bijection. Furthermore, we have $f^{\#}(u + \beta) = 0 \iff \langle \beta, u\rangle^o_{\nabla f} = 0$. Hence

denoting by $i : F \subset \widetilde{B}$ the inclusion, we have

$$\langle \beta, u \rangle^{o}_{\nabla f} = i_*^{-1}[i_*\beta\ ,\ u]^{o}_{\nabla f}$$

$$= i_*^{-1}(f^*(u + i_*\beta) - f^*(u))$$

$$= i_*^{-1}(-\partial\widetilde{\pi}_f + (u + \beta)_* f - (-\partial\widetilde{\pi}_f + u_* f))$$

$$= f^{\#}(u + \beta) - f^{\#}(u) = f^{\#}(u + \beta),$$

here $f^{\#}(u) = 0$ since we assumed that u has a section extension u_o. □

Remark: Barcus-Barratt [5] and Rutter [106] have also investigated the isotropy groups of the $[SA, U]$-action on $[C_f, U]$, which we identify in (3.3.2). In essence, they use the homomorphism $\Sigma(f)\gamma$ in Rutter's notation $\Gamma(u, f)$. We have shown that this homomorphism can be expressed in terms of a partial suspension. It will turn out that for this reason the properties of $\Gamma(u, f)$ discussed below, can be demonstrated much more simply. For example the Whitehead product theorem, the proof of which occupies four pages (p. 70 - 73) in [5] for $A = S^n$, can be reduced by our methods in a few easy steps to the statement that the Whitehead product is bilinear. This formulation in terms of partial suspensions has the further advantage that it enables us to prove all the theorems for section homotopy sets as well. The reader will note that $\Gamma(u, f)$ has here been embedded in a long exact sequence, whereas in [5] and [106] only an end of this sequence is considered.

It follows immediately from properties of the partial suspension that the following theorems hold for the operators $\nabla^n(u, f)$ and $\widetilde{\nabla}^n(u, f)$.

(3.3.4) Composition theorem: Let $A \xrightarrow{f} B \xrightarrow{g} C$ be maps, where A and B are co-H-groups and C is a mapping cone or a co-H-space

Then for $u \in [C, U]$ we have

$$\nabla^n(u, gf) = \nabla^n(ug, f) \circ \nabla^n(u, g).$$

Suppose $\tilde{C} \to C$ is a fibration. Then for a section $u \in \langle C, \tilde{C}\rangle$ we have

$$\tilde{\nabla}^n(u, gf) = \tilde{\nabla}^n(g^*u, f) \circ \tilde{\nabla}^n(u, g)$$

where $g^*u \in \langle B, g^*\tilde{C}\rangle$ denotes the induced section.

Proof: We have

$$\nabla f = -f^*(\sigma_2^B) + f^*(\sigma_2^B + \sigma_1^B) \in [A, B \vee B]_2$$

$$\nabla g = -g^*(\sigma_2) + g^*(\sigma_2 + \sigma_1) \in [B, D \vee C]_2$$

where $D = C$ if C is a co-H-space, and $D = SD'$ if C is a mapping cone of $D' \to X$. We have further

(3.3.5) $$\nabla(gf) = -(gf)^*(\sigma_2) + (gf)^*(\sigma_2 + \sigma_1)$$

$$= -f^*g^*(\sigma_2) + f^*g^*(\sigma_2 + \sigma_1)$$

$$= -f^*(g^*\sigma_2) + f^*(g^*(\sigma_2) + \nabla g)$$

$$= (\nabla f)^*(\nabla g, g^*(\sigma_2))$$

$$= (\nabla g, \sigma_2 \circ g) \circ \nabla f.$$

It then follows from (3.1.6) that

$$E^n \nabla(gf) = (E^n \nabla g, \sigma_2 g) \circ E^n \nabla f.$$

A comparison of this with the definition of $[\ , u]^n_{\nabla f}$ and $\langle\ , u\rangle^n_{\nabla f}$ in (3.2.1) completes the proof. □

(3.3.6) Additivity theorem: Let A be an abelian co-H-group, and

C a co-H-space or a mapping cone. Let $f, g \in [A, C]$, and $u \in [C, U]$ or $u \in \langle C, \tilde{C}\rangle$. Then

$$\tilde{\nabla}^n(u, f + g) = \hat{\nabla}^n(u, f) + \tilde{\nabla}^n(u, g),$$

Proof: Let $\sigma_2 + \sigma_1 : C \to D \vee C$ be as in (3.3.5). Then in $[A, D \vee C]$ we have

$$(3.3.7) \quad \nabla(f + g) = -(f + g)^*(\sigma_2) + (f + g)^*(\sigma_2 + \sigma_1)$$

$$= -g^*\sigma_2 - f^*\sigma_2 + f^*(\sigma_2 + \sigma_1) + g^*(\sigma_2 + \sigma_1)$$

$$= \nabla f + \nabla g.$$

Since E^n is a homomorphism, (3.3.6) follows from definition (3.2.1). □

A special case of the relative Whitehead product in (0.3) is the following.

(3.3.8) Definition: Let $p : Y \to B$ be a fibration with fiber F. Given a space X, there is an isomorphism $\chi : \pi_2^X(Z_p, Y) \cong \pi_1^X(F)$ where Z_p is the mapping cylinder of p, cf. (0.1.11). The relative Whitehead product

$$[\ ,\ ] \quad : \pi_1^A(Z_p, Y) \times \pi_1^B(Y) \longrightarrow \pi_2^{A \wedge B}(Z_p, Y)$$

therefore gives us an operation

$$\langle\ ,\ \rangle \quad : \pi_1^A(F) \times \pi_1^B(Y) \longrightarrow \pi_1^{A \wedge B}(F)$$

by means of $\langle \alpha, \beta\rangle = \chi[\chi^{-1}\alpha, \beta]$. We call this the relative Whitehead product for the fibration p. If A and B are co-H-spaces, $\langle\ ,\ \rangle$ is a bilinear pairing of abelian groups.

This definition is in close analogy with the definition of the group operation Φ_n in (1.5.9). Denoting by $i : F \subset Y$ the inclusion, we have $i_* = \partial \circ \chi^{-1} : \pi_1^X(F) \cong \pi_2^X(Z_p, Y) \longrightarrow \pi_1^X(Y)$ and so the equation

$$(3.3.9) \qquad i_* \langle \alpha, \beta \rangle = [i_*\alpha, \beta] \in \pi_1^{A \wedge B}(Y)$$

follows from the universal property of the relative Whitehead product as as expressed in (0.3). Therefore this equation completely characterizes $\langle\ ,\ \rangle$ when i_* is injective. This is the case for instance, when $Y \twoheadrightarrow B$ has a section. The following special case will be important in the sequel. If $\widetilde{SB} \twoheadrightarrow SB$ is a fibration with fiber F, then we have the operation, see [60],

$$(3.3.10) \qquad \langle,\rangle : [SA, F] \times \langle SB, \widetilde{SB}\rangle \longrightarrow [SA \wedge B, F]$$

defined by $\langle \alpha, u \rangle = i_*^{-1}[i_*\alpha, u]$. Mc Carty calls this operator $\langle\ ,\ \rangle$ simply the <u>'brace' product</u>. It is in fact a special case of the relative Whitehead product. If the fibration $\widetilde{SB} = SB \times F$ is trivial, then $\langle \widetilde{SB}, SB \rangle = [SB, F]$ and the product $\langle\ ,\ \rangle$ is the Whitehead product. If B is a co-H-space then by (1.2.21) the map $\beta \mapsto u + \beta$ is a bijection $[SB, F] \to \langle SB, \widetilde{SB}\rangle$, and we find that

(3.3.11) <u>Lemma</u>: <u>If</u> B <u>is a co-H-space, then</u> $\langle \alpha, u + \beta \rangle = \langle \alpha, u \rangle + [\alpha, \beta]$.

<u>Proof:</u> By (1.2.20), $\Theta(u + \beta) = (\Theta u) + i_*\beta$ in $[SB, \widetilde{SB}]$. From the bilinearity of the Whitehead product we can then conclude that

$$\begin{aligned} \langle \alpha, u + \beta \rangle &= i_*^{-1}[i_*\alpha, u + \beta] \\ &= i_*^{-1}[i_*\alpha, (\Theta u) + i_*\beta] \\ &= i_*^{-1}[i_*\alpha, \Theta u] + [\alpha, \beta] \\ &= \langle \alpha, u \rangle + [\alpha, \beta] . \quad \square \end{aligned}$$

Thus, if for a single $u \in \langle SB, \widetilde{SB}\rangle$ (where B is a co-H-space) we can calculate the map $\langle\ ,u\rangle : [SA, F] \to [SA \wedge B, F]$, then by (3.3.11) we already have the relative Whitehead product $\langle\ ,\ \rangle$ within reach. Again, the bilinearity of the Whitehead product implies that $\langle\ ,u\rangle$ is a homomorphism when A is a co-H-space. In an appendix to this section we obtain a twisting map which induces the homomorphism $\langle\ ,u\rangle$.

The operators $[\ ,\ ]^n_\xi$ and $\langle\ ,\ \rangle^n_\xi$ of (3.3.1) can be expressed in terms of Whitehead products. To show this we use the lemma of Ganea (3.1.20) and the evaluation map $R_Y : S\Omega Y \to Y$.

(3.3.12) <u>Theorem</u>: <u>Let</u> $\xi = i_1\xi_B + [R_B, R_Y]\,\xi_o \in \pi^A_o(B \vee Y)_2$. <u>Then</u>

$$[\beta, u]^n_\xi = \beta(S^n\,\xi_B) + [\beta(S^n R_B), u\,R_X]\,(S^n\xi_o)$$

<u>where</u> $(\beta, u) \in [S^nB, U] \times [Y, U]$. <u>If</u> $\tilde{Y} \to Y$ <u>is a fibration with fiber</u> F, <u>then</u>

$$\langle\beta, u\rangle^n_\xi = \beta(S^n\,\xi_B) + \langle\beta(S^nR_B), R_X^* u\rangle\,(S^n\,\xi_o)$$

<u>where</u> $(\beta, u) \in [S^nB, U] \times \langle Y, \tilde{Y}\rangle$.

<u>Proof</u>: We have $E^n\xi = E^n(i_1\,\xi_B) + E^n((R_B \vee R_Y)\,w\,\xi_o)$ where w is the Whitehead product map. It follows from (3.1.3) and (3.1.5) that $E^n(i_1\,\xi_B) = i_1(S^n\,\xi_B)$ and $E^n((R_B \vee R_Y)\,w\,\xi_o) = (S^nR_B \vee R_Y)(E^nw)(S^n\xi_o)$, where $E^nw = w$ by (3.1.11). ☐

It follows from familiar properties of the Whitehead product that the $[\ ,\ ]^n_\xi$ product has the following triviality behavior.

(3.3.13) <u>Corollary</u>: <u>Let</u> $\xi_B : A \to B \vee Y \to B$ <u>be the composition of</u> ξ <u>with the retraction onto</u> B. <u>Then</u>

1) $[\beta, 0]^n_\xi = \beta(S^n \xi_B)$.

2) <u>If</u> u <u>is an H-space, then</u> $[\beta, u]^n_\xi = \beta(S^n \xi_B)$ <u>for all</u> u.

3) $S[\beta, u]^n_\xi = (S\beta)(S^{n+1} \xi_B)$.

The Whitehead product map $w = w_{A,B} : SA \wedge B \to SA \vee SB = S(A \vee B)$ maps into a co-H-space, so we can form ∇w as in (3.3.1). Given $u = (u_A, u_B) \in [SA, U] \times [SB, U]$, we obtain

$$\nabla^n(u, w) : [S^{n+1}A, U] \times [S^{n+1} B, U] \to [S^{n+1} A \wedge B, U].$$

Suppose that $\widetilde{SA \vee SB} \to SA \vee SB$ is a fibration with fiber F and that $u = (u_A, u_B) \in \langle SA \vee SB, \widetilde{SA \vee SB} \rangle$, where $u_A \in \langle SA, \widetilde{SA} \rangle$ and $u_B \in \langle SB, \widetilde{SB} \rangle$. Then

$$\tilde{\nabla}^n(u, w) : [S^{n+1} A, F] \times [S^{n+1} B, F] \to [S^{n+1} A \wedge B, F]$$

is defined as well. Denote by $\tau: S^{n+1} A \wedge B \to S^{n+1} B \wedge A$ the map exchanging A and B. Then

(3.3.14) <u>Whitehead product theorem:</u> <u>If</u> A <u>and</u> B <u>are co-H-spaces, then for</u> $n \geq 1$

$$\nabla^n(u, w)(\alpha, \beta) = [\alpha, u_B] - \tau^*[\beta, u_A],$$

$$\tilde{\nabla}^n(u, w)(\alpha, \beta) = \langle \alpha, u_B \rangle - \tau^* \langle \beta, u_A \rangle.$$

<u>Proof</u>: Let A_i, B_i $(i = 1, 2)$ be copies of A and B. Writing $w = w_{A,B}$, we can express $w : SA \wedge B \to (SA_1 \vee SB_1) \vee (SA_2 \vee SB_2)$ as

$$\begin{aligned} \nabla w &= -w^*(a_2, b_2) + w^*(a_2 + a_1, b_2 + b_1) \\ &= -[a_2, b_2] + [a_2 + a_1, b_2 + b_1] \end{aligned}$$

where a_i, b_i are the inclusions of SA_i, SB_i in $SA_1 \vee SB_1 \vee SA_2 \vee SB_2$. If A and B are co-H-spaces, then it follows from bilinearity of the Whitehead product that $\nabla w = [a_1, b_1] + [a_1, b_2] + [a_2, b_1]$

Since $S[a_1, b_1] = 0$, we conclude from (3.1.3) and (3.1.11) that for $n \geq 1$ $E^n \nabla w = [S^n a_1, b_2] - \tau^*[S^n b_1, a_2]$. The theorem then follows from (3.2.1). ☐

In accordance with (0.3), there is a homotopy equivalence $C_w \simeq SA \times SB$. It then follows from the classification theorem (3.3.2) that

(3.3.15) Corollary: There is a bijection

$$[SA \times SB, U] \approx \bigcup_u [S^2A \wedge B, U] \big/ \big(\mathrm{Im}[\ \ , u_B] + \mathrm{Im}\, \tau^*[\ , u_A]\big)$$

where the union is taken over all $u = (u_A, u_B) \in [SA, U] \times [SB, U]$ with $[u_A, u_B] = 0$.

This corollary is valid even when A and B are not co-H-spaces, as we proved with other methods in [11] From (3.3.3) we deduce

(3.3.16) Corollary: Let A and B be co-H-spaces, and $\widetilde{SA \times SB} \to SA \times SB$ a fibration with fiber F and section u. Then there is a bijection

$$\langle SA \times SB, \widetilde{SA \times SB} \rangle \approx \bigcup_{\alpha, \beta} [S^2A \wedge B, F] \big/ \big(\mathrm{Im}\langle\ \ , u_A + \alpha\rangle + \mathrm{Im}\langle\ \ , u_B + \beta\rangle\big)$$

u_A and u_B denote the restrictions of u to SA and SB. The union is taken over all $(\alpha, \beta) \in [SA, F] \times [SB, F]$ with $\langle \alpha, u_B\rangle - \tau^*\langle \beta, u_A\rangle + [\alpha, \beta] = 0$.

We will now use the Hilton-Hopf invariants to characterize $\nabla(u, f)$ and $\tilde{\nabla}(u, f)$. First we describe the induction step in the Hilton-Milnor theorem. Let $B^{(i)} = B \wedge \ldots \wedge B$ be the i-fold smash product, with $B^{(0)} = S^0$. We have a map

$$P = \bigvee_{i \geq 0} SX \wedge B^{(i)} \longrightarrow SX \vee SB$$

such that the restriction P_i of P to $SX \wedge B^{(i)}$ is the iterated

Whitehead product $P_i = [i_X, i_B^{(i)}] = [\dots[i_X, i_B],\dots, i_B]$. i_X and i_B are the inclusions of SX and SB in $SX \vee SB$ and $P_o = i_X$.

(3.3.17) <u>Theorem</u> (Hilton-Milnor): <u>If</u> X <u>is a suspension, then the sequence</u>

$$\bigvee_{i \geq 0} SX \wedge B^{(i)} \xrightarrow{P} SX \vee SB \xrightarrow{r} SB$$

<u>is a fiber sequence, where r is the retraction.</u>

The definition of fiber sequence is given in (0.1.10). The theorem follows from 4.2 and 4.4 in [15]. An induction over (3.3.17) leads to the full Hilton-Milnor theorem [22].

(3.3.18) <u>Definition:</u> Let B be a suspension. Given $f \in [SA, SB]$ we define the <u>Hilton-Hopf invariants</u> $H_i(f)$ as follows. By the Hilton-Milnor theorem there is an isomorphism $P_* : \pi_1^A(\bigvee_{i \geq 1} SB^{(i)}) = \pi_1^A(SB \vee SB)_2$. Therefore from f we obtain for $i \geq 1$

$$H_i(f) = r_{i*}P_*^{-1}(\nabla f) \in [SA, SB^{(i)}],$$

where $r_i : \bigvee SB^{(i)} \to SB^{(i)}$ is the retraction. Thus $H_1(f) = f$.

The Hilton-Hopf invariants $H_i(f)$ lead to a complete characterization of the homomorphisms

$$\nabla^n(u, f) : [S^{n+1}B, U] \to [S^{n+1}A, U], \quad \tilde{\nabla}^n(u, f) : [S^{n+1}B, F] \to [S^{n+1}A, F]$$

(3.3.19) <u>Theorem:</u> <u>Let</u> B <u>be a suspension and</u> $f \in [SA, SB]$. <u>For and</u> $u \in [SB, U]$ <u>we have</u> $(n \geq 1)$

$$\nabla^n(u, f)(\beta) = (S^n f)^*\beta + \sum_{i \geq 1} [\beta, u^{(i)}] \circ (S^n H_{i+1}(f)).$$

<u>If</u> $p : \widetilde{SB} \to SB$ <u>is a fibration with fiber</u> F, <u>then for any</u> $u \in \langle SB, \widetilde{SB}\rangle$ <u>we have</u>

$$\tilde{\nabla}^n(u, f)(\beta) = (S^n f)^* \beta + \sum_{i \geqslant 1} \langle \beta , u^{(i)} \rangle (S^n H_{i+1}(f))$$

<u>where</u> $\langle \beta , u^{(i)} \rangle = \langle \ldots \langle \beta , u \rangle, \ldots, u \rangle$ <u>is the i-fold iterated relative Whitehead product for</u> p.

<u>Proof</u>: Let $\xi = P_*^{-1}(f)$. Then for $n \geqslant 1$

$$S^n \xi = \sum_{i \geqslant 1} (S^n j_i)(S^n H_i(f))$$

where $j_i : SB^{(i)} \subset \bigvee SB^{(i)}$ is the inclusion. On the other hand it follows from $\nabla f = P_* \xi = \xi^* P$ and naturality (3.1.5) that $E^n \nabla f = E^n \xi^* P = (S^n \xi)^* E^n P$. We conclude from (3.1.11) that $E^n P_i = [S^n i_1, i_2^{(i)}]$. This is the statement of the theorem. □

The following example illustrates these results.

(3.3.20) <u>Example</u>: Let $f : S^3 \to S^2$ be the Hopf map. It is well-known that $H_2(f) = 1$, so it follows from (3.3.18) that $f^*(\sigma_1 + \sigma_2) = f^*\sigma_1 + f^*\sigma_2 + [\sigma_1, \sigma_2]$ in $\pi_3(S^2 \vee S^2)$, and therefore $\nabla f = f^*\sigma_1 + [\sigma_1, \sigma_2]$. The mapping cone $C_f = \mathbb{C}P_2$ is the complex projective plane. In accordance with (3.3.2) there is a bijection

$$[\mathbb{C}P_2, U] \approx \bigcup_u \pi_4(U)/\{\eta^*\alpha + [\alpha , u] \mid \alpha \in \pi_3(U)\}$$

where the union is taken over all $u \in \pi_2(U)$ with $f^*u \doteq 0$. We have denoted by $\eta = Sf \in \pi_4(S^3) = \mathbb{Z}_2$ a generating element. (If $U = S^3, S^2$ it follows that every map $\mathbb{C}P_2 \to S^3, S^2$ is null-homotopic). Given a fibration $\widetilde{\mathbb{C}P}_2 \to \mathbb{C}P_2$ with fiber F and section u_o, we have a bijection

$$\langle \mathbb{C}P_2, \widetilde{\mathbb{C}P}_2 \rangle \approx \bigcup_\beta \pi_4(F)/\{\eta^*\alpha + \langle \alpha , u \rangle + [\alpha , \beta] \mid \alpha \in \pi_3(U)\}$$

where $u = u_o | S^2$ and the union is taken over all $\beta \in \pi_2(F)$ with $f^*\beta + \langle \beta, u \rangle = 0$.

(3.3) Appendix: Classification of fibrations over a suspension. Twisting maps.

As we have seen, relative Whitehead products play the same role in the classification of sections as do Whitehead products in the classification of maps. As a result of the classification of fibrations over a suspension carried out here, we will be able to represent relative Whitehead products by means of twisting maps. For the classification we use an adjunction lemma for fibrations.

First we divide fibrations into equivalence classes.

(3.3.21) Definition: Given spaces F and X, we consider pairs (p, i) or triples (p, u, i), where $p : E \to X$ is a fibration, $i : F \to p^{-1}(*)$ is a homotopy equivalence, and u is a section of p. Two such pairs are defined to be equivalent when there is a homotopy equivalence $h : E \to E'$ such that $p'h = p$ and $hi \simeq i'$. If in addition $hu \simeq u'$, we call (p, u, i) and (p', u', i') equivalent. We define

$\mathfrak{F}(F, X)$ = set of equivalence classes of pairs (p, i),

$\mathfrak{F}_0(F, X)$ = set of equivalence classes of triples (p, u, i).

(3.3.22) Classification theorem: Let B be a connected CW-space and $i : F \subset F \times B$ the inclusion. There are canonical bijections p

$$\mathfrak{F}(F, SB) \overset{\varphi}{\approx} \{ \eta \in [F \times B, F] \mid i^*\eta = 1_F \},$$

$$\mathfrak{F}_0(F, SB) \overset{\varphi}{\approx} \{ \eta \in [F \rtimes B, F] \mid i^*\eta = 1_F \}.$$

(3.3.23) Note: If $F \times SB \to SB$ is trivial with fiber F, $i : F \subset F \times SB$, then $u \in [SB, F]$ represents an element $u = (pr, u, i)$ $\mathfrak{F}_0(F, SB)$ for which $\varphi(u) : F \rtimes SB \simeq F \times B \cup CB \xrightarrow{pr \cup \pi} F \vee SB \xrightarrow{(1,u)} F.$

Given $\xi \in \mathcal{F}_0(F, SB)$, we call $\varphi(\xi) \in [F \rtimes B, F]$ the <u>twisting map</u> of ξ because

(3.3.24) <u>Theorem:</u> <u>If</u> $(p, u, i) \in \xi \in \mathcal{F}_0(F, SB)$ <u>and</u> $\alpha \in [SA, F]$, <u>then the relative Whitehead product</u> $\langle \alpha , u \rangle = i_*^{-1}[i_* \alpha , u]$ <u>is given by</u> $\langle \alpha , u \rangle : SA \wedge B \subset SA \vee SA \wedge B \underset{\bar{\mu}}{\overset{\simeq}{\longleftarrow}} SA \rtimes B \xrightarrow[\alpha \rtimes 1]{} F \rtimes B \xrightarrow[\varphi(\xi)]{} F.$

The homotopy equivalence $\bar{\mu}$ is defined by $(pr \vee \pi)(\mu \rtimes 1_B)$ where μ is the comultiplication on SA. If F is a co-H-group or an H-group, then there is a bijection $\Psi : \mathcal{F}_0(F, SB) \approx [F \wedge B, F]$. From the Puppe sequence for i we extract the short exact sequence of groups

$$0 \to [F \wedge B, F] \xrightarrow{\pi^*} [F \rtimes B, F] \xrightarrow{i} [F, F] \to 0$$

in which $\Psi(\xi) = (\pi^*)^{-1}(-pr + \varphi(\xi))$. However, only when F is an H-group is it the case that $\langle \alpha , u \rangle = (\Psi\xi)(\alpha \wedge 1_B)$. If F is a co-H-group, we must use an expansion as in (3.3.19). This is the case in particular for spherical fibrations over spheres. If F and B are suspensions, we derive from (3.3.15) and (3.3.22) the classification

$$(3.3.25) \qquad \mathcal{F}(SF, S^2B) \approx \bigcup_\beta [S^2F \wedge B, SF]/(\mathrm{Im}\,[\ ,\beta] + \mathrm{Im}\,\tau^*[\ , 1_{SF}])$$

where $\beta \in [SB, SF]$ satisfies $[\beta , 1_{SF}] = 0$.

<u>Remark:</u> (3.3.25) for spherical fibrations over spheres expresses a classical result of James and J.H.C. Whitehead [61]. A special case of (3.3.24) was proved by James in § 2 of [60]. He gives the relative Samelson products as examples of brace products. Further examples can be found in Hardie and Porter [42].

<u>Proof of</u> (3.3.22) Let $p : E \to SB$ be a fibration with fiber $F_0 = p^{-1}(*)$, and $i : F \to F_0$ a homotopy equivalence with homotopy inverse $\bar{I}$. Let $\pi : I \times B \to SB$ be the quotient map. We then have a

commutative diagram

$$\begin{array}{ccccc} I \times F \times \{*\} \cup \{1\} \times F \times B & \xrightarrow{pr} & F & \xrightarrow{i} & E \\ \Big\downarrow & & & & \Big\downarrow p \\ I \times F \times B & \xrightarrow{pr} & I \times B & \xrightarrow{\pi} & SB \end{array}$$

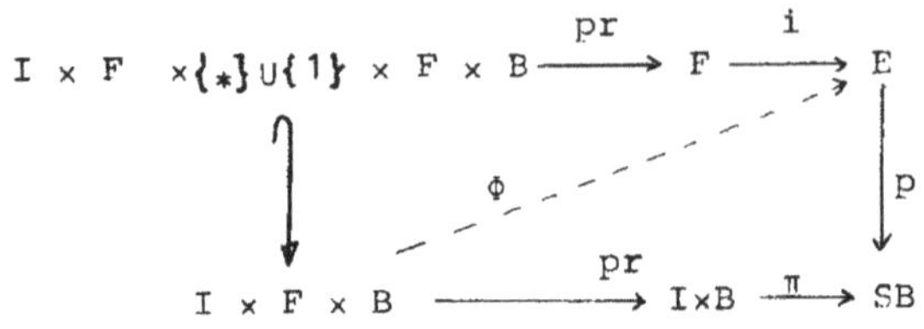

Since we require that F and B be well-pointed, there exists a pointed homotopy Φ which extends the diagram commutatively, see (0.1.5). Φ restricts to $\Phi_o : F \times B \to F_o$ with $\Phi_o\, i_F = i$. The element $\bar{i}\Phi_o \in [F \times B, F]$ for which $\bar{i}\Phi_o\, i_F = \bar{i}\, i = 1_F$ then represents the class (p, i). If $(p, i) \sim (p', i')$ then $\varphi(p, i) = \varphi(p', i')$: we can see this by again finding a relative lifting. If $u : SB \to E$ is a section of p, we can form the commutative diagram

$$\begin{array}{ccccc} I \times * \times B \cup (I \times F \times * \cup 1 \times F \times B) & & \xrightarrow{u\pi \cup (i\, pr)} & & E \\ \Big\downarrow & & & & \Big\downarrow p \\ I \times F \times B & \xrightarrow[pr]{} & I \times B & \xrightarrow[\pi]{} & SB \end{array}$$

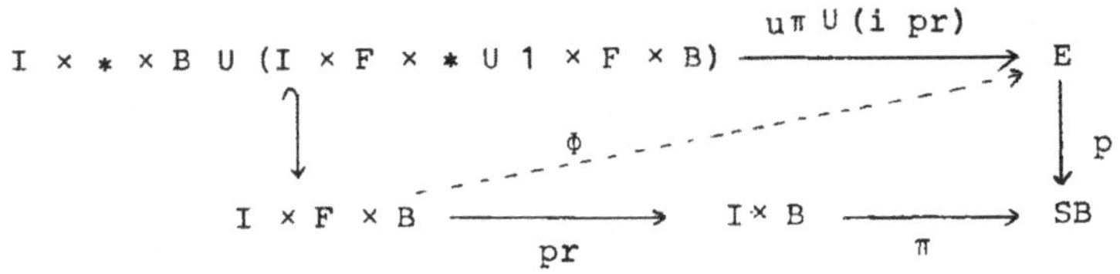

If Φ is a relative lifting, and $\Phi_o : F \times B \to F_o$ satisfies $\Phi_o\, i_F = i$ and $\Phi_o|B \times * = *$. The map $\bar{i}\Phi_o$ thus represents the class $\varphi(p, u, i) \in [F \rtimes B, F]$. If $(p, u, i) \sim (p', u', i')$, it follows as before from the existence of a relative lifting that $\varphi(p, u, i) = \varphi(p', u', i')$.

We now construct an inverse for φ, using the lemma cited immediately following this proof. If $f : F \times B \to F$ is a map with $f\, i_F \simeq 1_F$, we have a commutative diagram

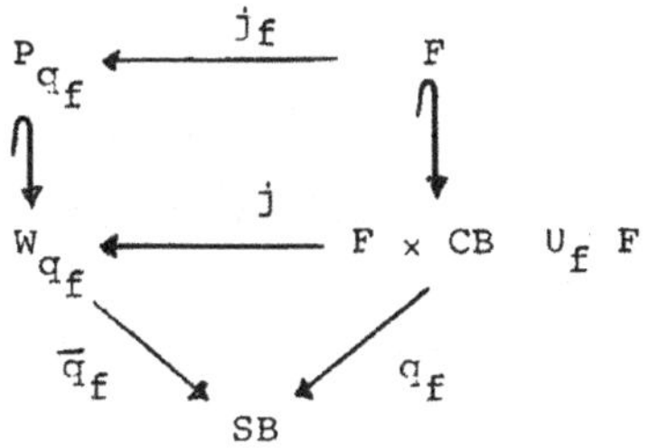

where q_f is defined to be the projection on $F \times CB$, that is $q_f(x, t, b) = (t, b) \in SB$ for $x \in F$ and $(t, b) \in CB$. By the lemma just mentioned, j and j_f are homotopy equivalences, see also (0.1.10). The equivalence class of the pair $((\bar{q}_f, j_f)$ in $\mathcal{F}(F, SB)$ is $\varphi^{-1}(f)$. If $f_{|B} = *$, then q_f has a canonical section $u : SB \to F \times CB \cup_f F$ defined by $u(t, b) = (*, t, b)$. Thus the equivalence class of the triple $(\bar{q}_f, ju, j_f)$ in $\mathcal{F}_o(F, SB)$ is $\varphi^{-1}(f)$. If f and f' represent the same element of $[F \times B, F]$ then $\varphi^{-1}(f) = \varphi^{-1}(f')$. We can see this as follows. Given a homotopy $H : f \simeq f'$, we let $\bar{F} = I \times F$ and take $g : \bar{F} \times B \to \bar{F}$ defined by $g(t, x, b) = (t, H(t, x, b))$. From the inclusion $i_o : F \subset \bar{F}$, $x \mapsto (x, 0)$ we pass to $i_o : F \times CB \cup_f F \subset \bar{F} \times CB \cup_g F$ defined by $\bar{q}_g\, i_o = \bar{q}_f$. It then follows from naturality of the construction that the map $\bar{i}_o : W_{q_f} \to W_{q_g}$ induced from i_o satisfies $\bar{q}_g\, \bar{i}_o = \bar{q}_f$. Since $\bar{i}_o$ is a homotopy equivalence, there is an equivalence $(q_f, j_f) \sim (q_g, j_g\, i_o)$. Similarly, there is an equivalence $(q_{f'}, j_{f'}) \sim (q_g, j_g\, i_1)$ where $i_1 : F \subset \bar{F}$, $x \mapsto (x, 1)$ denotes the inclusion. It is now easy to see that φ^{-1} is inverse to φ. □

(3.3.26) <u>Lemma</u>: <u>In the commutative diagram</u>

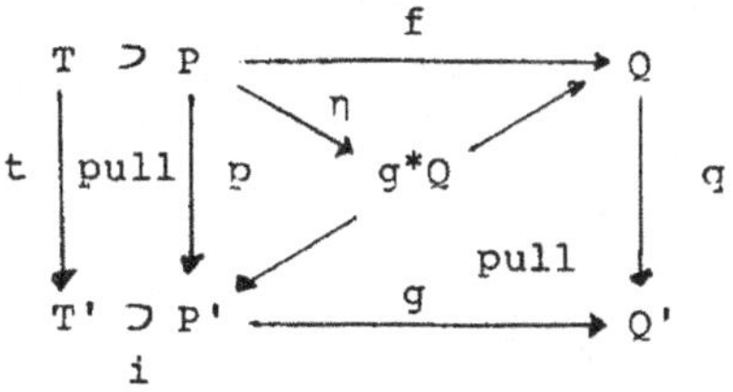

let t *and* q *be fibrations* , q *actually need only be an* h-*fibration, defined in* [24], *and* i *a closed cofibration. If the map* η *defined by* (p, f) *is a homotopy equivalence, then in the commutative diagram*

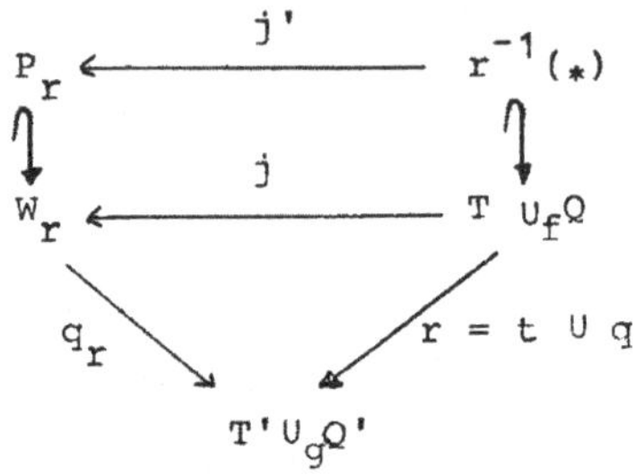

the maps j *and* j' *are homotopy equivalences.* W_r, P_r *and* j *are defined as in (0.1.10), and* j' *is the restriction of* j.

In proving (3.3.22) we considered the special case

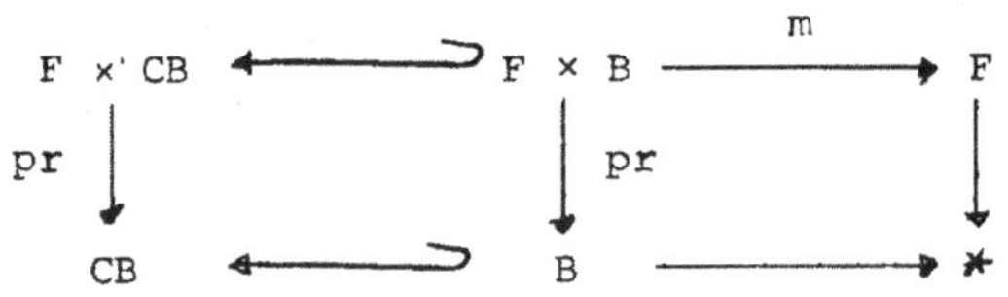

where $i_F{}^*(m) = 1_F$. We assumed that B is a CW-space, and so $\eta_m : F \times B \to F \times B$, $\eta_m(x, y) = (m(x, y), y)$ is a homotopy equivalence. This follows from 9.3 in [24].

Remark: Lemma (3.3.26) was proved in [43] using the methods of [24]: see in this connection [101]. K. A. Hardie has obtained a result similar to (3.3.26) for quasi-fibrations [41].

Proof of (3.3.24): Consider the homotopy commutative diagram

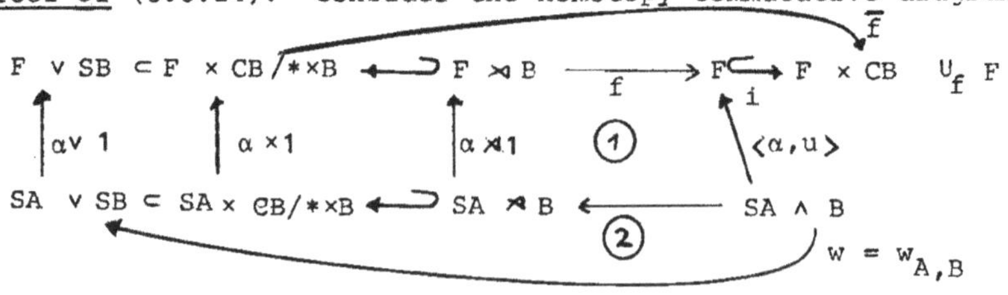

Since i_* is injective, it suffices for the proof of (3.3.24) to show that $i_*(f(\alpha \rtimes 1)\bar{\mu}) = i_* \langle \alpha , u \rangle = [i_* \alpha , u]$, where $f = \varphi(\xi)$ and $u : SB \subset F \times CB \cup_f F$ denotes the inclusion. Chasing around the outside of the diagram, we find that $i_* \langle \alpha , u \rangle = [i_* \alpha , u]$, since $\bar{f}$ is the quotient map. Therefore (3.3.24) follows from the homotopy commutativity of the above diagram. ☐

(3.4) The functional suspension. Some general suspension theorems

We will now define a functional operation called the functional suspension. One special case of it is the partial suspension in (3.1), thus it can be regarded as a very generalized Freudenthal suspension. Surprisingly enough, the Blakers-Massey excision results can be used to prove Freudenthal suspension theorems for the functional suspension as well. These general suspension theorems are an essential part of the proof that a CW-decomposition has a principal reduction, see (3.5.13).

Let Y be a space under D, that is suppose we have a map $d : D \to Y$. Further let B be an ex-cofiber space over D with projection ρ and section σ. Given a map $g : B \to Y$ under d, we can fit the mapping cone $C_D g$ into the commutative diagram

(3.4.1)

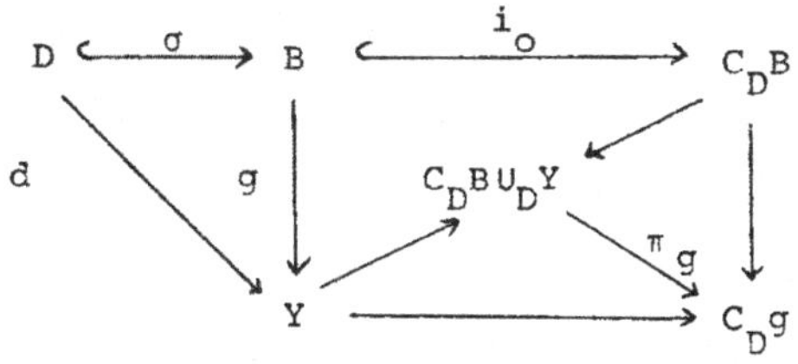

in which the square is cocartesian. We call π_g the identification map for $C_D g$. If $D = *$ is a point, then $\pi_g : CB \vee Y \to C_g$ is the usual identification map for the mapping cone. Consider the following diagram of homotopy groups.

$$\begin{array}{ccccccccc}
 & & 0 \to \pi_1^A(C_D B \cup_D Y,\ B \cup_D Y) & \xrightarrow{\partial} & \pi_0^A(B \cup_D Y) & \xrightarrow{\rho_*} & \pi_0^A(Y) \to 0 \\
 & & \downarrow{\scriptstyle \pi_{g*}} & & \downarrow{\scriptstyle (g \cup 1)_*} & & \\
\pi_1^A(Y) \xrightarrow{i} \pi_1^A(C_D g) & \xrightarrow{j} & \pi_1^A(C_D g,\ Y) & \xrightarrow{\partial} & \pi_0^A(Y) & &
\end{array}$$

Each row is a portion of an exact homotopy sequence of a pair. By (2.1.4), the projection $\rho \cup 1 : C_D B \cup_D Y \to Y$ is a homotopy equivalence under and over Y, therefore the map $\sigma : Y \subset B \cup_D Y$ induces a splitting σ_* of ρ_* and the upper row is exact. Generalizing the terminology of (3.1), we say that $\xi : A \to B \cup_D Y$ is <u>trivial on</u> Y when $\rho_*\xi = 0$, and we write

$$\pi_o^A(B \cup_D Y)_2 = [A, B \cup_D Y]_2 = \operatorname{Ker} \rho_*$$

for the subset (group) of all elements trivial on Y. If A is a co-H-group, then ∂ in the upper row of (3.4.2) is injective. The reader should give some thought to what the diagram says in case $D = *$. We can extract from diagram (3.4.2) a functional operation which we call the <u>functional suspension</u>:

(3.4.3) $$E_g : \pi_o^A(B \cup_D Y)_2 \cap \operatorname{Ker}(g \cup 1)_* \to \pi_1^A(C_D g)/\operatorname{Im} i, \quad E_g = j^{-1}\pi_{g*}\partial^{-1}.$$

We will say that an element $f \in \pi_o^A(C_D g)$ <u>is functional with respect to</u> g when $j(f) \in \operatorname{Im} \pi_{g*}$. This condition holds exactly when there is an element $\xi \in \pi_o^A(B \cup_D Y)_2$ with $f \in E_g(\xi)$. If A is a co-H-group, E_g is a homomorphism of groups. The important special case of E_g when $D = *$ will be discussed with examples in (3.6).

Now we generalize the partial suspension E of (3.1.1) so that we can use it on ex-cofiber spaces. If we take $g = d\rho : B \to D \to Y$ in diagram (3.4.2), then $C_D g = S_D B \cup_D Y$ and we have the maps

(3.4.4)

$$\begin{array}{ccc} & & \pi_1^A(C_D B \cup_D Y, B \cup_D Y) \xrightarrow[\cong]{\partial} \pi_o^A(B \cup_D Y)_2 \\ & & \big\downarrow (\pi_{d\rho})_* \\ \pi_1^A(S_D B \cup_D Y)_2 & \xrightarrow[\cong]{j} & \pi_1^A(S_D B \cup_D Y, Y) . \end{array}$$

If A is a co-H-group, then ∂ and j are group isomorphisms, and we define the <u>partial suspension</u>

(3.4.5) $$E : \pi_o^A(B \cup_D Y)_2 \to \pi_1^A(S_D B \cup_D Y)_2, \quad E = j^{-1}(\pi_{d\rho})_* \partial^{-1} .$$

When $D = *$ this is just the definition (3.1.1) of partial suspension. It is clear from diagram (3.4.2) that E_g and E are natural with respect to the appropriate maps. The reader may care to work out for himself other properties of E that generalize those of the partial suspension E in (3.1) .

We can generalize (3.1.2) to a procedure for finding representatives of a partial suspension. Let $\xi : A \to B \cup_D Y$ be trivial on Y, so that there exists an extension $\bar{\xi} : CA \to Y$ of $\rho\xi$. Let $\sigma : Y \subset B \cup_D Y$ denote the inclusion. Then

$$H_\xi : I \ltimes A \xrightarrow{1 \times \xi} I \ltimes (B \cup_D Y) \xrightarrow{\pi} S_D B \cup_D Y$$

is a self-homotopy of $\sigma\rho\xi$, where π is the union of the quotient map $I \ltimes B \to S_D B$ and the projection $I \ltimes Y \to Y$. The difference

(3.4.6) $$E\xi = d(\sigma\bar{\xi}, H, \sigma\bar{\xi}) \in [SA, S_D B \cup_D Y]_2$$

is then a representative of the partial suspension $E\xi$ (see the definition of difference in (1.2.5)). If A is a co-H-space, the homotopy class of $E\xi$ does not depend on which extension $\bar{\xi}$ of $\rho\xi$

we take, see (2.4.3). If A is not a co-H-space, we will mean by the partial suspension $E\xi$ an element as in (3.4.6), which thus is not uniquely defined. In the proof of (3.5.3) we will describe a representative of the functional suspension $E_g(\xi)$ that corresponds to the one found for $E\xi$ in (3.4.6). The Freudenthal suspension theorems have the following generalization for the functional suspension.

(3.4.7) Theorem: *Let* B *be an ex-cofiber space over* D *such that* (B, D) *is* $(b - 1)$-*connected. If* $g : B \to Y$ *is any map, then*

$$\pi_{g*}: \pi_n(C_D B \cup_D Y, B \cup_D Y) \longrightarrow \pi_n(C_D g, Y)$$

is an isomorphism for $n < 2b$, *and an epimorphism for* $n = 2b$.

(3.4.8) Corollary (General suspension theorem): *Let* B *be an ex-cofiber space over* D *such that* (B, D) *is* $(b - 1)$-*connected. Then*

$$E_{g*}: \pi_{n-1}(B \cup_D Y)_2 \cap \operatorname{Ker}(g \cup 1)_* \longrightarrow \pi_n(C_D g)/\operatorname{Im} i$$

and $$E \; : \pi_{n-1}(B \cup_D Y)_2 \longrightarrow \pi_n(S_D B \cup_D Y)_2$$

are isomorphisms for $n < 2b$ *and epimorphisms for* $n = 2b$.

A special case of these statements is (3.1.4). If $D = Y = *$ then (3.4.8) is just the Freudenthal suspension theorem.

Remark: In [14] we treated the special case where $D = *$. Because we used relative Whitehead products in investigating the image of the map $E' = \pi_{g*} : \pi_n(CB \vee Y, B \vee Y) \longrightarrow \pi_n(C_g, Y)$, the dimension conditions in [14] depended on the 'suspension degree' of Y. This was done by James [54] and Toda [126] for the case that $(C_g, Y) = (X^{b+1}, X^b)$ is the $(b + 1)$-skeleton of a CW-complex, that is when g is the attaching map of the $(b + 1)$-cells. Ganea [34] and Gray [37] studied the homotopy groups of a mapping cone as well. Ganea imbeds the homomorphism $E'' = \pi_{g*} : \pi_n(CB, B) \longrightarrow \pi_n(C_g, Y)$ in an exact EHP-sequence (see 5.3 of [34]). However, the length of this sequence depends on the connectivity of Y and B. Using E' instead of E'', we obtained in [15] an EHP-sequence whose length depends only on the connectivity of B. An additional requirement we made was that B and Y themselves be suspensions. However, the general suspension theorem above makes it seem probable that $(\pi_g)_*$ in (3.4.7) can always be embedded in an EHP-sequence generalizing the one in [15].

In proving (3.4.7) we use the excision theorem of Blakers and Massey, see [24] and [38] p. 144.

(3.4.9) <u>Theorem</u> (Blakers-Massey): <u>Let</u> $X = X_1 \cup X_2$, $A = X_1 \cap X_2$ <u>and assume</u> $A \subset X_1$ <u>and</u> $A \subset X_2$ <u>are cofibrations. Suppose</u> (X_1, A) <u>is</u> $(n - 1)$-<u>connected and</u> (X_2, A) <u>is</u> $(m - 1)$-<u>connected. Then</u> $i_* : \pi_r(X_1, A) \rightarrow \pi_r(X, X_2)$ <u>is an isomorphism for</u> $r < m + n - 2$ <u>and an epimorphism for</u> $r \leq m + n - 2$.

<u>Proof of</u> (3.4.7): Let $X = C_D g$ and define

$$X_1 = \{x \in C_D g \mid x \in Y \text{ or } x = (t, b) \in C_D B \text{ with } t \geq 1/2\},$$

$$X_2 = \{x \in C_D g \mid x \in Y \text{ or } x = (t, b) \in C_D B \text{ with } t \leq 1/2\}.$$

We can sketch the situation as

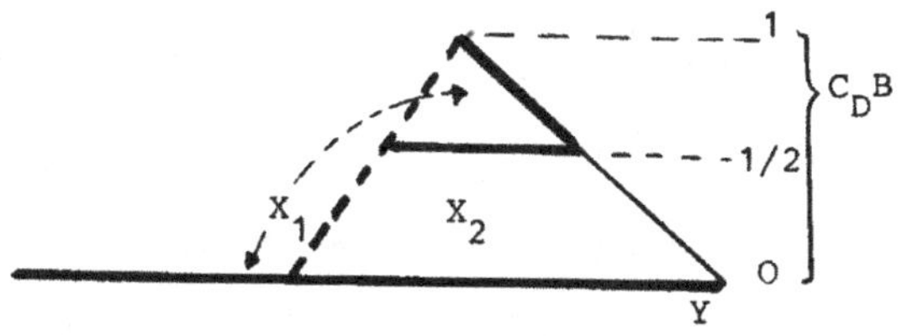

The heavy lines represent X_1, and the broken heavy line is identified to a point. Homeomorphisms $[0, 1/2] \approx I$ and $[1/2, 1] \approx I$ induce homeomorphisms

$$X_1 = C_D B \cup_D Y, \qquad X_2 = I_D \ltimes B \cup_g Y = Z_D g$$

where $Z_D g$ is the relative mapping cylinder of g. We then have $A = X_1 \cap X_2 = B \cup_D Y$, furthermore $A \subset X_1$ and $A \subset X_2$ are closed cofibrations. The map π_{g*} is the composite

$$\pi_{g*} : \pi_r(C_D B \cup_D Y, B \cup_D Y) \approx \pi_r(X_1, A) \xrightarrow{i_*} \pi_r(X, X_2) \approx \pi_r(C_D g, Y).$$

We now show that (X_1, A) and (X_2, A) are b-connected. (3.4.7) is therewith a consequence of the Blakers-Massey theorem. Consider the exact sequences

$$0 \to \pi_n(C_D B \cup_D Y, B \cup_D Y) \xrightarrow{\partial} \pi_{n-1}(B \cup_D Y) \underset{\sigma_*}{\overset{\rho_*}{\rightleftarrows}} \pi_{n-1}(Y) \to 0$$

$$0 \to \pi_n(Z_D g, B \cup_D Y) \xrightarrow{\partial} \pi_{n-1}(B \cup_D Y) \underset{\sigma_*}{\overset{(g \cup 1)_*}{\rightleftarrows}} \pi_{n-1}(Y) \to 0$$

These are portions of the long exact homotopy sequences of the indicated pairs since there are homotopy equivalences $\rho : C_D B \cup_D Y \to Y$ and $r : Z_D g \to Y$, and the inclusion $\sigma : Y \subset B \cup_D Y$ induces splitting morphisms. Since (B, D) is $(b - 1)$-connected by assumption, we know by (1.4.6) that we can take (B, D) to be a relative CW-complex with cells only in dimensions $\geq b$. The cellular approximation theorem (1.4.2) then implies that σ_* in the above sequences is surjective when $n \leq b$. It follows that ρ_* and $(g \cup 1)_*$ are isomorphisms when $n \leq b$, and so $\operatorname{Im} \partial = 0$. This proves that (X_1, A) and (X_2, A) are b-connected. ☐

From (3.4.7) we can derive an exact sequence, the use of which simplifies the computation of homotopy groups of a mapping cone $C_D g$ in a stable range, particularly when $D = *$.

(3.4.10) *Theorem: Let* B *be an ex-cofiber space over* D *and suppose* (B, D) *is* (b - 1)-*connected. If* $n \leqslant 2b - 1$ *then* E *and* j' *in the following commutative diagram are isomorphisms, and its row is exact.*

$$
\begin{array}{ccccccccc}
\pi_n(Y) & \xrightarrow{i_*} & \pi_n(C_D g) & \xrightarrow{\overline{E}_g} & \pi_{n-1}(B \cup_D Y)_2 & \xrightarrow{(g,1)_*} & \pi_{n-1}(Y) & \xrightarrow{i_*} & \cdots \\
 & & \Big\downarrow \mu_* & & \cong \Big\downarrow E & & & & \\
 & & & & \pi_n(S_D B \cup_D Y)_2 & & & & \\
 & & & & \cong \Big\downarrow j' & & & & \\
 & & \pi_n(S_D B \cup_D C_D g) & \xrightarrow{j} & \pi_n(S_D B \cup_D C_D g,\ C_D g) & & & &
\end{array}
$$

j' is induced by the inclusion $i : Y \subset C_D g$, and μ is the comultiplication. $\overline{E}_g$ can be regarded as the inverse of E_g, so with the notation of (3.4.2) we also have $\overline{E}_g = \partial(\pi_{g*})^{-1} j$. This exact sequence is dual to the one of E. Thomas that we treat in (6.4.8).

(3.5) Functional suspension and the principal reduction

It is well-known that the mapping cone of a suspension is equivalent to a suspension, that is $C_{Sf} = SC_f$. We will here show that the functional suspension behaves similarly. This is why, when we are given a functional suspension, obstruction sets turn to be cosets of subgroups. In the next chapter we will see for example that for CW-complexes, higher-order obstructions in a stable range are indeed cosets of subgroups. This section is mainly preparatory to the investigation of CW-complexes carried out in the next chapter. In the appendix we prove that CW-complexes have a principal reduction.

Consider a mapping cone C_f under a relative mapping cone $C_D g$, as in the diagram

(3.5.1)
$$\begin{array}{ccc} & & C_f \\ & & \uparrow \cup \\ SA & \xrightarrow{f} & C_D g \\ & & \uparrow \cup\, i \\ B & \xrightarrow{g} & X \end{array}$$

where B is an ex-cofiber space over D and g is an extension of $d : D \to X$. If $D = *$ we obtain a double mapping cone as discussed in (3.2).

(3.5.2) Theorem (Principal reduction of a double principal cofibration): *If f in (3.5.1) is functional with respect to g, that is if there exists an $\xi_0 \in \pi_0^A(B \cup_D X)_2$ such that $f \in E_g(\xi_0)$, then the double inclusion $X \subset C_D g \subset C_f$ is a principal cofibration relative to X.*

We will need the following somewhat more detailed version of this theorem.

(3.5.3) <u>Lemma</u>: <u>Suppose the map</u> $d = g|D$ in (3.5.1) <u>factors over a space</u> L <u>such that</u> $d : D \to L \xrightarrow{e} X$. <u>Further suppose</u> ξ: $A \to B \cup_D L$ <u>is a map trivial on</u> L <u>and such that</u> $f \in E_g(\xi_o)$, <u>where</u> $\xi_o = (1 \cup e)_* \xi$. <u>Then the double inclusion</u> $X \subset C_D g \subset C_f$ <u>is a principal cofibration relative to</u> L <u>with the property that there exists a retraction</u> $\bar{\xi} : C_\xi \to L$ (<u>making</u> C_ξ <u>an ex-cofiber space over</u> L) <u>and a map</u> $\xi_g : C_\xi \to X$ <u>extending</u> e, <u>for which</u> $C_f \simeq C_L\, \xi_g$ <u>under</u> X.

This theorem and the general suspension theorem (3.4.7) form the basis of the proof in (2.3.5) that a CW-decomposition has a principal reduction.

<u>Proof of</u> (3.5.3): The definition of E_g in the context of diagram (3.4.2) shows that $f \in E_g(\xi_o)$ can be represented as follows. There exist null-homotopies $\bar{\xi}$ and ξ_g making the diagram

(1) 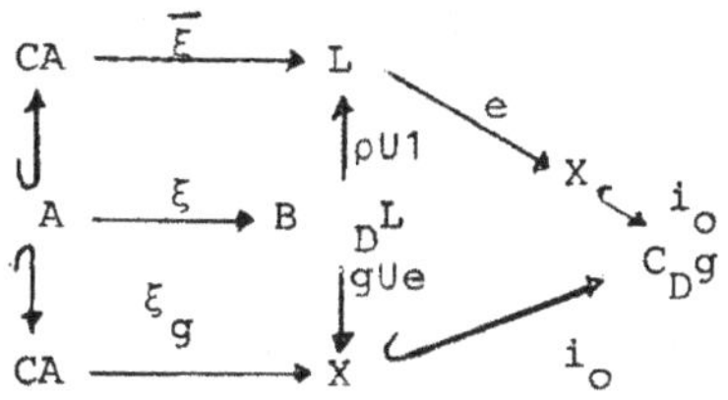

commute. There is also a homotopy

$$H^g_\xi : I \ltimes A \xrightarrow{1 \times \xi} I \ltimes (B \cup_D L) \xrightarrow{\pi} C_D B \cup_D L \xrightarrow{\pi_g} C_D g,$$

where π is the union of the quotient map $I \ltimes B \to C_D B$ and the projection $I \ltimes L \to L$. π_g as in (3.4.1) is an extension of e. It is not hard to see that $H^g_\xi : i_o (g \cup e)\xi \simeq i_o\, e\, (\rho \cup 1)$. The null-homotopies $\bar{\xi}$ and ξ_g can be chosen so that

(2) $$f = d(i_o \xi_g,\, H^g_\xi,\, i_o\, e\, \bar{\xi}) : SA \longrightarrow C_D g.$$

Because of (1), the difference is well-defined. This representation (2) of f generalizes (3.4.6). The mapping cone C_ξ of the map

$\xi : A \to B \cup_D L$ is an ex-cofiber space over L with inclusion $L \subset C_\xi$. The homotopy $\bar{\xi}$ in (1) yields a retraction $\bar{\xi} : C_\xi \to L$. Furthermore, the map ξ_g in (1) gives us a map

(3) $\xi_g : C_\xi \longrightarrow X$ extending $e : L \longrightarrow X$.

We will now show that the mapping cone $C_L \xi_g$ of ξ_g relative to L is such that $C_L \xi_g \simeq C_f$ under X, thus completing the proof of (3.5.3). Consider the following commutative diagram in which all squares are cocartesian.

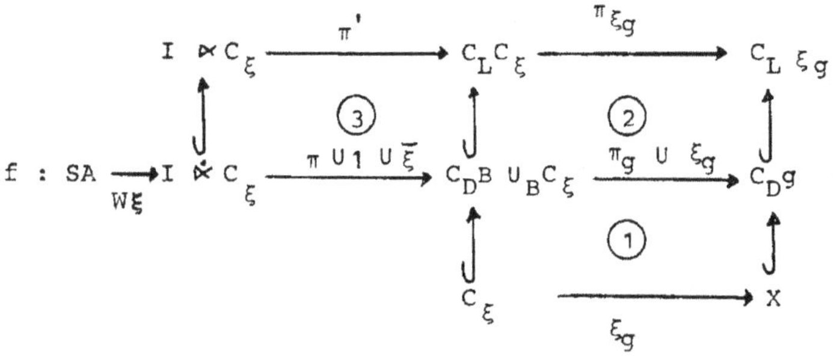

(1) is cocartesian by the definition of $C_D g$, and the union of (1) and (2) is cocartesian by the definition of $C_L \xi_g$, see (2.1.5). By (0.0.2) therefore (2) is also cocartesian. In (3), π' is the quotient map and $\pi \cup 1 \cup \bar{\xi}$ is the restriction of π' to $I \dot{\ltimes} C_\xi = I \ltimes (B \cup_D L) \cup \{0\} \times C_\xi \cup \{1\} \times C_\xi$, so (3) is cocartesian as well. It follows from all this that $C_D g \subset C_L \xi_g$ is a closed cofibration. Now let $W\xi$ be as in (1.2.3). Then by (2) we have

(4) $f = (\pi_g \cup \xi_g)(\pi \cup 1 \cup \bar{\xi})(W\xi)$.

It follows from (1.2.4) that $C_L \xi_g \simeq C_f$ under $C_D g$. $\square$

One consequence of (3.5.3) is that the mapping cone and the relative suspension commute with each other. That is, setting $L = X$ and

$g = d\rho$ in (3.5.3) we conclude

(3.5.4) Corollary: Let $\xi : A \to B \cup_D X$ be trivial on X and suppose $E\xi : SA \to S_D B \cup_D X$ is a partial suspension of ξ. Then a retraction $\xi : C_\xi \to X$ exists, and so C_ξ is an ex-cofiber space over X such that $C_{E\xi} \simeq S_X C_\xi$.

The relative suspension $S_X C_\xi$ is just the relative mapping cone of $g = d\rho$, see (2.1). If we set $D = X = *$, the corollary is equivalent to $C_{S\xi} = SC_\xi$, where ξ is any map $A \to B$.

We take up again now our investigation of the double mapping cone begun in (3.2). We saw that the element ∇f plays a significant role. There is a more general, relative version of ∇f: with the comultiplication $\mu : C_D g \to S_D B \cup_D C_D g$ as in (2.1.13), we obtain the element

$$(3.5.5) \qquad \nabla f = -f^*(\sigma) + f^*(\mu) \in \pi_1^A(S_D B \cup_D C_D g)_2$$

σ is the inclusion of $C_D g$. This ∇f is also trivial on $C_D g$, so we can form the partial suspensions $E^n \nabla f$ with E as in (3.4.5). An important property of the functional suspension is that

(3.5.7) Theorem: If $f \in E_g(\xi)$, then $\nabla f = i_*(E\xi)$, where $i : S_D B \cup_D X \subset S_D B \cup_D C_D g$ denotes the inclusion.

Proof: The map $\xi : A \to B \cup_D X$ is trivial on X. We set $L = X$ in the proof of (3.5.3), and see from (2) there that $f \in E_g(\xi)$ has the representation.

(1) $f = d(i_o \xi_g, H_\xi^g, i_o \bar{\xi})$. For the comultiplication μ then

(2) $\mu f = d(\sigma i_o \xi_g, \mu(H_\xi^g), \sigma i_o \bar{\xi})$ and

(3) $\mu H_\xi^g = \sigma H_\xi^g + i H_\xi$, with the homotopy H_ξ as in (3.4.6).

From (1), (2) and (3) and (1.2.6) we conclude that

$$\begin{aligned}
\nabla f &= -\sigma f + \mu f = \\
&= -\sigma d(i_o \xi_g, H^g_\xi, i_o\bar{\xi}) + d(\sigma i_o \xi_g, \sigma H^g_\xi + iH_\xi, \sigma i_o\bar{\xi}) \\
&= d(\sigma i_o\bar{\xi}, -\sigma H^g_\xi, \sigma i_o\xi_g) + d(\sigma i_o \xi_g, \sigma H^g_\xi + iH_\xi, \sigma i_o\bar{\xi}) \\
&= d(\sigma i_o\bar{\xi}, iH_\xi, \sigma i_o\bar{\xi}) \\
&= i\, d(\sigma\bar{\xi}, H_\xi, \sigma\bar{\xi}) = i\,(E\xi).
\end{aligned}$$

The last equation follows from the representation of $E\xi$ in (3.4.6). □

Now we generalize the operation (3.2.1) for the relative case. As before, let B be an ex-cofiber space over D, let $d : D \to Y \subset X$ be a map and A a co-H-group. If $\xi : A \to B \cup_D X$ is trivial on X, the partial suspensions $E^n\xi : S^nA \to S^n{}_DB \cup_D X$ are defined. Given a map $v : Y \to U$ with $w = vd$, or given a fibration $\tilde{X} \quad X$ with fiber F and section $v : Y \to \tilde{Y}$ with $w = d^*v$, we find that $E^n\xi$ induces for $n \geq 0$ the products

(3.5.8)
$$[\ ,\ ]^n_\xi : [S^n{}_DB, U]^w \times [X, U]^v \to [S^nA, U], \quad [\beta, u]^n_\xi = (E^n\xi)^*(\beta \cup_D u)$$

$$\langle\ ,\ \rangle^n_\xi : \langle S^n{}_DB, \rho^*\tilde{D}\rangle^w \times \langle X, \tilde{X}\rangle^v \to [S^nA, F], \quad \langle\beta, u\rangle^n_\xi = i_*^{-1}[\tilde{d}_*\beta, u]^n_\xi$$

Here $\tilde{d} : \tilde{D} = d^*\tilde{X} \to \tilde{X}$ is the map over d and $i : F \subset \tilde{X}$ is the inclusion of the fiber. For the same reasons as in (3.2.1), $\langle\ ,\ \rangle^n_\xi$ is well-defined. $[\ , u]^n_\xi$ and $\langle\ , u\rangle^n_\xi$ are homomorphisms of groups for $n \geq 1$. We derive directly from (3.5.7) the following statement about obstructions.

(3.5.9) <u>Corollary</u>: <u>Let</u> $f \in E_g(\xi)$ <u>in diagram</u> (3.5.1). <u>If</u> $u_o \in [C_Dg, U]^v$ <u>and</u> $u = u_o|X$, <u>then</u> $-f^*(u_o) + f^*(u_o + \beta) = [\beta, u]^1_\xi$. <u>If furthermore</u> $\tilde{C}_f \to C_f$ <u>is a fibration</u>, $u_o \in \langle C_Dg, \widetilde{C_DG}\rangle^v$ <u>and</u>

$u = u_o | X$, <u>then</u> $-f^{\#}(u_o) + f^{\#}(u_o + \beta) = \langle \beta, u \rangle^1_\xi$.

The operation $+\beta$ is defined as for the classification sequences (C) and (D) of (2.4). To prove the second equation of (3.5.9), the reader should consult the proof of (3.3.3).

If $u \in [X, U]^v$, or $u \in \langle X, \tilde{X} \rangle^v$, let

$$\mathcal{H}(u) = \{ f^{\#}(u_o) \mid u_o \text{ extends } u \text{ over } C_D g \}$$

denote the obstruction to extending u over C_f. From exactness of the classification sequences (C) and (D) we deduce

(3.5.10) <u>Corollary:</u> <u>If</u> $f \in E_g(\xi)$, <u>then</u> $\mathcal{H}(u)$ <u>is either empty or is a coset of the group</u> $\mathrm{Im}\,[\ ,\ u]^1_\xi$, <u>respectively of</u> $\mathrm{Im} \langle\ ,\ u \rangle^1_\xi$.

This corollary is essentially the reason why in CW-complexes higher-order obstructions in a stable range are cosets of subgroups. We will go into this in more detail in (4.3.11). As an application of corollary (3.5.10), we will in (3.6) describe second-order obstructions for products of spheres and for complex projective spaces.

We presented a classification formula in (3.2.5) for the double mapping cone $X \subset C_g \subset C_f$. If f is a functional suspension with respect to g, we have

(3.5.11) <u>Corollary</u> (Classification theorem): <u>Given</u> $X \subset C_g \subset C_f$, <u>let</u> $u : X \to U$ <u>be a map, or let</u> $u : X \to \tilde{X}$ <u>be a section of a fibration</u> $\tilde{C}_f \to C_f$ <u>with fiber</u> F, <u>and suppose</u> u <u>is extendable over</u> C_f. <u>If</u> $f \in E_g(\xi)$ <u>then there are bijections</u>

$$[C_f, U]^u \approx \mathrm{Ker}\,[\ ,\ u]^1_\xi \times \mathrm{Coker}\,[\ ,\ u]^2_\xi ,$$

$$\langle C_f, \tilde{C}_f \rangle^u \approx \mathrm{Ker} \langle\ ,\ u \rangle^1_\xi \times \mathrm{Coker} \langle\ ,\ u \rangle^2_\xi .$$

We have formulated the corollary only for double mapping cones $X \subset C_f \subset C_g$. The reader may care to work out for himself and prove the corresponding statements for the relative case, that is for $X \subset C_D g \subset C_f$ or even $X \subset C_D g \subset C_D f$. The classification theorem is dual to the classification theorem of James and Thomas for double principal fibrations, see (6.5.7).

<u>Proof of</u> (3.5.11): We content ourselves with a comparison of (3.5.11) and (3.2.5) from the map standpoint. Let u_o be an extension of u over C_g that can also be extended over C_f. Then $f^*(u_o) = 0$, and by (3.5.9) we have for $\beta \in [SB, U]$, $f^*(u_o + \beta) = 0 \Leftrightarrow [\beta, u]^1_\xi = 0$. Furthermore, by (3.5.7) $[\ , u_o + \beta]^1_{\nabla f} = [\ , u]^2_\xi$ so the claim follows from (3.2.5). $\square$

The principal reduction procedure of (3.5.2) allows us to interpret the classification formulas (3.5.11) more precisely. We know that when $f \in E_g(\xi)$, the inclusion $X \subset C_f$ is a relative principal cofibration with classifying map $C_\xi \to X$. By (2.1.11), there is a bijection $[S_X C_\xi, U]^u \approx [C_f, U]^u$ providing $[C_f, U]^u$ with a group structure. Therefore

$$0 \to \mathrm{Coker}\,[\ , u]^2_\xi \to [C_f, U]^u \to \mathrm{Ker}\,[\ , u]^1_\xi \to 0$$

is a short exact sequence of groups. There is a similar sequence in the case of section homotopy sets.

Applications of this classification will be discussed in the next section (3.6). Concluding this one, we will just mention the following general example for (3.5.7). Let f and g be as in (3.5.1) and let $Y \subset X$. Then the classifying maps

$$Wf : S^{n+1}A \to \Sigma^n_Y C_D g \cup C_f \subset \Sigma^n_Y C_f,$$
$$Wg : S^n_D B \to \Sigma^n_Y X \cup C_D g \subset \Sigma^n_Y C_D g$$

are defined, see (2.4.6) and (2.4.8). It can be shown that for $n \geqslant 1$

$$(-1)^{n+1}\, Wf \in E_{Wg}\,(i_* E^{n-1}\, \nabla f) \quad ,$$

i denotes the inclusion of $S_D^{n-1} B \cup_D C_D g$ in $S_D^{n-1} B \cup_D (\Sigma_Y^n X \cup C_D g)$, and ∇f is defined as in (3.5.5). It follows from theorem (3.5.7) that

$$\text{(3.5.12)} \qquad \begin{array}{ccc}
\Sigma_Y^n C_D g \cup C_f & \xrightarrow{\mu \cup 1} & S_D^{n+1} B \cup_D (\Sigma_Y^n C_D g \cup C_f) \\
 & & \big\downarrow (-1)^{n+1} \cup \rho \\
\big\uparrow Wf & & S_D^{n+1} B \cup_D C_f \\
 & & \big\uparrow j' \\
S^{n+1} A & \xrightarrow{E^n \nabla f} & S_D^{n+1} B \cup_D C_D g
\end{array}$$

is homotopy commutative. μ is the comultiplication on $^{n}_{nY} C_D g$ $C_D Wg$ as in (2.1.13), ρ is the retraction and j' is the inclusion. When $D = *$ this diagram is exactly the diagram in the proof of (2.4.6). The reader may take this opportunity to formulate for himself the relative version of (2.4.6). That (3.5.12) is homotopy commutative follows from the next calculations. The retraction $\rho : \Sigma_Y^n C_D g \cup C_f \to C_f$ can be extended over $\Sigma_Y^n C_f$, so $(Wf)^* \rho = 0$. By the definition of ∇Wf and setting $\bar{\mu} = (\mu \cup 1)^*((-1)^{n+1} \cup \rho)$, $(Wf)^*(\bar{\mu}) = (\nabla Wf)^*((-1)^{n+1} \cup \bar{\rho}) = (E\, i_*\, E^{n-1} \nabla f)^*(1 \cup \rho) = j'_*(E^n \nabla f)$. The second equation holds by (3.5.7), the third because $(1 \cup \rho) i = j'$.

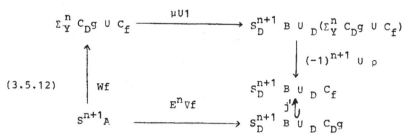

Appendix: The principal reduction of a CW-decomposition

Theorem (2.3.5) on the principal reduction of a CW-decomposition can now be proved by induction. The general suspension theorem (3.4) and the principal reduction procedure of (3.5.3) are essential to the proof. The principal reduction of CW-complexes yields relative principal cofibrations. Thus we can apply the obstruction theorems (3.5.9) and the classification theorem (3.5.10) to CW-complexes. This will be described in more detail in (4.4) with the aid of a spectral sequence.

(3.5.13) *Theorem: Let* (X, Y) *be a relative CW-complex with skeletons* X_n. *Let* Y *be path-connected and* $X_{r-1} = Y$ $(r \geq 1)$, *so that* $X - Y$ *has cells only in dimensions* $\geq r$. *Suppose that* $d : D \to Y$ *is an* $(r - 1)$*-connected map. Then there exists an ex-cofiber space* Q *over* D *and a map* $f : Q \to X_n$ *extending* $d : D$ $D \to Y \subset X_n$, *such that under* X_n

$$X_{n+r} \simeq C_D f$$

are homotopy equivalent. $Q = S_D^{n-r+1} Q'$ *is an* $(n - r + 1)$*-fold relative suspension of an ex-cofiber space* Q' *over* D.

The ex-cofiber space Q' over D to be constructed in the proof a arises inductively by attaching cells to D, so (Q', D) is a relative CW-complex. It is easy to see from the definitions of relative suspension S_D and a cone C_D that $(S_D^m Q', D)$ and $(C_D S_D^m Q', S_D^m Q')$ are also relative CW-complexes, the cells of which correspond exactly to those of $Q' - D$ by a dimensional shift of m and $m + 1$ respectively.

Note: (Q', D), and therefore $(C_D Q, Q)$ and $(C_D f, X^n)$ in the theorem, are relative CW-complexes. The cells of $C_D Q - Q = C_D f - X_n$ are in one-to-one correspondence with those of $X_{n+r} - X_n$, so the homotopy equivalence $h : C_D f \simeq X_{n+r}$ is obtained simply by homotopically altering the attaching maps of these cells, see (1.2.13).

Proof: By (1.4.4) we can assume in the proof that (X, Y) is strictly pointed. We now construct inductively the skeletons (Q_j, D) of the relative CW-complex (Q, D) and maps $f_j : Q_{j-1} \to X_n$ extending $d : D \to Y \subset X_n$, such that

$$(1) \qquad X_j \simeq C_D f_j \quad \text{under } X_n \text{ with } n + 1 \leq j \leq n + r.$$

At each step Q_{j-1} is as subspace of Q an ex-cofiber space over D. In fact $Q_{j-1} = S_D^{n-r+1} Q'_{j-1}$ is the $(n - r + 1)$-fold relative suspension of an ex-cofiber space Q'_{j-1} over D. Let $\varphi : S^n Z_{n+1} \to X_n$ We set $Q_n = D_n \vee S^n Z_{n+1}$, $Q'_n = D \vee S^{r-1} Z_{n+1}$ and $f_{n+1} = (d, \varphi)$, with the result that $X_{n+1} = C_\varphi = C_D f_{n+1}$. We now assume that for $n + 1 < j < n + r$ the ex-cofiber spaces $B = Q_{j-1}$ and $B' = Q'_{j-1}$ over D (with $B = S_D^m B'$, $m = n - r + 1$) have been constructed, and also that $g = f_j : B \to X_n$ has been sconstructed satisfying (1). We let

$$(2) \qquad f : S^j A = S^j Z_{j+1} \to X_j \simeq C_D f_j$$

be the attaching map of the $(j + 1)$-cells of $X_{j+1} - X_j$. This gives us a homotopy equivalence $C_f \simeq X_{j+1}$ under X_n. In (2) we have $j \leq n + r - 1$, and since $X_{r-1} = Y$ we can suppose that $r - 1 \leq n$, so that $j \leq 2n$. Now (B, D) is $(n - 1)$-connected, so the general suspension theorem (3.4.7) with f as in (2) says that there exists

$$(3) \qquad \xi_o \in \pi^A_{j-1}(B \cup_D X_n)_2 \qquad \text{with} \qquad f \in E_g(\xi_o).$$

We will show that the map

$$(4) \qquad i_* \;:\; \pi_{j-1}(B)_2 \longrightarrow \pi_{j-1}(B \cup_D X_n)_2$$

induced by the inclusion $i : B \subset B \cup_D X_n$ is surjective. It follows from this that there is an $\xi \in \pi_{j-1}(B)_2$ such that $i_*\xi = \xi_o$ in (3). We can then conclude from (3.5.3) that $X_n \subset C_D g \subset C_f$ is a principal cofibration relative to D and therefore that $X_n \subset X_{j+1}$ is also. As in the proof of (3.5.3) we set $Q_j = C_\xi$ and $f_{j+1} = \xi_g$. Since (B', D) is $(r - 2)$-connected, it also follows from the suspension theorems that

$$(5) \qquad E^m : \pi_{j-1-m}(B')_2 \longrightarrow \pi_{j-1}(B)_2$$

is surjective when $j \leqslant n + r - 1$, where $m = n - r + 1$. Then there exists a ξ' with $E^m\xi' = \xi$. Setting $Q_j' = C_{\xi'}$, we conclude from (3.5.4) that $S_D^m Q_j' = Q_j$. This completes the induction step and also the proof.

The surjectivity of i_* in (4) can be derived from the diagram with exact rows $(k = j - 1)$

$$\begin{array}{ccccc}
\pi_k(B) & \xrightarrow{\;i_*\;} & \pi_k(B \cup_D X_n) & \xrightarrow{\;j\;} & \pi_k(B \cup_D X_n, B) \\
\sigma_* \uparrow \downarrow \rho_* & & \sigma_*' \uparrow \downarrow \rho_*' & & \sigma_*'' \uparrow \downarrow \rho_*'' \\
\pi_k(D) & \longrightarrow & \pi_k(X_n) & \longrightarrow & \pi_k(X_n, D)
\end{array}$$

as follows. We can assume that $D \subset Y$ is a cofibration. Then (B, D) is $(n - 1)$-connected and (X_n, D) is $(r - 1)$-connected. By the excision theorem (3.4.8), σ_*'' is surjective for $k = j - 1 \leqslant n + r - 2$. Since σ_*'' splits ρ_*'', σ_*'' is in fact an isomorphism. If $\xi_o \in \pi_k(B \cup_D X_n)$ with $\rho_*'(\xi_o) = 0$, then $j(\xi_o) = 0$ and so there exists an $\eta \in \pi_k(B)$ with $i_*(\eta) = \xi_o$. Setting $\xi = \eta - \sigma_*\rho_*\eta$ we see that $i_*\xi = \xi_o$ and $\rho_*\xi = 0$, so i_* in (4) is surjective. $\square$

(3.6) Examples of secondary homotopy classifications, Toda brackets, triple Whitehead products, and secondary obstructions in complex projective spaces.

We here discuss some typical kinds of functional suspension. The first example involves co-extensions, which occur in the definition of the well-known Toda brackets. Then, using the functional suspension, a direct generalization of the Toda brackets will be described. This is a secondary homotopy operation based on certain relations in which Whitehead products and map compositions appear. An instance of this secondary homotopy operation is the triple Whitehead product. Finally, we show that Hopf maps for complex projective spaces are functional suspensions.

Given a map $g : B \to X$, we see by setting $D = *$ in (3.4.3) that the functional suspension

$$E_g : \pi_o^A(B \vee X)_2 \cap \operatorname{Ker}(g, 1)_* \to \pi_1^A(C_g)/i\,\pi_1^A(X)$$

is defined, where A is a co-H-group. We want to look more closely at the domain of origin of E_g. From Ganeas isomorphism Θ in (3.1.20) we can form the composite

$$(3.6.1) \qquad \Theta_g : \pi_o^A(B) \times \pi_o^A(S\Omega B \wedge \Omega Y) \cong \pi_o^A(B \vee X)_2 \xrightarrow{(g,1)_*} \pi_o^A(X),$$

for which $\Theta_g(\xi_B, \xi_o) = g_*(\xi_B) + [gR_B, R_Y]_*\,(\xi_o)$. We can distinguish three kinds of element ξ for which $E_g(\xi)$ is defined:

(A) $\xi = \Theta(\xi_B)$ with $g_*(\xi_B) = 0$

(B) $\xi = \Theta(\xi_o)$ with $[gR_B, R_Y]_*\,(\xi_o) = 0$

(C) $\xi = \Theta(\xi_B, \xi_o)$ with $g_*(\xi_B) + [gR_B, R_Y]_*\,(\xi_o) = 0$.

We call (A) a composition relation, (B) a Whitehead product relation, and (C) with $\xi_B \neq 0$ and $\xi_o \neq 0$ a mixed relation. We now consider functional suspensions of these three kinds of relation. The functional suspension of a composition relation (A) is identical with the well-known co-extension.

(3.6.2) <u>Definition</u>: Given maps $A \xrightarrow{\xi} B \xrightarrow{g} X$, let $v : CA \to X$ be a null-homotopy of $g\xi$. Then $\xi_v : SA \to C_g$ with $\xi_v(t, a) = v(1 - 2t, a) \in X$, for $0 \leq t \leq 1/2$ and $= (2t - 1, \xi a)$ for $1/2 \leq t \leq 1$ is called a <u>co-extension</u> of ξ, see [127].

<u>Theorem</u>: Let $\xi \in \pi_o^A(B) \subset \pi_o^A(B \vee X)_2$ <u>and</u> $g_*\xi = 0$. <u>Then</u> $E_g(\xi) \subset \pi_1^A(C_g)$ <u>is the set of co-extensions of</u> ξ.

The Toda brackets or secondary compositions which play an important role in stable homotopy theory, are obtained from co-extensions.

(3.6.3) <u>Definition</u>: Given maps $U \xleftarrow{u} X \xleftarrow{g} B \xleftarrow{\xi} A$ with $ug \simeq 0$ and $g\xi \simeq 0$, we call the set $\{u, g, \xi\} \subset [SA, U]$, consisting of all composites $U \xleftarrow{u'} C_g \xleftarrow{f} SA$ with u' an extension of u and f a co-extension of ξ, the <u>Toda bracket</u> of (u, g, ξ).

The preceding theorem suggests the following more general concept of secondary composition.

(3.6.4) <u>Definition</u>: Given maps $U \xleftarrow{u} X \xleftarrow{(g,1)} B \vee X \xleftarrow{\xi} A$ with $ug \simeq 0$, $(g, 1)\xi \simeq 0$ and ξ trivial on X, we call the set $\langle u, g, \xi \rangle \subset [SA, U]$, consisting of all composites $U \xleftarrow{u'} C_g \xleftarrow{f} SA$ with u' an extension of u and $f \in E_g(\xi)$, the <u>secondary composition</u> of (u, g, ξ).

The indeterminancy of the Toda bracket $\{u, g, \xi\}$ is the subgroup $u_* [SA, X] + (S\xi)^* [SB, U] \subset [SA, U]$. More generally, it follows from (3.5.10) that

(3.6.5) Theorem: The secondary composition $\langle u, g, \xi \rangle \subset [SA, U]$ is a coset of the subgroup $u_* \pi_1^A(u) + [\pi_1^B(U), u]_\xi^1$.

(3.6.6) Theorem: $S\langle u, g, \xi\rangle \subset -\{Su, Sg, S\xi_B\}$

The bracket $\langle u, g, \xi \rangle$ can therefore be used to desuspend the Toda bracket $\{Su, Sg, S\xi_B\}$ in cases where $g\xi_B$ is not null-homotopic and so $\{u, g, \xi_B\}$ is not defined.

Proof of (3.6.6): We have $SE_g(\xi) \subset -E_{Sg}(S\xi_B)$ because the diagram used to define E_g in (3.4.2) commutes up to a sign with S. $\square$

We now give a typical example of a Whitehead product relation (B), namely the Jacobi identity ofor Whitehead products, see (0.3). The appropriate secondary composition then gives us the so-called second-order Whitehead product, or triple Whitehead product. Let $T = ($ (S_1, S_2, S_3) be a triple of spheres $S_i = S^{n_i}$, $n_i \geq 2$. The cellular decomposition $S_i = e_i \cup *$ produces a CW-decomposition of the product $T^3 = S_1 \times S_2 \times S_3$ with cells $e = e_1 \times e_2 \times e_3$, $e_{ij} = e_1 \times e_j$ $(i < j)$, e_i and $*$. In virtue of this cellular decomposition, T^3 is a double mapping cone

$$\begin{array}{ccccl}
 & & T^3 & = & S_1 \times S_2 \times S_3 \\
 & & \uparrow & & \\
S^{n-1} & \xrightarrow{W} & T^2 & = & S_1 \times S_2 \cup S_1 \times S_3 \cup S_2 \times S_3 \\
 & & \uparrow & & \\
\mathcal{J} & \xrightarrow{\underline{W}} & T^1 & = & S_1 \vee S_2 \vee S_3 \xrightarrow[(\alpha,\ \beta,\ \gamma)]{} U
\end{array}$$

where $\mathcal{J} = S^{n(12)-1} \vee S^{n(13)-1} \vee S^{n(23)-1}$, $n(ij) = n_i + n_j$ and $n = n_1 + n_2 + n_3$. The cells of T^3 have characteristic maps $W \in \pi_n(T^3, T^2)$, $W_{ij} \in \pi_{n(ij)}(T^2, T^1)$ which we normalize as follows. Denoting by $p : T^3 \to T^3/T^2 = S^n$ the identification map, we require $p_*W \in \pi_n(S^n)$ to represent the identity. Similarly, denoting by

(3.6.7)

$$p_{ij} : T^2 \to T^2/T^1 = \bigvee_{i<j} S^{n(ij)} \longrightarrow S^{n(ij)}$$

the identification maps, we want $p_{ij*}W_{ij}$ to represent the identity. For the inclusions $\xi_i : S_i \subset T^1$ we find that

(3.6.7) $\partial W_{ij} = (-1)^{n_i} [\xi_i, \xi_j], \qquad 1 \leq i < j \leq 3,$

is the Whitehead product (0.3), up to sign. The attaching maps w and $\underline{w}$ in the diagram are given by $w = \partial W$ and $\underline{w} = (\partial W_{12}, \partial W_{13}, \partial W_{23})$. The map w is called a 2nd order Whitehead product map. Given a map $(\alpha_1, \alpha_2, \alpha_3) : T^1 \to U$ with $\alpha_i \in \pi_{n_i}(U)$, we call the secondary obstruction

(3.6.8) $[\alpha_1, \alpha_2, \alpha_3] = \{w^*(u') \mid u' \text{ extends } (\alpha, \beta, \gamma) \text{ over } T^2\}$

$$\subset \pi_{n-1}(U)$$

the triple Whitehead product of $(\alpha_1, \alpha_2, \alpha_3)$. By (3.6.7) it is defined exactly when all the Whitehead products $[\alpha_i, \alpha_j] = 0$ vanish.

(3.6.9) Theorem: Nakaoka-Toda [89]: Denoting by $j : \pi_{n-1}(T^2) \to \pi_{n-1}(T^2, T^1)$ the homomorphism from the long homotopy sequence, we have

$$j(w) = (-1)^r [W_{12}, \xi_3] + (-1)^s [W_{13}, \xi_2] + (-1)^t [W_{23}, \xi_1]$$

where $[\ ,\ ]$ is the relative Whitehead product and where $r = n_1 + n_2$, $s = n_1 + n_3 + n_2 \cdot n_3$, $t = n_2 + n_3 + n_1 n_2 + n_1 n_3$.

Since $\partial j = 0$, this theorem implies the Jacobi identity for Whitehead products. Together with (3.1.12), the definition of functional suspension implies that the last theorem is equivalent to the

(3.6.10) Corollary: Let $\xi_{ij} : S^{n(ij)-1} \subset \mathcal{Y}$ be the inclusions and $\mathcal{J} \in \pi_{n-2}(\mathcal{Y} \vee T^1)_2$ be such that

$$\mathfrak{J} = (-1)^r [\xi_{12}, \xi_3] + (-1)^s [\xi_{13}, \xi_2] + (-1)^t [\xi_{23}, \xi_1].$$

Then $w \in E_{\underline{w}}(\mathfrak{J})$.

The Jacobi identity for Whitehead products is equivalent to the relation $(\underline{w}, 1)_* \mathfrak{J} = 0$. The latter is a Whitehead product relation of type (B) in (3.6.1). Thus the suspension $E_{\underline{w}}(\mathfrak{J})$ can be defined without the use of (3.6.9). It follows directly from (3.3.11), (3.6.10) and (3.5.10) that the indeterminancy of the triple Whitehead product is given by

(3.6.11) Corollary (Hardie [39]): Let $\alpha_i \in \pi_{n_i}(U)$ be such that $[\alpha_i, \alpha_j] = 0$. Then the triple Whitehead product $[\alpha_1, \alpha_2, \alpha_3] \subset \pi_{n-1}(U)$ is not empty, and is in fact a coset of $[\pi_{n_1+n_2}(U), \alpha_3] + [\pi_{n_1+n_3}(U), \alpha_2] + [\pi_{n_2+n_3}(U), \alpha_1]$

We now compare the triple Whitehead product with the secondary composition (3.6.4). Given maps

$$U \xleftarrow{\alpha} T^1 \xleftarrow{(\underline{w},1)} \mathfrak{f} \vee T^1 \xleftarrow{\mathfrak{J}} S^{n-2}$$

with $\alpha = (\alpha_1, \alpha_2, \alpha_3)$, suppose $[\alpha_i, \alpha_j] = 0$. Then the secondary composition $\langle \alpha, \underline{w}, \mathfrak{J} \rangle \subset \pi_{n-1}(U)$ is defined, and we have by (3.6.5)

(3.6.12) $$\langle \alpha, \underline{w}, \mathfrak{J} \rangle = [\alpha_1, \alpha_2, \alpha_3] + \alpha_* \pi_{n-1}(T^1).$$

This shows that the indeterminancy of the secondary composition is greater than that of the triple Whitehead product. This greater degree of indeterminancy is generated by the various null-homotopies $(\underline{w}, 1)\mathfrak{J} \simeq 0$ which the Jacobi identity for Whithead products may have. The triple Whitehead product is obtained by selecting one particular null-homotopy for the Jacobi identity. This null-homotopy is defined from a product of spheres with the aid of the Nakaoka Toda theorem (3.6.9). The next classification result is a consequence of (3.6.9) and (3.6.11).

(3.6.13) Corollary. Let $\alpha = (\alpha_1, \alpha_2, \alpha_3) : S_1 \vee S_2 \vee S_3 \to U$ be a map which can be extended over $S_1 \times S_2 \times S_3$. Then

$$[S_1 \times S_2 \times S_3, U]^\alpha \approx \text{Ker } \mathcal{J}_0(\alpha) \times \text{Coker } \mathcal{J}_1(\alpha)$$

is a bijection for the homotopy set of extensions, where for $\varepsilon = 0, 1$ the homomorphism

$$\mathcal{J}_\varepsilon(\alpha) : \pi_{n_1 + n_2 + \varepsilon} \times \pi_{n_1 + n_3 + \varepsilon} \times \pi_{n_2 + n_3 + \varepsilon} \to \pi_{n + \varepsilon - 1}$$

with $\pi_k = \pi_k(U)$ is defined to be

$$\mathcal{J}_\varepsilon(\alpha)(\xi_{12}, \xi_{13}, \xi_{23}) = (-1)^r [\xi_{12}, \alpha] + (-1)^s [\xi_{13}, \alpha_2] + (-1)^t [\xi_{23}, \alpha_1]$$

Remark: In Massey's list of problems [75], Zeeman asked about the indeterminancy of the triple Whitehead product. Hardie later obtained the result (3.6.11) in [39] using methods other than those here. Our characterisation of the indeterminancy is an easy consequence of the formula of Nakaoka and Toda. In [16] we showed that the Nakaoka-Toda formula is just a special case of a large number of similar formulas for products of spheres. It is possible to obtain considerably more general results on higher-order Whitehead products and the homotopy classification of maps on products of spheres, by proceeding as above using the formulas we exhibited in [16].

We now give an example of a Whitehead product relation that is not derived from the Jacobi identity. Let $\sigma = [i_2, i_2] : S^3 \to S^2$ be the Whitehead product of the identity $i_2 \in \pi_2(S^2)$. Then $[i_2, i_2] = 2h$ is twice the Hopf map. Since S^3 is a covering group of $SO(3)$, we have an operation $S^2 \times S^3 \to S^2$ which in fact extends (i_2, h). Therefore $[h, i_2] = 0$ and so $[[i_2, i_2], i_2] = 0$. With $\xi = [i_3, i_2] : S^4 \to S^3 \vee S^2$, we see that

$$E_g(\xi) \subset \pi_5(S^2 \cup_g E^4) \xrightarrow{\ i\ } \pi_6(S^3)$$

$$E_h(\xi) \subset \pi_5(S^2 \cup_h E^4) = \pi_5(\mathbb{C}P_2) = \mathbb{Z}$$

are well-defined. We have $i(\alpha) = r_*(S\alpha)$, because $Sg \simeq 0$ ensures that a retraction $r : SC_g \to S^3$ exists. Then

(3.6.14) Theorem : $i\,E_g(\xi) \subset \pi_6(S^3) \cong Z_{12}$ contains exactly one element, which has order three. $E_h(\xi) \subset Z$ contains exactly one element, a generator in fact.

Thus for complex projective space $\mathbb{C}P_2 = C_h$, $E_h(\xi)$ is just the Hopf map $h_3 : S^5 \to \mathbb{C}P_2$. In connection with this theorem we refer the reader to the example on p. 521 of [15] and to the calculation of the Hopf invariants in [18]. We now go on to prove a corresponding statement for the higher Hopf maps.

(3.6.15) Corollary: Let $\alpha \in \pi_6(S^3) = Z_{12}$ be a generator. The suspension of the Hopf map h_3 makes

$$\begin{array}{ccc} S^6 & \xrightarrow{\ Sh_3\ } & S\mathbb{C}P_2 \\ & {\scriptstyle\alpha}\searrow \quad \nearrow{\scriptstyle j} & \\ & S^3 & \end{array}$$

homotopy commutative.

Proof of (3.6.15): Since ξ is unstable

$$Sh_3 \in SE_h(\xi) \subset -E_{Sh}(S\xi) = -E_{Sh}(0) = j_*(\pi_6(S^3))$$

(see the proof of (3.6.6)). Therefore there exists an element $\beta \in \pi_6(S^3)$ such that $j_*\beta = Sh_3$. Since $2h = [i_2, i_2] = g$ there is a map $F : C_g \to C_h$ extending the identity on S^2. For any $\gamma \in E_g(\xi)$ we have $F_*(\gamma) = 2\,E_h(\xi) = 2\,h_3$. Thus the diagram

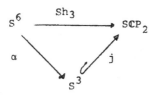

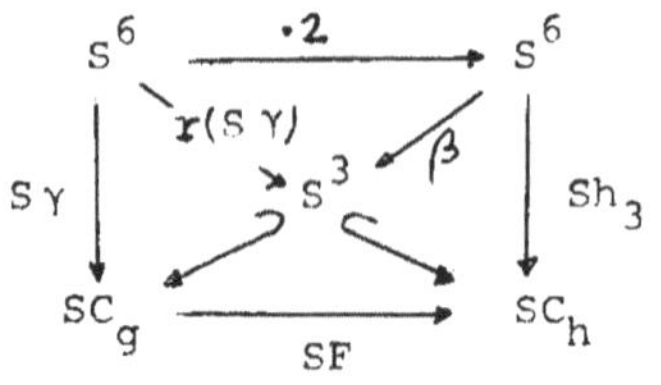

is homotopy commutative, so $2j_*(\beta) = j_* r_*(S\gamma)$. Since $r_*(S\gamma)$ is by (3.6.14) an element of order 3, so also is $2j_*(\beta)$. Therefore β has order at least 6 in $\pi_6(S^3) = Z_{12}$. Since $\ker j_* = Z_2$, it is in fact possible to choose β to be an element of order 12. ▭

A consequence of (3.6.15) for complex projective space is the

(3.6.16) <u>Corollary</u>: <u>The homotopy sets</u> $[\mathbb{C}P_3, S^2]$ <u>and</u> $[\mathbb{C}P_3, S^3]$ <u>each contain exactly two elements.</u>

<u>Proof</u>: By (3.3.20), $[\mathbb{C}P_2, S^i] = 0$ for $i = 2, 3$. It follows from the Puppe sequence that $[\mathbb{C}P_3, S^i] = \pi_6(S^i)/\mathrm{Im}\,(Sh_3)^*$, where $\mathrm{Im}\,\{(Sh_3)^* : [S\mathbb{C}P_2, S^i] \to \pi_6(S^i) = Z_{12}\}$ is equal to Z_6 by (3.6.15) and since $2i_3 : S^3 \to S^3$ can be extended over $S\mathbb{C}P_2$ (Sh is an element of order 2). Then because S^3 is an H-space $\alpha \circ (2i_3) = 2\alpha \in \mathrm{Im}\,(Sh_3)^*$. ▭

The reader should notice the following phenomenon. We call $E_g(\xi)$ <u>unstable</u> when $0 \in SE_g(\xi)$. Since the Whitehead product map $w \in E_{\underline{w}}(\mathfrak{J})$ in (3.6.10) is unstable, that is $Sw \simeq 0$, $E_{\underline{w}}(\mathfrak{J})$ is also unstable. In this case $\underline{w}$ and $\mathfrak{J}$ are themselves unstable. But it cannot be deduced from $Sg = S\xi = 0$ that $E_g(\xi)$ is unstable, as example (3.6.14) shows. For this reason it is not quite clear under what conditions $E_g(\xi)$ is unstable. The 'kind' of null-homotopy $(g, 1)\xi \simeq 0$ chosen must have some significance. For instance, the null-homotopy $(g, 1)\xi = [[i_2, i_2], i_2] \simeq 0$ came from the operation $S^2 \times S^3 \to S^2$, and we had $0 \notin SE_g(\xi)$. The Jacobi identity yields a null-homotopy $(g, 1)(3\xi) =$

$3[[i_2, i_2], i_2] \simeq 0$ for which $0 \in SE_g(3\xi)$.

We will now show that the higher Hopf maps are functional suspensions too. This provides us with mixed relations of type (C) in (3.6.1). Further rather complicated examples of mixed relations are the relations between Hopf invariants and Whitehead products which we described in [15], and the Barcus-Barratt formula [5],[15].

Let CP_n be complex projective space of real dimension $2n$ and let $h_n : S^{2n-1} \to \mathbb{C}P_{n-1}$ be the Hopf map, that is a generator of $\pi_{2n-1}(\mathbb{C}P_{n-1}) = Z$. $\mathbb{C}P_n$ arises from $\mathbb{C}P_{n-1}$ by attaching a $2n$-cell by h_n, $\mathbb{C}P_n = \mathbb{C}_{h_n}$ is a mapping cone. Let $\tilde{h}_n : (E^{2n}, S^{2n-1}) \to (\mathbb{C}P_n, \mathbb{C}P_{n-1})$ be the characteristic map of this cell. By (3.4.7)

$$(\tilde{h}_n, 1)_* : \pi_{2n+1}(E^{2n} \vee \mathbb{C}P_{n-1}, S^{2n-1} \vee \mathbb{C}P_{n-1}) \to \pi_{2n+1}(\mathbb{C}P_n, \mathbb{C}P_{n-1})$$

is an isomorphism. Therefore the Hopf map h_{n+1} is functional with respect to h_n. Let $i_2 : S^2 = \mathbb{C}P_1 \subset S^{2n-1} \vee \mathbb{C}P_{n-1}$ and $i_{2n-1} : S^{2n-1} \subset S^{2n-1} \vee \mathbb{C}P_{n-1}$ be the inclusions, and let $\eta \in \pi_m(S^{m-1}) = Z_2$ be a generator for $m \geq 4$.

(3.6.17) <u>Theorem:</u> <u>There are generators</u> $h_n \in \pi_{2n-1}(CP_{n-1})$, $n \geq 2$, <u>for which</u>

$$h_{n+1} = \begin{cases} E_{h_n}([i_{2n-1}, i_2]) & \text{if } n \text{ is even,} \\ E_{h_n}([i_{2n-1}, i_2] + i_{2n-1}\eta) & \text{if } n \text{ is odd.} \end{cases}$$

If n is odd, $[h_n, i_2] + h_n \circ \eta$ is a mixed relation for which h_{n+1} is defined. (If n is even, it follows as before that the suspension Sh_{n+1} factors through $S\mathbb{C}P_{n-1}$, see the proof of (3.6.15). We do not know whether Sh_{n+1} factors through an $S\mathbb{C}P_k$ for $k < n - 1$ as well. This question is related to the James numbers.)

Proof of (3.6.17): By (3.4.7), there is an isomorphism $\pi_{2n+1}(CP_n, CP_{n-1}) \cong Z \oplus Z_2$ for $n \geq 2$, where Z is generated by the relative Whitehead product $[\tilde{h}_n, i_2]$ and Z_2 by $\eta' = (\tilde{h}_n, 1)_* \partial^{-1}(i_{2n-1}\eta)$. We need to show that

$$(1) \qquad j(h_{n+1}) = \begin{cases} [\tilde{h}_n, i_2] & \text{if } n \text{ is even,} \\ [\tilde{h}_n, i_2] + \eta' & \text{if } n \text{ is odd.} \end{cases}$$

From the long homotopy sequence of the pair $(\mathbb{C}P_n, \mathbb{C}P_{n-1})$ we extract the short exact sequence

$$0 \to \pi_{2n+1}(\mathbb{C}P_n) \xrightarrow{j} \pi_{2n+1}(\mathbb{C}P_n, CP_{n-1}) \xrightarrow{\partial} \pi_{2n}(\mathbb{C}P_{n-1}) \to 0$$

We conclude from it that

(2) either $j(h_{n+1}) = [\tilde{h}_n, i_2]$ or $j(h_{n+1}) = [\tilde{h}_n, i_2] + \eta'$

If $\pi: (\mathbb{C}P_n, \mathbb{C}P_{n-1}) \to (S^{2n}, *)$ denotes the pinching map, then by (2)

(3) $j(h_{n+1}) = [\tilde{h}_n, i_2] \iff \pi_* j(h_{n+1}) = 0$.

Because of (2) and (3), it suffices to prove the equivalence

(4) $\pi_*(h_{n+1}) = 0 \iff n$ even

in order to prove (1). Now there is a map $\bar{\pi} : \mathbb{C}P_{n+1} \to S^{2n} \cup_{\pi h_{n+1}} E^{2n+2} = K$ which induces isomorphisms in cohomology in dimensions $2n$ and $2n + 2$. Let $x \in H^{2n}(K, Z_2)$ and $y \in H^{2n+2}(K, Z_2)$ be generators. Then

(5) $\pi_*(h_{n+1}) = 0 \iff Sq^2 x = 0$

because the generator of $\pi_{2n+1}(S^{2n}) = Z_2$ is detected by Sq^2. On the other hand, for a generator $a \in H^2(\mathbb{C}P_{n+1}, Z_2)$ we have that $\bar{\pi}^*(x) = a^n$ and $\pi^*(y) = a^{n+1}$. Since $Sq^2(a^n) = na^{n+1}$, see [121] Lemma 2.5, we conclude that $Sq^2(a^n) = 0$ exactly when n is even. □

(3.5.10) shows that secondary obstructions in complex projective spaces can be described as follows. Let $u : \mathbb{C}P_{n-1} \to U$ be a map, $n \geq 2$ and set $\mathcal{H}(u) = \{h^*_{n+1}(u') \mid u'$ extends u over $\mathbb{C}P_n\}$.

Then u can be extended over $\mathbb{C}P_{n+1}$ exactly when $0 \in \mathcal{H}_2(u)$. Writing $\pi_k = \pi_k(U)$, we have the

(3.6.18) <u>Corollary</u>: <u>The obstruction</u> $\mathcal{H}_2(u) \subset \pi_{2n+1}(U)$ <u>is, when not empty, a coset of</u> $[\pi_{2n}, u_*i_2]$ <u>if</u> n <u>is even, of</u> $\{[\alpha, u_*i_2] + \eta^*\alpha \mid \alpha \in \pi_{2n}\}$ <u>if</u> n <u>is odd</u>.

It follows moreover from (3.5.11) with $u_2 = u_*i_2 \in \pi_2(U)$ that

(3.6.19) <u>Corollary</u>: <u>If</u> $0 \in \mathcal{H}_2(u)$, <u>then there are bijections</u>.

n <u>even</u>, $[\mathbb{C}P_{n+1}, U]^u \approx \{\alpha \in \pi_{2n} \mid [\alpha, u_2] = 0\} \times \left(\pi_{2n+2} / [\pi_{2n+1}, u_2]\right)$

n <u>odd</u>, $[\mathbb{C}P_{n+1}, U]^u \approx \{\alpha \in \pi_{2n} \mid [\alpha, u_2] = \eta^*\alpha\} \times \left(\pi_{2n+2} / \{[\alpha, u_2] + \eta^*\alpha \mid \alpha\}\right)$

This corollary extends (3.3.20). The reader may formulate this corollary for sections of fibrations. Corollary (3.6.16) can also be obtained as follows. If $u : \mathbb{C}P_1 \to * \in S^3$ is the trivial map, then by (3.6.19)

$$[\mathbb{C}P_3/\mathbb{C}P_1, S^3] = [\mathbb{C}P_3, S^3]^u = \pi_4(S^3) \times \pi_6(S^3) = Z_2 \times Z_{12}.$$

The Puppe sequence for the inclusion $i : \mathbb{C}P_1 \subset \mathbb{C}P_3$ then looks like

$$\begin{array}{ccccccccc} [S\mathbb{C}P_3, S^3] & \xrightarrow{(Si)^*} & [S^3, S^2] & \to & [\mathbb{C}P_3/\mathbb{C}P_1, S^3] & \to & [\mathbb{C}P_3, S^3] & \to & 0 \\ \downarrow & & \| & & \| & & \| & & \\ 12Z & \hookrightarrow & Z & \to & Z_2 \times Z_{12} & \to & Z_2 & \to & 0 \end{array}$$

where $\text{Im}\,(Si)^* = 12\,\mathbb{Z}$ by (3.6.15).

CHAPTER 4: OBSTRUCTION THEORY FOR CW-COMPLEXES

(4.1) Cohomology with local coefficients

Classical obstruction theory describes secondary obstructions for CW-complexes as cohomology classes in cohomology groups with local coefficients (see (4.2)). Given a CW-complex, we define cohomology wi with local coefficients by means of the cellular chain complex of the universal covering. We then give an alternative characterization of this chain complex in terms of the attaching maps f_n, using to this end the partial suspensions $E^i(\nabla f_n)$ of the difference elements ∇f_n defined in (3.2). An appendix contains some examples of local coefficients used in the further course of this chapter.

Let X be a connected CW-complex with skeletons X^n, $n > 0$ and let $\hat{X} \xrightarrow{p} X$ be the universal covering of X (see [116]). Given $\alpha \in \pi = \pi_1(X)$, let $\cdot\alpha : \hat{X} \to \hat{X}$ be the covering transformation induced by α. $H_n(A, B)$ denotes the singular homology of the pair (A, B) with coefficients in Z. Let $L \subset X$ be a sub-complex and let $X_n = L \cup X^n$, $\hat{X}_n = p^{-1}(X_n) \subset \hat{X}$. The covering transformations induce an operation from the right of π on $H_n(\hat{X}_n, \hat{X}_{n-1})$ such that the boundary operator

$$(4.1.1) \qquad \partial = \partial_{n+1} : H_{n+1}(\hat{X}_{n+1}, \hat{X}_n) \to H_n(\hat{X}_n, \hat{X}_{n-1})$$

of the triple $(\hat{X}_{n+1}, \hat{X}_n, \hat{X}_{n-1})$ is a π-equivariant homomorphism.

Remark: The universal covering $\hat{X}$ of X has a natural CW-decomposition with skeletons $\hat{X}^n = p^{-1}(X^n)$. When $L = \phi$ is empty, the map ∂ in (4.1.1) is the boundary operator of the cellular chain complex of this CW-decomposition.

Let G be an abelian group on which $\pi = \pi_1(X)$ operates. The group

G together with this operation is called a local group and is denoted by $\hat{G}$, see (0.4). Given the chain groups $H_n(\hat{X}_n, \hat{X}_{n-1})$, we define the cochain group

(4.1.2) $$C^n(X, L; \hat{G}) = \mathrm{Hom}_\pi (H_n(\hat{X}_n, \hat{X}_{n-1}), \hat{G})$$

to be the group of π-equivariant homomorphisms. We obtain a cochain complex by introducing the boundary operator

(4.1.3) $\delta = \delta^n : C^n(X, L; \hat{G}) \to C^{n+1}(X, L; \hat{G})$ with $\delta^n = \mathrm{Hom}_\pi (\partial_{n+1}, 1_G)$. From this cochain complex we derive for $n \geq 0$

(4.1.4) $$H^n(X, L; \hat{G}) = \mathrm{Ker}\, \delta^n / \mathrm{Im}\, \delta^{n-1},$$

the cohomology of the pair (X, L) with local coefficients $\hat{G}$. If $L = \phi$ is empty, we write $H^n(X; \hat{G}) = H^n(X, \phi; \hat{G})$. Let Y be another connected CW-complex with K as subcomplex. A cellular map $f : (Y, K) \to (X, L)$ of pairs (cellular means $f(Y^n) \subset X^n$) induces a homomorphism

(4.1.5) $$f^* : H^n(X, L; \hat{G}) \to H^n(Y, K; f^*\hat{G})$$

of cohomology groups with local coefficients, as follows. By composing the operation of $\pi = \pi_1(X)$ of gG with the homomorphism $f_* : \pi_1(Y) \to \pi_1(X)$, we get an operation on $\pi_1(Y)$ on G. This local group in Y is denoted by $f^*\hat{G}$. The map f lies under a unique basepoint-preserving map $\hat{f} : \hat{Y} \to \hat{X}$ between the universal coverings that is simultaneously cellular and equivariant. Regarding the operation of the covering transformation, we have $\hat{f}(x \cdot \alpha) = \hat{f}(x) \cdot f_*(\alpha)$. Thus f induces a map of cochain complexes

$$\hat{f}^* = \mathrm{Hom}_\pi (f_*, 1) : C^*(X, L; \hat{G}) \to C^*(Y, K; f^*\hat{G})$$

which then induces f^* in (4.1.5).

As usual, a homomorphism $\gamma : \hat{G} \to \hat{H}$ of local groups in X induces a homomorphism

(4.1.6) $$\gamma_* : H^n(X, L; \hat{G}) \to H^n(X, L; \hat{H}).$$

γ_* is defined on the cochain complexes to be $Hom_\pi(1, \gamma)$.

<u>Remark</u>: A definition of singular cohomology with local coefficients for arbitrary pairs (A, B) is given in Spanier p. 282 [116]. Using this singular definition, one can show that the groups defined in (4.1.4) are independent of the cellular decomposition of X. When the coefficients are simple, there is the familiar proof that the homology of the cellular chain complex of a CW-complex is the same as its singular homology. This proof is easily extended to the case of local coefficients. Using the singular definition, one can define for any continuous map $f : (Y, K) \to (X, L)$ an induced homomorphism between homologies with local coefficients.

If the subcomplex $L(i : L \subset X)$ is connected, the sequence of cochain complexes

$$0 \to C^*(X, L; \hat{G}) \to C^*(X, \hat{G}) \to C^*(L, i^*\hat{G}) \to 0$$

is exact. We then obtain in the usual way a <u>long exact cohomology sequence</u>

(4.1.7) $$\to H^n(X, L; \hat{G}) \xrightarrow{j} H^n(X; \hat{G}) \xrightarrow{i^*} H^n(L, i^*\hat{G}) \xrightarrow{\delta} H^{n+1}(X, L; \hat{G}) \to$$

j is induced by the inclusion $(X, \phi) \subset (X, L)$.

We will now show that the boundary operator (4.1.1) in the cellular chain complex of the universal covering $\hat{X}$ can be described in terms of the attaching maps in X. We may assume that X is a strictly pointed CW-complex and L a subcomplex, possibly empty (see (1.4.4)).

Then $X_n = X^n \cup L = C_{f_n}$ is the mapping cone of the attaching map

$$f_n : S^{n-1} Z_n = \bigvee_{Z_n^*} S^{n-1} \to X^{n-1} \subset X_{n-1} .$$

Here Z_n^* is the set of n-cells in $X - L$ and is possibly empty. $Z_n = Z_n^* + \{*\}$ is the disjoint union of Z_n^* with a basepoint $*$. Since $X_{n-1} = C_{f_{n-1}}$, the element

$$\nabla f_n \in [S^{n-1} Z_n, S^{n-1} Z_{n-1} \vee X_{n-1}]_2 , \quad n \geq 2$$

is defined as in (3.2.2). By (3.1.4) the partial suspensions $E^i \nabla f_n$ are also defined for negative i with $i \geq 3 - n$. We thus obtain the sequence of maps $(k \geq 2)$

(4.1.8) $$\ldots \to S^k Z_n \vee X_2 \xrightarrow{(\partial_n^k,1)} S^k Z_{n-1} \vee X_2 \to \ldots \xrightarrow{(\partial_2^k,1)} S^k Z_1 \vee X_2$$

where $\partial_n^k = E^{k-n+1}(\nabla f_n)$ for $n \geq 2$, see (3.2.16). We have here used the fact that ∇f_n depends only on the 2-skeleton X_2. We call the sequence of maps (4.1.8) the <u>differential mapping complex</u> of the CW-complex (X, L), see (3.2.17). There are π-equivariant isomorphisms $(k \geq 2)$

(4.1.9) $$\tau : H_n(\hat{X}_n, \hat{X}_{n-1}) \underset{\lambda}{\cong} \bigoplus_{Z_n^*} Z[\pi] \underset{\sigma}{\cong} \pi_k(S^k Z_n \vee X_2)_2$$

which arise in the following way. σ is determined by the inclusions $i_e : S^k \wedge \{*, e\} \subset S^k Z_n$, $e \in Z_n^*$ and by σ of (3.1.27). The CW-decomposition of X canonically induces a CW-decomposition of the universal covering $\hat{X}$ of X as follows. If $\hat{X} \xrightarrow{p} X$ is the projection and $e \in Z_n^*$ an n-cell with $e \subset X$, then $p^{-1}(e)$ consists only of n-cells. There is exactly one n-cell $\hat{e}$ in $p^{-1}(e)$ whose closure contains the basepoint $* \in \hat{X}$, since X is strictly pointed. All other cells in $p^{-1}(e)$ are of the form $\hat{e} \cdot \alpha$ for some $\alpha \in \pi$. Thus p is a cellular map. Again with $e \in Z_n^*$, let $\{\hat{e}\} \in H_n(\hat{X}_n, \hat{X}_{n-1})$ be the homology class represented by the oriented cell e. Then $\lambda\{\hat{e}\} = e$ defines a π-equivariant isomorphism, and we have $\tau\{\hat{e}\} = i_e$. We can now describe the boundary operator ∂_{n+1} of the cellular chain complex of X in terms of the isomorphism τ.

(4.1.10) Theorem: The isomorphism τ yields a commutative diagram $(n \geq 1,\ k \geq 2)$:

$$\begin{array}{ccc} H_{n+1}(\hat{X}_{n+1}, \hat{X}_n) & \xrightarrow{\partial_{n+1}} & H_n(\hat{X}_n, \hat{X}_{n-1}) \\ \cong \downarrow \tau & & \cong \downarrow \tau \\ \pi_k(S^k Z_{n+1} \vee X_2)_2 & \xrightarrow{(\partial^k_{n+1}, 1)_*} & \pi_k(S^k Z_n \vee X_2)_2 \end{array}$$

If $\pi = \pi_1(X) = 0$ then (4.1.10) expresses a well-known property of the boundary operator of the cellular chain complex, namely that $\partial : H_{n+1}(X^{n+1}, X^n) \to H_n(X^n, X^{n-1})$ is also induced by $S^n Z_{n+1} \to X^n \to X^n/X^{n-1} = S^n Z_n$. By (4.1.10), we can describe cohomology with local coefficients using solely the differential mapping complex (4.1.8). Since $\partial\partial = 0$ in the cellular chain complex, (4.1.10) implies

(4.1.11) Corollary: In the differential mapping complex, $(\partial^k_n, 1) \circ \partial^k_{n+1}$ is null-homotopic.

This corollary was proved in (3.2.15) without the use of the cellular chain complex of $\hat{X}$.

Proof of (4.1.10): It suffices to show that the diagram commutes for generators $\{\hat{e}\} \in H_{n+1}(\hat{X}_{n+1}, \hat{X}_n)$ and for $n = k$. To this end we consider the following commutative diagram for $n \geq 2$, where h again denotes the Hurewicz homomorphism.

$$\begin{array}{ccccc}
\pi_n(S^nZ_n \vee X_n, X_n) & \overset{j_o}{\underset{\cong}{\longleftarrow}} & & & \pi_n(S^nZ_n \vee X_n)_2 \\
\uparrow j' & & & & \\
\pi_n(S^nZ_n \vee X_n, X_{n-1}) & & & & \cong \big\downarrow \tau \\
\uparrow \mu_* & & & & \\
\pi_n(X_n, X_{n-1}) & \overset{p_*}{\underset{=}{\longleftarrow}} & \pi_n(\hat{X}_n, \hat{X}_{n-1}) & \xrightarrow{h} & H_n(\hat{X}_n, \hat{X}_{n-1}) \\
\big\uparrow j & & & & \\
\pi_n(X_n) & & & &
\end{array}$$

Denoting by $f_e \in \pi_n(X_n)$ the attaching map of the cell $e \in Z^*_{n+1}$, we have

(1) $\quad hp_*^{-1}j(f_e) = \partial_{n+1}\{\hat{e}\}.$

On the other hand, by the definition of ∇f_e

(2) $\quad j'\mu_* j(f_e) = j_o(\nabla f_e)$

where μ is the cooperation on $X_n = C_{f_n}$.

The commutativity of the preceeding diagram then implies that

(3) $\quad \tau\, \partial_{n+1}\{\hat{e}\} = \nabla f_e = (\partial^n_{n+1}, 1)_*\, \tau^{-1}\{\hat{e}\}.$

This proves (4.1.10). ▭

We now describe the coboundary operator δ in (4.1.3) in terms of twisted products, as defined in (3.2.1). Let $\tilde{X} \to X$ be a fibration with fiber F and let $L \subset X$ be a subcomplex and $X_n = X^n \cup L$, as before. Let $u : X_1 \to \tilde{X}_1$ be a section which can be extended over X_2. By composing the operation of $\pi_1(\tilde{X})$ on $\pi_k(F)$, defined in (1.5.9), with the homomorphism

$$\theta : \pi = \pi_1(X) \cong \pi_1(X_2) \xrightarrow{u_*} \pi_1(\tilde{X}_2) \cong \pi_1(\tilde{X})$$

we get the local group $u^*\pi_k(F)$ in X for $k \geq 2$. This is studied in more detail in the appendix to this section.

The boundary operator $\partial^k_{n+1} : S^k Z_{n+1} \to S^k Z_n \vee X_2$ of the differential mapping complex is trivial on X_2 and so by (3.2.1) induces for $k \geq 2$ a homomorphism of abelian groups

$$(4.1.12) \quad \langle\ ,u\rangle^n_k : [S^k Z_n, F] \to [S^k Z_{n+1}, F]$$

by $\langle\alpha, u\rangle^n_k = i_*^{-1}\, \partial^{k*}_{n+1} (i_* \alpha, u')$. Here $i : F \subset \tilde{X}_2$ is the inclusion of the fiber and $u' : X_2 \to \tilde{X}_2$ is a section extending u.

(4.1.13) Corollary: The map λ of (4.1.9) induces an isomorphism λ^* of abelian groups which makes the diagram

$$\begin{array}{ccc} C^n(X, L; u_\bullet \pi_k(F)) & \xrightarrow{\ \delta^n_k\ } & C^{n+1}(X, L; u_\bullet \pi_k(F)) \\ \lambda^* \downarrow \cong & & \cong \downarrow \lambda^* \\ [S^k Z_n, F] & \xrightarrow{\ \langle\ ,u\rangle^n_k\ } & [S^k Z_{n+1}, F] \end{array}$$

commute. The coboundary δ^n_k is defined as in (4.1.3) for $n \geq 1$, $k \geq 2$.

We define λ^* by

$$\begin{aligned} \lambda^* : C^n(X, L; u_\bullet \pi_k(F)) &= \mathrm{Hom}_\pi (H_n(\hat{X}_n, \hat{X}_{n-1}), u_\bullet \pi_k(F)) \\ &\overset{\lambda^*}{=} \mathrm{Hom}_\pi \Big(\bigoplus_{Z_n^*} Z[\pi], u_\bullet \pi_k(F)\Big) \\ &= \mathrm{Hom} \Big(\bigoplus_{Z_n^*} Z, \pi_k(F)\Big) \\ &= \bigtimes_{Z_n^*} \pi_k(F) = [S^k Z_n, F] \end{aligned}$$

Proof of (4.1.13): For $\alpha \in [S^k Z_{n+1}, F]$ let $\alpha|e \in \pi_k(F)$ be the restriction of α to $i_e : S^k\{*, e\} \subset S^k Z_{n+1}$, where $e \in Z^*_{n+1}$. Then λ^* is defined by $\lambda^*(c)|e = c\{\hat{e}\}$ for c $C^{n+1}(X, L; u_\bullet \pi_k(F))$.

We have

$$
\begin{aligned}
(\lambda^* \circ \delta_k^n)(c)|e &= \lambda^*(c \circ \partial_{n+1})|e \\
&= c \circ \partial_{n+1}\{\hat{e}\} \\
&= c\,\tau^{-1}\,(\partial_{n+1}^k,\ 1)_*\,\tau\{\hat{e}\}, \quad \text{cf. } (4.1.10) \\
&= c\,\tau^{-1}\,(\partial_{n+1}^k \circ i_e).
\end{aligned}
$$

On the other hand,

$$
\begin{aligned}
(\langle\ ,\ u\rangle_k^n \circ \lambda^*)(c)|e &= \langle \lambda^*(c),\ u\rangle_k^n|e \\
&= i_*^{-1}((i_*\,\lambda^*(c),\ u') \circ \partial_{n+1}^k)|e, \quad \text{cf. } (4.1.12) \\
&= i_*^{-1}((i_*\,\lambda^*(c),\ u')_*(\partial_{n+1}^k \circ i_e)).
\end{aligned}
$$

The corollary now follows from the diagram

$$
\begin{array}{ccccc}
H^n(X_n,\ X_{n-1}) & \xrightarrow{\ c\ } & \pi_k(F) & \xrightarrow{\ i_*\ } & \pi_k(\widetilde{X}_2) \\
\cong \big\downarrow \tau & & & & \big\uparrow (i,\ 1)_* \\
\partial_{n+1}^k \circ i_e \in \pi_k(S^k Z_n \vee X_2)_2 & & \xrightarrow[(\lambda^* c \vee u')_*]{} & & \pi_k(F \vee X)_2
\end{array}
$$

which commutes because $c(\hat{e}) = (\lambda^* c)|e$ and all maps are π-equivariant.

□

(4.1) <u>Appendix:</u> <u>Examples of local groups.</u>

A local group $\hat{G}$ in a space X is an abelian group together with a group operation of $\pi = \pi_1(X)$ on G. We will write this operation in an exponential fashion as $\xi^\alpha \in G$, where $\xi \in G$ and $\alpha \in \pi$. Let $\Theta : \pi_1(Y) \to \pi_1(X)$ be a homomorphism. A local group G can be pulled back to a local group in Y:

(4.1.14) $\qquad \Theta^*\hat{G}$ with $\xi^\beta = \xi^{\Theta(\beta)}$ for $\beta \in \pi_1(Y)$.

If $f : Y \to X$ is a map and $\Theta = f_*$, we will also write $f^*\hat{G} = \Theta^*\hat{G}$. If Y is a CW-complex, a map $f : Y^1 \to X$ which can be extended over the 2-skeleton Y^2 induces a homomorphism $\Theta = f_* : \pi_1(Y) = \pi_1(Y^2) \to \pi_1(X)$, so that here also we can pull back to the local group $f^*\hat{G}$.

The fundamental group $\pi_1(A)$ acts on the homotopy groups $\pi_n(A)$ and $\pi_{n+1}(X, A)$ for $n \geq 1$, giving us local groups in A. The boundary operator $\partial : \pi_{n+1}(X, A) \longrightarrow \pi_n(A)$ is a homomorphism of local groups, that is $\partial(\xi^\alpha) = (\partial\xi)^\alpha$. All the local groups which we will consider are derived from such local groups $\pi_n(A)$ and $\pi_{n+1}(X, A)$.

For instance, given a fibration $p: \tilde{X} \to X$ with fiber $i : F \subset \tilde{X}$ we define the local group $\pi_n(F)^p$ in X as follows (see (1.5.9) in this connection). There is an isomorphism $\chi : \pi_{n+1}(Z_p, \tilde{X}) \cong \pi_n(F)$ where Z_p is the mapping cylinder of p, see (0.1.11). We define the local group

(4.1.15) $\pi_n(F)^p$ by $\xi^\alpha = \chi((\chi^{-1}\xi)^\alpha)$ for $\xi \in \pi_n(F)$, $\alpha \in \pi_1(\tilde{X})$,

which has the characteristic property $i_*(\xi^\alpha) = (i_*\xi)^\alpha$ because $i_* = \partial\chi^{-1} : \pi_n(F) \to \pi_n(\tilde{X})$. (see the similar definition in (3.3.8)). When X is a CW-complex, the local group $\pi_n(F)^p$ in $\tilde{X}$ depends only on the restriction $p' : \tilde{X}_2 \to X_2$ of p to the 2-skeleton. This is because the isomorphism $\pi_1(\tilde{X}_2) = \pi_1(\tilde{X})$ induces an isomorphism $\pi_n(F)^p = \pi_n(F)^{p'}$.

The local group $\pi_n(F)^p$ is natural for maps $f : Y \to X$. If $f^*p : f^*\tilde{X} \to Y$ is the fibration induced by f, and $\tilde{f} : f^*\tilde{X} \to \tilde{X}$ is the corresponding map lying over f, then by definition (4.1.14)

$$\tilde{f}^*(\pi_n(F)^p) = \pi_n(F)^{f^*p}. \tag{4.1.16}$$

The local group $\pi_n(F)^p$ can be regarded as a particular instance of the following local group (see (4.1.20)). Let $p : \tilde{X} \to X$ be a fibration with fiber $i : F \subset \tilde{X}$ and section $u : X \to \tilde{X}$. We define the local group

(4.1.17) $u \cdot \pi_n(F)$ in X by $\xi^{\alpha} = i_*^{-1}((i_*\xi)^{u_*\alpha})$ for $\xi \in \pi_n(F)$, $\alpha \in \pi_1(X)$

Since p has a section, $i_* : \pi_n(F) \to \pi_n(\tilde{X})$ is injective. With this notation and keeping (4.1.14) in mind, we have

(4.1.18) $u \cdot \pi_n(F) = \theta^*(\pi_n(F)^p)$

for $\theta = u_* : \pi_1(X) \to \pi_1(\tilde{X})$. This follows from the remark after (4.1.15) since i_* is injective. A map $f : Y \to X$ induces a section $f^*u : Y \to f^*\tilde{X}$, and the naturality of the local group $u \cdot \pi_n(F)$ is expressed by

(4.1.19) $f^*(u \cdot \pi_n(F)) = (f^*u) \cdot \pi_n(F)$.

<u>Proof</u>: For f as in (4.1.16) we have $\tilde{f}^*(f^*u) = uf$. It follows from (4.1.18) that $f^*(u \cdot \pi_n(F)) = f^* \theta^*(\pi_n(F)^p) = \underline{\theta}^* \tilde{f}^*(\pi_n(F)^p) = \underline{\theta}^*(\pi_n(F)^{f^*p}) = (f^*u) \cdot \pi_n(F)$, where $\underline{\theta} = (f^*u)_* : \pi_1(Y) \to \pi_1(f^*\tilde{X})$. $\square$

In the fibration $p^*\tilde{X} \to \tilde{X}$ pulled back from $p : \tilde{X} \to X$ we have the canonical diagonal section $d : \tilde{X} \to p^*\tilde{X}$ with $d(x) = (x, x) \in p^*\tilde{X} \subset \tilde{X} \times \tilde{X}$. Then

(4.1.20) $d \cdot \pi_n(F) = \pi_n(F)^p$.

<u>Proof</u>: The map $\tilde{p} : p^*\tilde{X} \to \tilde{X}$ lying over p satisfies $\tilde{p}d = 1$. Therefore $d \cdot \pi_n(F) = d^*(\pi_n(F)^{p^*p}) = d^*\tilde{p}^*(\pi_n(F)^p) = \pi_n(F)^p$ by (4.1.18) and (4.1.16). $\square$

(4.1.21) <u>Lemma</u>: <u>For</u> $i_* : \pi_1(F) \to \pi_1(X)$ <u>we have the equation</u> $\xi^{\beta} = \xi^{i_*\beta}$ <u>with</u> $\xi \in \pi_n(F)$ <u>and</u> $\beta \in \pi_1(F)$. ξ^{β} <u>is given by the</u>

<u>local group</u> $\pi_n(F)$ <u>and</u> $\xi^{i_*\beta}$ <u>is given by</u> $\pi_n(F)^p$.

<u>Proof:</u> We have $\pi_n(F)^p = d \cdot \pi_n(F)$, therefore

(1) $$\xi^{i_*\beta} = j_*^{-1}((j_*\xi)^{d_*i_*\beta})$$

for the inclusion $j : F \subset p^*\tilde{X}$ (see (4.1.17)). The inclusions $j, j' : F \subset F \times F \subset p^*\tilde{X}$ are given by $j(x) = (x, *)$ and $j'(x) = (*, x)$. It follows that

(2) $$d_*i_*\beta = j_*\beta + j'_*\beta \in \pi_1(p^*\tilde{X})$$

and therefore

$$\begin{aligned} \xi^{i_*\beta} &= j_*^{-1}((j_*\xi)^{j_*\beta + j'_*\beta}) \\ &= j_*^{-1}((j_*\xi^{j_*\beta})^{j'_*\beta}) \\ &= j_*^{-1}((j_*(\xi^\beta))^{j'_*\beta}) \qquad (3) \\ &= j_*^{-1}j_*(\xi^\beta) = \xi^\beta \end{aligned}$$

Equation (3) is due to the fact that $j'_*\beta$ operates trivially on $\operatorname{Im} j_* \subset \pi_n(F \times F)$. □

A consequence of (4.1.21) is

(4.1.22) <u>Theorem:</u> <u>Let</u> F <u>be path-connected and</u> n-<u>simple, that is let</u> $\pi_1(F)$ <u>operate trivially</u> on $\pi_n(F)$. <u>If</u> $p : X \to X$ <u>is a fibration with fiber</u> F, <u>then there is exactly one local group</u> $\pi_n(F)_p$ <u>in</u> X <u>such that</u> $p^*\pi_n(F)_p = \pi_n(F)^p$.

<u>Proof:</u> Since F is path-connected, $p_* : \pi_1(\tilde{X}) \to \pi_1(X)$ is surjective. There can thus be at most one local group $\pi_n(F)_p$. (4.1.21) implies that one exists: given $\beta', \beta'' \in p_*^{-1}(\alpha)$, we see that $\xi^\beta = \xi^{i_*\gamma} = \xi^\gamma = \xi$ for $\beta = \beta'\beta''^{-1} = i_*(\gamma)$ because $\pi_1(F)$ operates trivially on $\pi_n(F)$. Therefore $\xi^{\beta'} = \xi^{\beta''}$ in $\pi_n(F)^p$, and the

operation $\xi^{\alpha} := \xi^{\beta}$ with $\beta \in p_*^{-1}(\alpha)$ yields a well-defined local group $\pi_n(F)_p$. ▭

The local group $\pi_n(F)_p$ of (4.1.22) is natural for maps $f : Y \to X$ that is

(4.1.23) $$f^*(\pi_n(F)_p) = \pi_n(F)_{f^*p}.$$

Moreover

(4.1.24) Theorem: If F is path-connected and n-simple and $u : X \to \tilde{X}$ is a section, then the local group $u \cdot \pi_n(F)$ is independent of u, and $u \cdot \pi_n(F) = \pi_n(F)_p$.

Proof: We have $u \cdot \pi_n(F) = u^*(\pi_n(F)^p) = u^*p^*(\pi_n(F)_p)$ and $u^*p^* = 1$. ▭

(4.1.25) Note: Let $\mathcal{E}(F)$ be the group of (non-pointed) homotopy equivalences of F. Any fibration $\tilde{X} \to X$ with fiber F determines a homomorphism $\psi : \pi_1(X) \to \mathcal{E}(F)$ as follows. An element $\alpha \in \pi_1(X)$ leads to a diagram

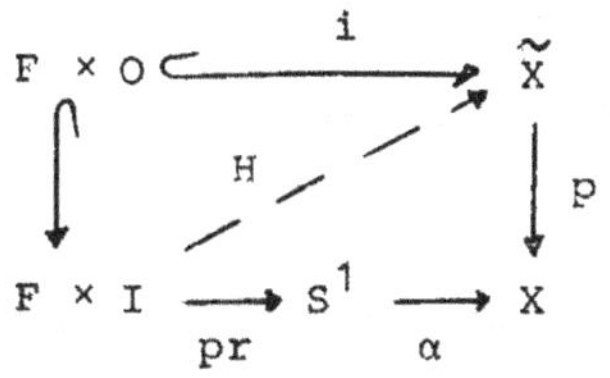

in which H is a lifting. We have set $pr(x, t) = t \in I/\partial I = S^1$. The map $H_1 : F \to F$ is then a homotopy equivalence unique up to free homotopy, therefore $\psi(\alpha) = H_1 \in \mathcal{E}(F)$. If F is path-connected and n-simple, $\psi(\alpha)$ induces an isomorphism $\psi(\alpha)_* : \pi_n(F) \to \pi_n(F)$. the operation ξ^{α} in $\pi_n(F)_p$ is such that $\xi^{\alpha} = \psi(-\alpha)_*(\xi)$ for $\xi \in \pi_n(F)$.

Remark: Let $(p : \tilde{X} \to X) = \mathcal{B}$ be a fiber bundle in Steenrod's sense, with path-connected and n-simple fiber F. The preceding note shows that the local group $\pi_n(F)_p$ induces exactly the coefficient bundle $\mathcal{B}(\pi_n(F))$ defined by Steenrod. Thus his obstruction theory, presented in [120], is a particular case of the obstruction theory we will develop in the sequel for fibrations. This more general situation was first treated by Barcus in [4]. In the classical treatments of obstruction theory by Eilenberg [31], Hu [52] and Steenrod [120], the fiber is always assumed to be n-simple in order that the fibration p should by itself suffice to determine unique local coefficients. This condition can be dropped by considering obstruction theory with local coefficients depending on a section , as in (4.1.17). Such coefficients were also used by Barcus in [4], however he does not bring out the connection between (4.1.18) and (4.1.20). On the other hand, coefficients as in (4.1.15) have been used by Olum [95] and Hill [46]. It thus seemed worthwhile to discuss in this appendix the relationships among the various coefficients.

(4.2) The obstruction cocycle and the difference cochain

In this section we will characterize the classical obstruction cocycle and difference cochain as elements in a cochain complex with local coefficients. We then compare these classical definitions with the primary obstructions and differences of section (1.2). As an easy consequence, we prove the obstruction theorem (4.2.9) which expresses essential properties of the obstruction cocycle and the difference cochain. After that, we state a classification theorem which implies certain classification theorems of Hopf, Eilenberg, Olum and Steenrod.

We assume in the following that X is a strictly pointed CW-complex. Thus X together with the skeletal filtration X^n is an iterated mapping cone, and the obstruction theory of the first three chapters can be applied.

<u>Remark</u>: Because of (1.4.4) and (1.1.6), there is no essential difference in examining extension problems for strictly pointed CW-complexes instead of for CW-complexes. Classical obstruction theory implicitly requires such strictly pointed CW-complexes, since it assumes that the images of the cellular attaching maps are connected by paths to the basepoint. See in this connection footnote 4 in [133] and § 12 of [132]. By taking X to be strictly pointed, we can dispense with the necessity of choosing connecting paths.

We now define the obstruction cocycle and the difference cochain for maps as well as for sections.

(4.2.1) <u>The obstruction cocycle for maps:</u>

Let $L \subset X$ be a subcomplex, possibly empty. We set $X_n = X^n \cup L$. Let a map $u : X_n \to U$ be given. For $n \geq 2$ let $\Theta : \pi_1(X) = \pi_1(X_n) \xrightarrow{u_*} \pi_1(U)$ be the induced map. Then the obstruction cocycle $c(u) \in C^{n+1}(X, L; \Theta^* \pi_n(U))$ is the $\pi_1(X)$-equivariant homomorphism making the diagram

$$\begin{array}{ccc}
\pi_{n+1}(X_{n+1}, X_n) & \xrightarrow{\partial} & \pi_n(X_n) \\
\cong \uparrow q_* & & \\
\pi_{n+1}(\hat{X}_{n+1}, \hat{X}_n) & & \downarrow u_* \\
\downarrow h & & \\
H_{n+1}(\hat{X}_{n+1}, \hat{X}_n) & \xrightarrow{c(u)} & \pi_n(U)
\end{array}$$

commute. Here the Hurewicz homomorphism h is an isomorphism for $n > 2$. For $n = 1$, let $\pi_1(U)$ be abelian: then $c(u)$ is given by

(4.2.2)

$$c(u) : H_2(\hat{X}_2, \hat{X}_1) \xrightarrow{q_*} H_2(X_2, X_1) \xrightarrow{\partial} H_1(X_1) \xrightarrow{u_*} H_1(U) \overset{h}{\cong} \pi_1(U)$$

$q : \hat{X} \to X$ is the projection of the universal covering $\hat{X}$ of X.

(4.2.2) <u>The obstruction cocycle for sections</u>:

Let $\tilde{X} \to X$ be a fibration with path-connected fiber F, and let $u : X_n \to \tilde{X}_n$ be a section with $n \geq 1$. Let the local group $\hat{\pi}_n(F)$ be $u_*\pi_n(F)$ for $n \geq 2$ as in (4.1.17), and for $n = 1$ let $\pi_1(F)$ be abelian and $\hat{\pi}_1(F)$ be equal to $\pi_1(F)_p$ as in (4.1.22). The obstruction cocycle $c(u) \in C^{n+1}(X, L; \hat{\pi}_n(F))$ is the $\pi_1(X)$-equivariant homomorphism making the diagram

$$\begin{array}{ccc}
\pi_{n+1}(\tilde{X}_{n+1}, \tilde{X}_n) & \xrightarrow{\partial} & \pi_n(\tilde{X}_n) \\
\cong \downarrow p_* & & \\
\pi_{n+1}(X_{n+1}, X_n) & & \downarrow u_{\#} \\
\cong \uparrow q_* & & \\
\pi_{n+1}(\hat{X}_{n+1}, \hat{X}_n) & & \\
\downarrow h & & \\
H_{n+1}(\hat{X}_{n+1}, \hat{X}_n) & \overset{c(u)}{\dashrightarrow} & \pi_n(F)
\end{array}$$

commute. $u_{\#}$ is defined by $u_{\#}(\alpha) = i_*^{-1}(-\alpha + u_* p_*(\alpha))$ where $p : \tilde{X}_n \to \tilde{X}_n$ is the projection. Since u is a section of p, $i_* : \pi_n(F) \to \pi_n(\tilde{X}_n)$ is injective. The Hurewicz homomorphism h is an isomorphism for $n \geq 2$ and is surjective for $n = 1$, therefore $c(u)$ is well-defined by the diagram. (The reader may check that $c(u)$ is equivariant, and use the Hurewicz isomorphism theorem to see that $u_{\#}\partial p_*^{-1} q_*$ factors over h in the diagram when $n = 1$.)

The obstruction cocycle for maps is just the special case of the obstruction cocycle for sections where we take the trivial fibration.

(4.2.3) <u>Theorem</u>: $c(u)$ <u>is a cocycle</u>.

Proof: We consider the commutative diagram

$$\begin{array}{ccccc} \pi_{n+2}(\tilde{X}_{n+2}, \tilde{X}_{n+1}) & \xrightarrow{\bar{\partial}} & \pi_{n+1}(\tilde{X}_{n+1}, \tilde{X}_n) & \xrightarrow{\partial} & \pi_n(\tilde{X}_n) \\ \downarrow \bar{h} & & \downarrow \bar{h} & & \downarrow u_{\#} \\ H_{n+2}(\hat{X}_{n+2}, \hat{X}_{n+1}) & \xrightarrow{\partial_{n+1}} & H_{n+1}(\hat{X}_{n+1}, \hat{X}_n) & \xrightarrow{c(u)} & \pi_n(F) \end{array}$$

where $\bar{h} = hq_*^{-1}p_*$ as in (4.2.2). Since $\partial\bar{\partial} = 0$ it follows that $c(u) \circ \partial_{n+1} = 0$. Definition (4.1.3) then implies that $\delta^n(c(u)) = c(u) \circ \partial_{n+1} = 0$. □

(4.2.4) The difference homomorphism:

As before, let $L \subset X$ and $X_n = X^n \cup L$. Also let $n \geq 2$. The exact homotopy sequence of the triple $(I \times X_n, I \dot{\times} X_n, I \times X_{n-1})$ where $I \dot{\times} X_{n-1} = I \times X_{n-1} \cup \{0, 1\} \times X_n$ gives us the exact sequence

$$0 \to \pi_{n+1}(I \times X_n, I \dot{\times} X_n) \xrightarrow{\bar{\partial}} \pi_n(I \dot{\times} X_n, I \times X_{n-1}) \xrightarrow{j} \pi_n(X_n, X_{n-1}) \to 0$$

Let $i_o, i_1 : (X_n, X_{n-1}) \to (I \dot{\times} X_n, I \times X_{n-1})$ be the inclusions $i_\tau(x) = (\tau, x)$ for $x \in X_n$, $\tau \in I$. Then $ji_{o*} = ji_{1*} = 1$. We thus obtain the map

$$\Delta : \pi_n(X_n, X_{n-1}) \xrightarrow{\Delta'} \pi_{n+1}(I \times X_n, I \dot{\times} X_n) \xrightarrow{\partial} \pi_n(I \dot{\times} X_n)$$

with $\Delta'(\alpha) := \bar{\partial}^{-1}(-i_{o*}(\alpha) + i_{1*}(\alpha))$.

Lemma: For $n \geq 2$, Δ is a homomorphism and is equivariant with respect to the operation of $\pi_1(X_{n-1})$.

Proof: The lemma is clear for $n > 2$. When $n = 2$, it must be shown that $\eta = -i_{o*}(\beta) + \bar{\partial}(\xi) + i_{o*}(\beta) = \bar{\partial}(\xi)$ with $\beta \in \pi_2(X_2, X_1)$ and $\xi \in \pi_3(I \times X_2, I \dot{\times} X_2)$. Let $\hat{X}_2$ be the universal covering of X_2. The homomorphism $\bar{h} : \pi_2(I \dot{\times} X_2, I \times X_1) \cong \pi_2(I \dot{\times} \hat{X}_2, I \times \hat{X}_1) \xrightarrow{h} H_2(I \dot{\times} \hat{X}_2, I \times \hat{X}_1)$ is injective on Im ∂. Since $\eta \in \text{Im}\,\partial$ and $\bar{h}(\eta) = \bar{h}(\bar{\partial}\xi)$, the equation $\eta = \bar{\partial}\xi$ follows. □

(4.2.5) The difference cochain for maps :

Let u_o, $u_1 : X_n \to U$ and let $H : u_o|_{X_{n-1}} \simeq u_1|_{X_{n-1}}$ be a homotopy

Then we have a map

$$u_o \cup u_1 \cup H : I \overset{\bullet}{\times} X_n = \{0, 1\} \times X_n \cup I \times X_{n-1} \to U.$$

Let $\theta : \pi_1(X) = \pi_1(X_n) \xrightarrow{u_{o*}} \pi_1(U)$ for $n \geq 2$ be the induced

homomorphism. Then the difference cochain $\Delta(u_o, H, u_1) \in$

$C^n(X, L; \theta^*\pi_n(U))$ is the $\pi_1(X)$-equivariant homomorphism making

$$
\begin{array}{ccc}
\pi_n(X_n, X_{n-1}) & \xrightarrow{\quad \Delta \quad} & \pi_n(I \overset{\bullet}{\times} X_n) \\
\cong \uparrow q & & \downarrow (u_o \cup u_1 \cup H)_* \\
\pi_n(\hat{X}_n, \hat{X}_{n-1}) & & \\
\downarrow h & \xrightarrow{\Delta(u_o, H, u_1)} & \\
H_n(\hat{X}_n, \hat{X}_{n-1}) & - - - - - \to & \pi_n(U)
\end{array}
$$

commute. Again, h is an isomorphism for $n > 2$ and is surjective

for $n = 2$. When $n = 2$, $(u_o \cup u_1 \cup H)_* \Delta q_*^{-1}$ factors over h,

so $\Delta(u_o, H, u_1)$ is well-defined in this case too. If $u_o|_{X_{n-1}} =$

$u_1|_{X_{n-1}} = u$, and $H = u \circ pr$ is the trivial homotopy, then we set

$\Delta(u_o, u_1) = \Delta(u_o, upr, u_1)$ in analogy with (1.2.5). For $n = 1$,

let $\pi_1(U)$ be abelian. We have $X_1 = L \vee \bigvee_e S^1$ where e runs

over the 1-cells of $X - L$. Let α, β be the restrictions of

u_o, u_1 to $\bigvee_e S^1$. Then the difference cochain is defined by

$$\Delta(u_o, u_1) : H_1(\hat{X}_1, \hat{L}) \xrightarrow{q_*} H_1(X_1, L) = H_1(\bigvee_e S^1) \xrightarrow{\beta_* - \alpha_*} H_1(U) \overset{h}{\cong} \pi_1$$

(4.2.6) The difference cochain for sections:

Let $p : \tilde{X} \to X$ again be a fibration with path-connected fiber F.

Let u_o, $u_1 : X_n \to \tilde{X}_n$ be sections, and let $H : u_o|_{X_{n-1}} \equiv u_1|_{X_{n-1}}$

be a section homotopy. We then have in the fibration $1 \times p :$

$I \times \tilde{X} \to I \times X$ the section $u_o \cup u_1 \cup \bar{H} : I \dot{\times} X_n \to I \dot{\times} \tilde{X}_n$, see (1.2.16) The local group $\hat{\pi}_n(F)$ is defined to be $u_o \cdot \pi_n(F)$ for $n \geq 2$ as in (4.1.17). The difference cochain $\Delta(u_o, H, u_1) \in C^n(X, L; \hat{\pi}_n(F))$ is then the $\pi_1(X)$ -equivariant homomorphism making

$$\begin{array}{ccc}
\pi_n(\tilde{X}_n, \tilde{X}_{n-1}) & \xrightarrow{\quad\Delta\quad} & \pi_n(I \dot{\times} \tilde{X}_n) \\
\cong \big\downarrow p_* & & \\
\pi_n(X_n, X_{n-1}) & & \\
\cong \big\uparrow q_* & & \big\downarrow (u_o \cup u_1 \cup \bar{H})_{\#} \\
\pi_n(\hat{X}_n, \hat{X}_{n-1}) & & \\
\big\downarrow h & & \\
H_n(\hat{X}_n, \hat{X}_{n-1}) & \xrightarrow{\Delta(u_o, H, u_1)} & \pi_n(F)
\end{array}$$

commute. $(u_o \cup u_1 \cup \bar{H})_{\#}$ is defined as in (4.2.2) (The reader may again check that $\Delta(u_o, H, u_1)$ is equivariant and that $(u_o \cup u_1 \cup H)_{\#} \Delta p_*^{-1} q_*$ factors over h in the diagram when $n = 2$.) For $n = 1$, let $\pi_1(F)$ be abelian and let the local group $\hat{\pi}_1(F) = \pi_1(F)_p$ as in (4.1.20). Then we define the difference cochain $\Delta(u_o, u_1) \in C^1(X, L; \hat{\pi}_1(F))$ as follows. We have first of all the exact sequence $0 \to \pi_1(F) \xrightarrow{i_*} \pi_1(\tilde{X}_1) \xrightarrow{p_*} \pi_1(X_1) \to 0$. From the inclusions $\alpha_e : S^1 \subset L \vee \bigvee_e S^1 = X^1$, where e runs over the 1-cells of $X - L$, we get the element

$$\Delta_e(u_o, u_1) = i_*^{-1}(u_{1*}\alpha_e - u_{o*}\alpha_e) \in \pi_1(F).$$

We now define $\Delta(u_o, u_1) : H_1(\hat{X}_1, \hat{L}) \to \pi_1(F)$ to tbe the equivariant extension of $\hat{e} \to \Delta_e(u_o, u_1)$. Here $\hat{e} \in H_1(\hat{X}_1, \hat{L})$ is represented by that 1-cell in $\hat{X}_1$ lying over e whose closure contains the basepoint in $\hat{X}_1$.

Again the difference cochain for maps is just the special case of the difference cochain for sections where we replace $\tilde{X}$ by $X \times U$, the trivial fibration with fiber U. If $u_o|X_{n-1} = u_1|X_{n-1}$ and we take the trivial section homotopy $u \circ pr$, then again $\Lambda(u_o, u_1) = \Delta(u_o, upr, u_1)$.

The following additive property of difference cochains follows from the definitions, or can be deduced from (1.2.17) with the aid of (4.2.8).

(4.2.7) Theorem: Let $u_o, u_1\ u_2 : X_n \to X_n$ be sections and let $u_o|X_{n-1} \overset{H}{\equiv} u_1|X_{n-1} \overset{G}{\equiv} u_2|X_{n-1}$ be section homotopies. Then

$$\Delta(u_o, H, u_1) + \Delta(u_1, G, u_2) = \Delta(u_1, H + G, u_2).$$

We now compare the definitions of obstruction cocycle and difference cochain above with the definitions of primary obstruction and primary difference in (1.2.11) and (1.2.16). With the isomorphism λ^* as in (4.1.13), we have

(4.2.8) Theorem: Let $n \geq 2$.

(a) If $u : X_n \to X_n$ is a section, then $\lambda^*(c(u)) = f^{\#}_{n+1}(u)$ in $[S^n Z_{n+1}, F]$.

(b) If $u_o, u_1 : X_n \to \tilde{X}_n$ are sections and $H : u_o|X_{n-1} \equiv u_1|X_{n-1}$ is a section homotopy, then $\lambda^*\Delta(u_o, H, u_1) = d(u_o, H, u_1)$ in $[S^n Z_n, F]$.

Proof: It is sufficient to show that

$$c(u)\ \{\hat{e}\} = f^{\#}_{n+1}(u)\,|e \quad \text{for} \quad e \in Z^*_{n+1} \quad \text{and}$$

$$\Delta(u_o, H, u_1)\{\hat{e}\} = d(u_o, H, u_1)\,|e \quad \text{for} \quad e \in Z^*_n.$$

These equations follow from the respective definitions without too much trouble. For (b) the reader may compare the definition of Δ' in (4.2.4) and the definition of w_f in (1.2.3). $\square$

The following fundamental theorem of classical obstruction theory is formulated here specifically for sections of fibrations. Taking the trivial fibration, we obtain the classical theorem on the obstruction cocycle and difference cochain for maps. As before, let $p : \tilde{X} \to X$ be a fibration with path-connected fiber F and let $X_n = X^n \cup L$ and $\tilde{X}_n = p^{-1}(X_n)$.

(4.2.9) Obstruction theorem (for primary obstructions): *Let* $n \geq 1$ *where if* $n = 1$ *then* $\pi_1(F)$ *must be abelian. Let* $u : X_n \to \tilde{X}_n$ *be a section, so that the obstruction cocycle* $c(u) \in C^{n+1}(X, L; \hat{\pi}_n(F))$ *is defined. Then*

c1) $\delta c(u) = 0$,

c2) $c(u) = 0$ *iff* u *can be extended to a section* $u' : X_{n+1} \to \tilde{X}_{n+1}$.

Let $u_o, u_1 : X_n \to \tilde{X}_n$ *be sections with* $u_o|X_{n-1} = u_1|X_{n-1}$, *so that the difference cochain* $\Delta(u_o, u_1) \in C^n(X, L; \pi_n(F))$ *is defined. Then*

Δ1) $\delta\Delta(u_o, u_1) = c(u_1) - c(u_o)$,

Δ2) $\Delta(u_o, u_1) = 0$ *iff* u_o *and* u_1 *are section homotopic rel* X_{n-1}

(4.2.10) Corollary: *Let* $u_o, u_1 : X_n \to \tilde{X}_n$ *be sections and* $H : u_o|X_{n-1} \simeq u_1|X_{n-1}$ *be a section homotopy. Then the difference cochain* $\Delta(u_o, H, u_1) \in C^n(X, L; \hat{\pi}_n(F))$ *is defined, and we have in generalization of* Δ1) *and* Δ2)

Δ'1) $\delta\Delta(u_o, H, u_1) = c(u_1) - c(u_o)$,

$\Delta'2)$ $\Delta(u_o, H, u_1) = 0$ iff the section homotopy H can be extended to a section homotopy $u_o \equiv u_1$.

Proof of (4.2.10): By the general homotopy extension principle (0.1.5), there exists a section extension $u_1' : X_n \to \tilde{X}_n$ of $u_o|X_{n-1}$ and a section homotopy $H' : u_1' \equiv u_1$ extending H. Then by (4.2.7) we have $\Delta(u_o, u_1') + \Delta(u_1', H, u_1) = \Delta(u_o, H, u_1)$, and it follows from $\Delta 2)$ that $\Delta(u_1', H, u_1) = 0$. Since $c(u_1)$ depends only on the homotopy class of u_1 (see Definition (4.2.2)) we have $c(u_1) = c(u_1')$. Thus $\Delta'1)$ follows from $\Delta 1$. Using the existence of H', $\Delta'2)$ can be deduced from $\Delta 2)$ as well. □

Proof of (4.2.9): c1) has already been proved in (4.2.3). c2) and $\Delta 2)$ follows from (4.2.8). The familiar coboundary formula $\Delta 1)$ is a special case of the equation in (3.5.9). Let $\beta = d(u_o, u_1) \in [S^n Z_n, F]$. Then $u_1 = u_o + \beta$, and by (3.5.9)

$$-f_{n+1}^{\#}(u_o) + f_{n+1}^{\#}(u_o + \beta) = \langle \beta, u \rangle_{\nabla f_{n+1}}$$

Here we use the facts that $f_{n+1} \in E_{f_n}(\xi_{n+1})$ is a functional suspension and $E\xi_{n+1} = \nabla f_{n+1}$, see (3.5.7) and the following lemma (4.2.11). Using (4.1.13) and (4.2.8), we see that equation (3.5.9) is equivalent to $\Delta 1)$ because of the isomorphism λ^*, $n \geq 2$. The case $n = 1$ must be checked separately. □

Let $\pi : E^n \wedge Z_n \to X_n$ be the characteristic map of the n-cells of $X - L$. We consider the homomorphisms $\pi_n(X_n) \xrightarrow{j} \pi_n(X_n, X_{n-1}) \xleftarrow{\pi_*} \pi_n((E^n, S^{n-1}) \wedge Z_n \vee X_{n-1})$ where $(E, S) \wedge Z \vee X = (E \wedge Z \vee X, S \wedge Z \vee X)$.

(4.2.11) Lemma: $\operatorname{Im} j \subset \operatorname{Im} \pi_*$ for $n > 2$.

It follows that $f_{n+1} \in [S^n Z_{n+1}, X_n]$ is functional with respect to f_n.

That is, for $n \geq 2$ there exists a $\xi_{n+1} \in [S^{n-1}Z_{n+1}, S^{n-1}Z_n \vee X_{n-1}]_2$ such that

(4.2.12) $\quad f_{n+1} \in E_{f_n}(\xi_{n+1})$, that is $j(f_{n+1}) = \pi_* \partial^{-1}\xi_{n+1}$.

Here E_{f_n} is a special case of the functional suspension of (3.4), see (3.6).

Proof of (4.2.11): By (3.4.7), π_* is epimorphic for $n > 3$. Using the universal cover of X_2 and the Hurewicz homomorphism, one can check the case $n = 2$. ☐

As an easy corollary of the obstruction theorem (4.2.9) we get the

(4.2.13) Obstruction theorem (for cohomology obstructions): Let $n > 1$, where if $n = 1$, then $\pi_1(F)$ must be abelian.

1) Let $u : X_{n-1} \to \tilde{X}_{n-1}$ be a section which can be extended to $u' : X_n \to X_n$ Then

$$\underline{c}(u) = \{c(u') \mid u' \text{ extends } u\} \in H^{n+1}(X, L; \hat{\pi}_n(F))$$

is a full cohomology class. $\underline{c}(u) = 0$ if and only if u can be extended to a section $u'' : X_{n+1} \to X_{n+1}$.

2) Let $u_o, u_1 : X_n \to X_n$ be sections which can be extended over X_{n+1} for $n > 2$. Let $H : u_o|X_{n-2} \simeq u_1|X_{n-2}$ be a section homotopy which can be extended to a section homotopy $H' : u_o|X_{n-1} \simeq u_1|X_{n-1}$ Then

$$\underline{\Delta}(u_o, H, u_1) = \{\Delta(u_o, H', u_1) \mid H' \text{ extends } H\} \in H^n(X, L; \hat{\pi}_n(F))$$

is a full cohomology class. $\underline{\Delta}(u_o, H, u_1) = 0$ if and only if H can be extended to a section homotopy $H'' : u_o \simeq u_1$.

That $\underline{\Delta}(u_o, H, u_1)$ is a full cohomology class follows from (3.2.4) and

(4.2.7). That $\Delta(u_o, H', u_1)$ is a cocycle follows from $\Delta'1)$ in (4.2.10).

In the next section we will use the above obstruction theorem to define primary cohomology obstructions.

(4.2.14) Classification theorem:

Let $n \geq 2$ *and let* $u : X_{n-1} \to \tilde{X}_{n-1}$ *be a section. The set* $\langle X_k, \tilde{X}_k \rangle^u$ *of section homotopy classes relative* u *has the following properties for* $k = n, n+1$.

(1) *If* u *can be extended to a section* u_o *over* X_n, *then there is a bijection*

$$\langle X_n, \tilde{X}_n \rangle^u \approx C^n(X, L; u_* \pi_n(F))$$

given by $u_1 \to \Delta(u_o, u_1)$.

(2)) *If* u *can be extended to a section* u_o *over* X_{n+1}, *then there is a bijection*

$$\langle X_{n+1}, \tilde{X}_{n+1} \rangle^u \approx \ker \delta_n^n \times \operatorname{coker} \delta_{n+1}^n$$

where $\delta_k^n : C^n(X, L; u_* \pi_k(F)) \to C^{n+1}(X, L; u_* \pi_k(F))$ *is the coboundary operator of (4.1.3).*

By (4.2.12) and (4.1.13), this is just a special case of the classification in (3.5.10). It contains the essential parts of classical classification theorems (see (4.4)). In section (4.4) we will extend this type of classification by using the spectral sequence of (3.2).

(4.3) 2-extendability of a section. Primary cohomology obstructions.

We first consider the question of whether a fibration has a section over the 2-skeleton of the base. We will give a criterion in terms of the fundamental-group functor for answering this existence question. If the fundamental group of the fiber is abelian, the criterion is a cohomological one. We then derive a fundamental theorem on primary cohomology obstructions that implies and unifies classical results of Hopf, Eilenberg, Olum, Hu, Steenrod and Postnikov. We use it to define the fundamental class of a CW-space and the characteristic class of a fibration.

(4.3.1) Theorem (Criterion for 2-extendability): Let (X, L) be a relative CW-complex, and $\tilde{X} \to X$ a fibration with fiber $F \subset \tilde{X}$. Further let X, L and F be path-connected. A section $u : L \to \tilde{L}$ can be extended to a section u' over the 2-skeleton $X_2 = L \cup X^2$ exactly when $i_* : \pi_1(F) \to \pi_1(\tilde{X})$ is injective and when there is a homomorphism Θ making the diagram

$$\begin{array}{ccccc} \pi_1(\tilde{L}) & \longrightarrow & \pi_1(\tilde{X}) & & \\ u_* \uparrow & & \Theta \uparrow & \searrow^{p_*} & \\ & & & & \pi_1(X) \\ \pi_1(L) & \longrightarrow & \pi_1(X) & \nearrow_{1} & \end{array}$$

commute. u' can be chosen so that $u'_* = \Theta$.

This criterion for 2-extendability is significant for obstruction theory since, if a section over the 2-skeleton is given, we know from (4.2.13) that all obstructions to extending this section over the n-skeletons, $n > 2$, can be expressed in cohomological terms. A much more complicated criterion was given by Barcus in 5.1 of [4]. If the fibration is trivial, that is $\tilde{X} = X \times U$, the above theorem expresses the well-known criterion for 2-extendability of maps that we use in the proof of (4.3.1) (see [53]).

(4.3.2) <u>Corollary</u>: <u>Let</u> $u_o, u_1 : X_2 \to \widetilde{X}_2$ <u>be section extensions of</u> $u : L \to \widetilde{L}$ <u>with</u> $u_{o*} = u_{1*} = \Theta$. <u>Then there exists a section homotopy</u> $H : u_o|X_1 \simeq u_1|X_1$ <u>relative</u> u.

<u>Proof</u>: From the fibration $X_2 \widetilde{\times} I \to X_2 \times I = X'$ induced by $\widetilde{X}_2 \to X_2$ and using u_o, u_1 and the constant section homotopy $\bar{u} : L \times I \to L \widetilde{\times} I$, we obtain a section

$$u' = (u_o, \cup u_1 \cup \bar{u}) : L' = X_2 \times \{0, 1\} \cup L \times I \to \widetilde{L}'.$$

By (4.3.1), u' can be extended to a section u'' over

$$X_2' = L' \cup (X_2 \times I)^2 = X_2 \times \{0, 1\} \cup X_1 \times I$$

since $i_* : \pi_1(F) \to \pi_1(X \widetilde{\times} I)$ is injective and Θ determines a homomorphism Θ' making

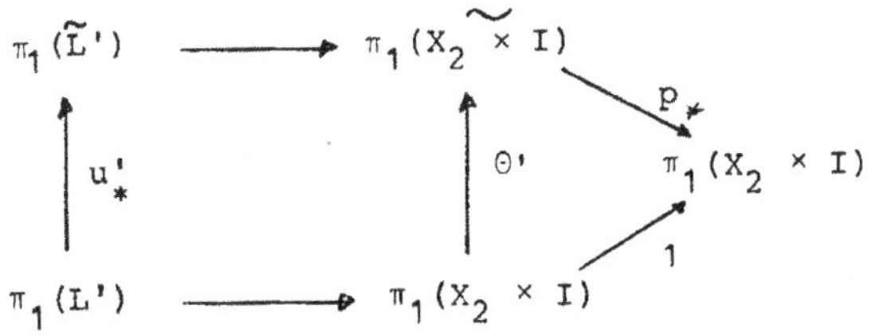

commute. The section u'' is equivalent to a section homotopy H as in (4.3.2). ☐

<u>Proof of</u> (4.3.1): By (1.4.4) and (1.1.6), it suffices to prove (4.3.1) for strictly pointed CW-complexes (X, L), so we assume $X_1 = L \vee \bigvee_e S^1$ where e runs over the 1-cells of $X - L$. Given Θ in (4.3.1), we want to construct u'. For the inclusions $\alpha_e : S^1 \subset X_1 \subset X_2$ we have maps $\beta_e : S^1 \to \widetilde{X}_2$ with $\beta_e \in \Theta\,\alpha_e$. Since $p_*\Theta = \mathrm{id}$ there are homotopies $H_e : p\,\beta_e \simeq \alpha_e$. Thus we obtain the commutative diagram

(1)
$$\begin{array}{ccc} I \times L \cup \{0\} \times (\bigvee_e S^1) & \xrightarrow{u \circ pr \cup \beta} & \tilde{X}_2 \\ \downarrow & \overset{\bar{H}}{\nearrow} & \downarrow p \\ I \times (L \vee \bigvee_e S^1) & \xrightarrow{i \circ pr \cup H} & X_2 \end{array}$$

Here β and H are unions of the β_e and H_e, $i : L \subset X_2$ is the inclusion and $pr : I \times L \to L$ is the projection. By (0.1.5) there exists a relative lifting $\bar{H}$ leaving the diagram commutative. This map gives us a section $u_1 = H_1 : X_1 \to \tilde{X}_1 \subset \tilde{X}$ extending u and making the diagram

(2)
$$\begin{array}{ccccc} \pi_1(\tilde{L}) & \longrightarrow & \pi_1(\tilde{X}_1) & \longrightarrow & \pi_1(\tilde{X}_2) \cong \pi_1(\tilde{X}) \\ \uparrow u_* & & \uparrow u_{1*} & & \uparrow \Theta \\ \pi_1(L) & \longrightarrow & \pi_1(X_1) & \longrightarrow & \pi_1(X_2) \cong \pi_1(X) \end{array}$$

commute. Notice that $\pi_1(X_1) = \pi_1(L \cup X^1) = \pi_1(L) * \pi_1(\vee S^1)$ is a free product of groups; we therefore need only show commutativity on the elements $\alpha_e \in \pi_1(X_1)$. Since Θ is a homomorphism, there exists by (2) a map v inducing Θ that fits into the commutative diagram

(3)
$$\begin{array}{ccc} X_1 & \subset\!\!\longrightarrow & X_2 \\ \uparrow p & & \uparrow p \\ \tilde{X}_1 & \subset\!\!\longrightarrow & \tilde{X}_2 \\ \uparrow u_1 & & \uparrow v \\ X_1 & \subset\!\!\longrightarrow & X_2 \end{array} .$$

Since pu_1 is the identity, we can form $\alpha = d(pv, 1_{X_2}) \in [X_2/X_1, X_2]$ as in (1.2.5). Here $X_2/X_1 = \bigvee_e S^2$, where e runs over the 2-cells of $X - L$. Since $i_* : \pi_1(F) \to \pi_1(\tilde{X}_2)$ is injective and consequently $p_* : \pi_2(\tilde{X}_2) \to \pi_2(X_2)$ is surjective, there exists an $\alpha' \in [X_2/X_1, \tilde{X}_2]$ with $p_* \alpha' = \alpha$. Now let $v' = v + \alpha'$. In accordance with (1.2.8) then, $pv' \simeq 1_{X_2}$ rel X_1. Using the relative lifting of this homotopy, we obtain a section $u_2 : X_2 \to \tilde{X}_2$ which extends u_1 and induces Θ. We have thus proved one direction in (4.3.1). The other is trivial. $\square$

Suppose (X, L) is a relative CW-complex and $\tilde{X} \to X$ is a fibration with fiber F, and let X, L and F be path-connected. If Θ is a homomorphism extending diagram (4.3.1), we call it a <u>splitting of</u> $p_* = \pi_1(p) : \pi_1(\tilde{X}) \to \pi_1(X)$ compatible with the section $u : L \to \tilde{L}$. The set of all such Θ is denoted by $\overline{\text{Split}\ \pi_1}(p)^u$. If $i_* : \pi_1(F) \longrightarrow \pi_1(\tilde{X})$ is injective, we see from (4.3.1) and (4.3.2) that the canonical map

$$\text{(4.3.3)} \qquad \overline{\text{Split}\ \pi_1}(p)^u \xleftarrow[\chi]{\approx} \text{Im}\ \{ \langle X_2, \tilde{X}_2 \rangle^u \xrightarrow{r} \langle X_1, \tilde{X}_1 \rangle^u \}$$

is a bijection, where r is the restriction and $\chi(r(v)) = v_*$. A cellular map $f : (X', L') \to (X, L)$ induces a map

$$\text{(4.3.4)} \qquad f^* : \overline{\text{Split}\ \pi_1}(p)^u \longrightarrow \overline{\text{Split}\ \pi_1}(f^*p)^{f^*u} \quad \text{with} \quad f^*(\Theta) = \chi f^* \chi^{-1} \Theta$$

where $f^*p : f^*\tilde{X} \to X'$ is the induced fibration and $f^*u : L' \to f^*\tilde{L}$ the induced section. If $\Theta \in \overline{\text{Split}\ \pi_1}(p)^u$ we can define as in (4.1.18) the local group $\Theta.\pi_n(F) = \Theta^*(\pi_n(F)^p)$, for which

$$f^*(\Theta.\pi_n(F)) = (f^*\Theta) . \pi_n(F)$$

with $f^*\Theta$ as in (4.4.4) (see (4.1.19)). (The map f^* of (4.3.4) is also defined for maps $f : (X', L') \to (X, L)$ between relative CW-spaces by taking CW-models (see (1.4).)

(4.3.5) <u>Note</u>: Suppose $i_* : \pi_1(F) \to \pi_1(\tilde{X})$ is injective. The set $\overline{\text{Split}\ \pi_1}(p)^u$ can be classified by using crossed homomorphisms. If $\Theta \in \overline{\text{Split}\ \pi_1}(p)^u$, we call a function $\varphi : \pi_1(X) \to \pi_1(F)$ with

$$\varphi(\alpha + \beta) = \varphi(\alpha)^{\Theta\beta} + \varphi(\beta)$$

a Θ-crossed homomorphism, where for $\xi \in \pi_1(F)$ and $\eta \in \pi_1(\tilde{X})$ the operation ξ^η is given by $\pi_1(F)^p$, see (4.1.15). Let $\overline{\text{Cross}}_\Theta\ (\pi_1(X),\ \pi_1(F))_L$ be the set of all such φ with $\varphi j_* = 0$, where $j_* : \pi_1(L) \to \pi_1(X)$ is induced by the inclusion. Then there is a bijection

$$\Delta_\Theta : \overline{\text{Split}\ \pi_1}(p)^u \approx \overline{\text{Cross}}_\Theta\ (\pi_1(X),\ \pi_1(F))_L$$

defined by $\Delta_\Theta(\Theta')(\alpha) = i_*^{-1}(-v_*(\alpha) + v'_*(\alpha))$. $v, v' : X_2 \to \tilde{X}_2$ are sections with $rv = \chi^{-1}\Theta$ and $rv' = \chi^{-1}\Theta'$, see (4.4.3). By (4.1.18) $\varphi = \Delta_\Theta(\Theta')$ is a crossed homomorphism with $\varphi j_* = 0$.

(4.3.6) Note: If the fundamental group $\pi_1(F)$ of the fiber is abelian, we can formulate a cohomology criterion for 2-extendability as follows. For a fibration $\tilde{X} \to X$ with fiber F, let the local group $\pi_1(F)_p$ be defined as in (4.1.22). Let $L \subset X$ be a subcomplex and X, L, F be path-connected. Then

A section $u : L \to \tilde{L}$ determines a cohomology class $\underline{c}(u) \in H^2(X, L;\ \pi_1(F)_p)$. $\underline{c}(u) = 0$ exactly when u can be extended as a section over $X_2 = X^2 \cup L$. In this case there is a bijection

$$\Delta : H^1(X, L;\ \pi_1(F)_p) \approx \operatorname{Im}\{\langle X_2, \tilde{X}_2\rangle^u \xrightarrow{r} \langle X_1, \tilde{X}_1\rangle^u\}$$

This statement can be derived as a special case of the obstruction theorem (4.2.13) and the classification theorem (4.2.14). □

We will formulate the next theorem not just for CW-complexes but for CW-spaces generally. We call (X, L) a CW-pair when X and L are CW-spaces and the inclusion $L \subset X$ is a closed cofibration. Given a CW-pair (X, L) let $\dim(X - L)$ be the smallest integer $n < \infty$ for which there exists a relative CW-complex (X', L') such that $(X, L) \simeq (X', L')$ and $\dim(X' - L') = n$.

(4.3.7) Theorem: (primary cohomology obstructions):

Let (X, L) be a CW-pair and let $p : \tilde{X} \to X$ be a fibration with fiber F. Let X, L and F be path-connected and $i_* : \pi_1(F) \to \pi_1(\tilde{X})$ be injective.

(*) Let q be the smallest integer > 1 with $\pi_q(F) \neq 0$, that is with $\pi_i(F) = 0$ for $1 < i < q$.

(1) A section $u : L \to L$ and a splitting $\Theta \in \overline{\text{Split}}\ \pi_1(p)^u$ determine a cohomology class $\tilde{c}_\Theta(u) \in H^{q+1}(X, L; \Theta_\bullet\pi_q(F))$ with the property that

(**) If $\pi_j(F) = 0$ for $q < j < \dim (X - L)$, then

$\tilde{c}_\Theta(u) = 0 \Leftrightarrow$ u can be extended over X as a section and compatibly with Θ.

(2) Two section extensions $u_0, u_1 : X \to \tilde{X}$ of u compatible with Θ determine a cohomology class $\tilde{\Delta}_\Theta(u_0, u_1) \in H^q(X, L; \Theta_\bullet\pi_q(F))$ with the property that

(***) If $\pi_j(F) = 0$ for $q < j \leq \dim (X - L)$, then the mapping

$$\langle X, \tilde{X}\rangle^{u,\Theta} \approx H^q(X, L; \Theta_\bullet\pi_q(F)), \quad u_1 \mapsto \tilde{\Delta}_\Theta(u_0, u_1)$$

is a bijection. $\langle X, \tilde{X}\rangle^{u,\Theta}$ is the set of all $\bar{u} \in \langle X, \tilde{X}\rangle^u$ compatible with Θ, that is with $\bar{u}_* = \Theta$.

(3) The primary cohomology obstructions $\tilde{c}_\Theta(u)$ and $\tilde{\Delta}_\Theta(u_0, u_1)$ are natural: given a map $f : (X', L') \to (X, L)$ between path-connected CW-pairs, we have

$$f^*\tilde{c}_\Theta(u) = \tilde{c}_{f^*\Theta}(f^*u)$$

$$f^*\tilde{\Delta}_\Theta(u_0, u_1) = \tilde{\Delta}_{f^*\Theta}(f^*u_0, f^*u_1)$$

where

$$f^* : H^*(X, L;\ \Theta.\pi_q(F)) \to H^*(X', L';\ f^*\Theta.\pi_q(F))$$

is the homomorphism induced by f.

(4) $\tilde{\Delta}_\Theta(u_o, u_1) + \tilde{\Delta}_\Theta(u_1, u_2) = \tilde{\Delta}_\Theta(u_o, u_2)$.

(5) Let $j : L \subset X$ be the inclusion and let

$$\delta : H^q(L, j^*\Theta\ \pi_q(F)) \to H^{q+1}(X, L;\ \Theta\ \pi_q(F))$$

be the connecting homomorphism of (4.1.7). If $u', u'' : L \to L$ are two sections compatible with $j^*\Theta$, then

$$\delta\tilde{\Delta}_{j*\Theta}(u', u'') = \tilde{c}_\Theta(u'') - \tilde{c}_\Theta(u').$$

When $\pi_1(F)$ operates trivially, we leave Θ out of the formulas.

Note: Conditions (*), (**) and (***) in (4.3.7) can be replaced by

(*) $H^{j+1}(X, L;\Theta.\pi_j(F)) = 0$ and $H^j(X, L;\Theta\pi_j(F)) = 0$ for $1 < j < q$.

(**) $H^{j+1}(X, L;\Theta.\pi_j(F)) = 0$ for $j > q$

(***) $H^{j+1}(X, L;\Theta.\pi_j(F)) = 0$ and $H^j(X, L;\Theta\pi_j(F)) = 0$ for $j > q$.

In (*), (1) and (2) it is permissible to have $q = 2$, $\dim(X - L) = q + 1$ and $\dim(X - L) = q$ respectively. L can be empty in (2), but then it should be remembered that we are working exclusively with basepoint-preserving maps, see (0.0) . □

From the theorem we obtain the following classification result. If conditions (*) and (***) in (4.3.7) hold and if $\langle X, \tilde{X}\rangle^u$ is non-empty, then there is a bijection

(4.3.8) $$\langle X, \tilde{X}\rangle^u \approx \coprod_{\Theta \in \overline{\mathrm{Split}\ \pi_1(p)}^u} H^q(X, L;\ \Theta.\pi_q(F))$$

where the set $\overline{\mathrm{Split}\ \pi_1(p)}^u$ can be replaced by the isomorphic sets mentioned in (4.3.5) and (4.3.6). For example, when $\pi_1(F)$ is abelian and operates trivially on $\pi_1(F)$ there is a bijection

$$(4.3.9) \quad \langle X, \tilde{X} \rangle^u \approx H^1(X, L; \pi_1(F)_p) \times H^q(X, L; \pi_q(F)_p)$$

where the local groups $\pi_1(F)_p$ and $\pi_q(F)_p$ are determined by the fibration p as described in (4.1.22). An example of (4.3.8) arises by considering maps into a lens space, or sections of a fibration with a lens space as fiber. In [94] Olum gave a classification of maps between lens spaces that corresponds to (4.3.8). For real projective space RP_n $(n > 2)$, the fundamental group $\pi_1(RP_n) = Z_2$ operates trivially on $\pi_n(RP_n) = Z$ for n odd and by multiplication with -1 for n even.

Proof of Theorem (4.3.7): We first note that if statements (1) through (5) of (4.3.7), hold for pairs of strictly pointed CW-complexes, then they also hold for CW-pairs of path-connected spaces. This is because under the assumption that $\tilde{c}$ and $\tilde{\Delta}$ have already been defined for strictly pointed CW-complexes and satisfy properties (1) - (5), we can define $\tilde{c}$ and $\tilde{\Delta}$ for CW-pairs (X, L) as follows. Let $(h, h_2) : (X', L') \to (X, L)$ be a homotopy equivalence of pairs, where X' is a strictly pointed CW-complex and L' a subcomplex, see (1.4.4). We define

$$\tilde{c}_\Theta(u) = h^{*-1}\, \tilde{c}_{h^*\Theta}\, (h_2^* u), \quad \tilde{\Delta}_\Theta(u_o, u_1) = h^{*-1}\, \tilde{\Delta}_{h^*\Theta}\, (h^* u_o, h^* u_1).$$

It is clear from (3) that these definitions do not depend on h, so we can use (1.1.6) to show that the elements satisfy properties (1) - (5).

We now prove (4.3.7) for strictly pointed CW-complexes.

Proof of (1): By (4.3.1), the section $u : L \to \tilde{L}$ compatible with $\Theta \in \overline{\text{Split}\ \pi_1}(p)^u$ can be extended to a section $u' : X_2 \to \tilde{X}_2$. It follows by repeated application of (4.2.13) under condition (*) that $u'|X_1$ also can be extended to a section $u_o : X_q \to \tilde{X}_q$. The cocycle $c(u_o)$ of (4.2.2) then represents the cohomology class

(a) $\tilde{c}_\Theta(u) = \underline{c}(u_o|X_{q-1}) \in H^{q+1}(X, L; \Theta.\pi_q(F))$.

It follows from (4.2.13) that $\tilde{c}_\Theta(u) = 0$ exactly when u can be extended compatibly with Θ over X_{q+1}. Again by repeated application of (4.2.13) under condition (**) we see that u can be extended over X if and only if $\tilde{c}_\Theta(u) = 0$. It remains to show that $\tilde{c}_\Theta(u)$ depends only on u and Θ. Let $u_1 : X_q \to \tilde{X}_q$ be another section extending u compatibly with Θ. Then by (4.3.2) there exists a section homotopy $H' : u_o|X_1 \simeq u_1|X_1$. It follows by repeated application of (4.2.13) under condition (*) that there exists a section homotopy $H : u_o|X_{q-1} \simeq u_1|X_{q-1}$. By 1') of (4.2.10) we have

(b) $\delta\Delta(u_o, H, u_1) = c(u_o) - c(u_1)$.

Therefore $c(u_o)$ and $c(u_1)$ represent the same cohomology class $c_\Theta(u)$ in (a).

Proof of (2): For u_o and u_1 in (2) there exists a section homotopy $H : u_o|X_{q-1} \simeq u_1|X_{q-1}$ as in (b). The cochain $\Delta(u_o, H, u_1)$ of (4.2.6) is a cocycle and represents

(c) $\tilde{\Delta}_\Theta(u_o, u_1) = \tilde{\Delta}(u_o, G, u_1) \in H^q(X, L; \Theta . \pi_q(F))$

where $G = H|X_{q-2}$, By (4.2.13) and (***), G can be extended over X if and only if $\tilde{\Delta}_\Theta(u_o, u_1) = 0$. By (4) then, the function $u_1 \to \tilde{\Delta}_\Theta(u_o, u_1)$ in (2) is injective. It is also surjective, since for a cocycle $\alpha \in C^q(X, L; \Theta.\pi_q(F))$ the section $u_o|X_q + \lambda_*^*\alpha$ can be extended to a section $u_\alpha : X \to \tilde{X}$ by (4.2.9) and (***), so we have $\tilde{\Delta}_\Theta(u_o, u_\alpha) = \{\Delta(u_o|X_q, u_o|X_q + \alpha)\} = \{\alpha\}$ (see (1.2.21) and (4.2.8)).

Proof of (3) - (5): $\tilde{c}_\Theta$ and $\tilde{\Delta}_\Theta$ are natural with respect to cellular maps $f : (X', L') \to (X, L)$ between strictly pointed CW-complexes, because of definitions (a) and (b) and the fact that the diagrams used to define the obstruction cocycle and the difference cochain in (4.2.2) and (4.2.6) are natural in this respect. Equation (4) of (4.4.7) follows easily from (4.2.7). Equation (5) is a consequence of (a) and (c) and also $\Delta 1'$) of (4.2.10) by the definition of the connecting homomorphism δ in (4.1.7). ☐

(4.3.10) Remark: Theorem (4.3.7) on primary cohomology obstructions developed gradually in the literature over a long period of time. In its present form it covers classical results first derived in the following special cases. The reader will see for himself how to formulate the theorem so as to apply to these cases.

A) Let $\tilde{X} = X \times F \to X$ be the trivial fibration and $\pi_1(F) = 0$. The theorem then gives a classification of maps $X \to F$, first derived by S. Eilenberg in his fundamental work [31]. The result was also treated by H. Hopf [51].

B) Let $\tilde{X} \to X$ be a fiber bundle with fiber F and suppose $\pi_1(F) = 0$. The statement of the theorem in this case was proved in the third section of N.E. Steenrod's book [120] see p. 177-186. In this case the primary cohomology obstruction is called the characteristic class of the fiber bundle.

C) Let $\tilde{X} \to X$ be the trivial fibration. The theorem then states results on the classification of maps $X \to F$ when the fundamental group of F does not vanish. This generalization of the initial results of S. Eilenberg was carried out by M. M. Postnikov [99], S. Hu [52] and P. Olum [92]. Olum also deals with the case where the fundamental group $\pi_1(F)$ does not operate trivially on $\pi_n(F)$.

The results in A), B) and C) are unified in theorem (4.3.7). W. Barcus [4] also arrived at a cohomology obstruction as in (4.2.13) for sections of fibrations, but the full theorem on primary cohomology obstructions, which implies B) and C) as well, does not appear. □

The theorem enables us to make some definitions.

(4.3.11) <u>Definition</u>: Let X be a CW-space and let q be the smallest integer with $\pi_q(X) \neq 0$, and let $\pi_1(X)$ be abelian if $q = 1$. Then (4.3.7) (2) defines, for the trivial map $0 : X \to * \in X$ and the identity $1_X : X \to X$, the primary difference $\Delta(X) = \Delta(1_X, 0) \in H^q(X, \pi_q(X))$. This is called the <u>fundamental class</u> of X.

<u>Remark</u>: The fundamental class $\Delta(X)$ can be obtained as in (0.5.4) from the isomorphism

$$H^q(X, \pi_q(X)) \overset{k}{=} \operatorname{Hom}(H_q(X, Z), \pi_q(X)) \overset{h}{=} \operatorname{Hom}(\pi_q(X), \pi_q(X)),$$

where k comes from the universal coefficient theorem and h is the Hurewicz homomorphism. The fundamental class $\Delta(X)$ is the class mapped onto the identity. It follows immediately from (4.3.7) (2) and (3) that the γ of (0.5.5) is an isomorphism.

(4.3.12) <u>Definition</u>: Let X be a CW-space with $\pi_q(X) = H_q(X, Z)$ as the first non-vanishing homotopy group. We use the fundamental class $\Delta(X)$ to define for an arbitrary space Y the <u>mapping degree</u> $\deg : [Y, X] \to H^q(Y, H_q(X, Z))$, $\deg(f) = f^*(\Delta(X))$.

If Y is a CW-space, $\deg(f) = \Delta(f, 0)$ by (4.3.7) (3) and therefore it follows from (4.3.7) (2) that

(4.3.13) <u>Theorem</u> (S. Eilenberg): <u>If</u> Y <u>is a CW-space and</u>

dim $Y \leq q$ <u>then</u> deg : $[Y, X] \to H^q(Y, H_q(X, Z))$ <u>is a bijection</u>.

Taking X to be a sphere S^q, we arrive at the well-known Hopf classification theorem.

(4.3.14) <u>Corollary</u> (H. Hopf): <u>If</u> dim $Y \leq q$ <u>then</u> deg : $[Y, S^q] \longrightarrow H^q(Y, Z)$ <u>is a bijection, where</u> deg (f) = $f^*(e_q)$ <u>with</u> $e_q \in H_q(S^q, Z)$ <u>a generator</u>.

In connection with case B) of theorem (4.3.7), we can make the

(4.3.15) <u>Definition</u> : Let X be a CW-space and $p : \tilde{X} \to X$ a fibration with fiber F. Let X and F be path-connected and suppose q is the smallest integer with $\pi_q(F) \neq 0$, where $\pi_1(F)$ is to be abelian if $q = 1$. The <u>characteristic cohomology class</u> $\bar{c}(p) \in H^{q+1}(X, \pi_q(F)_p)$ of the fibration p is obtained as follows. By (4.3.7) (1) when $L = \{*\}$ we have the primary cohomology obstruction $\tilde{c}(0)$ to extending the trivial section $0 : * \to F$. We set $\bar{c}(p) = \tilde{c}(0)$. The characteristic cohomology class of a fibration is thus the primary cohomology obstruction to the existence of a section.

From here one can go on to study characteristic classes of vector bundles (Stiefel-Whitney classes, Chern classes, Euler class). These can also be obtained as primary cohomology obstructions, just as in (4.3.14). In this connection see § 38 [120] and [87].

(4.4) The spectral sequence for classifying maps and sections on CW-complexes.

In (3.2) we constructed, for any given iterated principal cofibration $Y \subset X$, a spectral sequence to be used in classifying maps and sections defined on X. In this section we study the spectral sequence for the case that (X, Y) is a relative CW-complex. The E_2-term can then be described in terms of cohomology with local coefficients. As an application we show that results of Hopf, Pontrjagin and Steenrod can be interpreted as the calculation of certain differentials in this spectral sequence. We then discuss the effect of principal reduction on the spectral sequence. In the course of this discussion we prove a general theorem on the linear structure of higher-order obstructions, see (4.4.11).

(4.4.1) General classification theorem for sections: Let (X, Y) be a relative CW-complex with skeletons X_n, and let $X_o = Y$ be path-connected. Let $p : \tilde{X} \to X$ be a fibration with fiber F and $\pi_i = \pi_i(F)$. Suppose $\underline{u} \in u \in \langle X, \tilde{X} \rangle$, $v = \underline{u}|Y$ and $u_n = \underline{u}|X_n$, and let $j_n : \langle X_n, \tilde{X}_n \rangle^v \to \langle X_{n-1}, \tilde{X}_{n-1} \rangle^v$ be the restriction. Then

Claim A): $\operatorname{Im} j_2 \approx \overline{\text{Split}\ \pi_1(p)^v}$ (see (4.3.3))

Claim B): If $n \geq 2$ then $J_n(X, Y; u) = j_n^{-1}(u_{n-1})$ has an abelian group structure with u_n as zero element, and there is a spectral sequence

$$\left\{ E_r^{**} = E_r^{**}(X, Y, u),\ d_r = d_r(u) \right\}_{r \geq 1}$$

where d_r has bidegree $(r, -1)$ and $E_r^{p,q} = 0$ for $p < 1$ or $q < 0$, such that

$$E_2^{p,q} = H^p(X_{p+q}, Y; u_\bullet \pi_{p+q}) \qquad \text{for } p + q \geq 2,\ p \geq 1$$

$$E_n^{n,0} = J_n(X, Y; u) \qquad \text{for } n \geq 2.$$

The local group $u_\bullet \pi_i$ is defined in (4.1.17).

Claim C): The spectral sequence in B) is natural for maps. That is, if in the diagrams

(4.4.2)

$$\begin{array}{ccc} \tilde{X} & \xrightarrow{g} & \tilde{\tilde{X}} \\ & {\scriptstyle p}\searrow \quad \swarrow{\scriptstyle p'} & \\ & X & \end{array} \qquad\qquad \begin{array}{ccc} \tilde{X}' & \longrightarrow & \tilde{X} \\ \downarrow & \text{pull} & \downarrow \\ X' & \xrightarrow{h} & X \end{array}$$

g <u>is a map over</u> X <u>between fibrations, and</u> $h : (X', Y') \to (X, Y)$ <u>is a cellular map between relative CW-complexes (where</u> $X'_o = Y'$ <u>is also path-connected), then</u> g <u>and</u> h <u>induce homomorphisms</u>

$$g_* : E_r^{**}(X, Y; u) \longrightarrow E_r^{**}(X, Y; g_* u)$$

$$h^* : E_r^{**}(X, Y; u) \longrightarrow E_r^{**}(X', Y'; h^* u)$$

<u>between the spectral sequences. When</u> $r = 2$, g_* <u>is induced by the coefficient homomorphism</u> $g_* : \pi_i(F) \to \pi_i(F')$ <u>where</u> F' <u>is the fiber of</u> p', <u>and</u> h^* <u>is the induced map in cohomology with local coefficients.</u>

<u>Proof</u>: We assume that (X, Y) is a relative strictly pointed CW-complex. Then the skeletal filtration $X_o = Y \subset X_1 \subset \ldots \subset X$ is an iterated principal cofibration with attaching maps $f_n : S^{n-1}Z_n \to X_{n-1}$, where $n \geq 1$ with Z_n^* the set of all n-cells in $Z_n = Z_n^* + \{*\}$, see (1.4.3) and (1.4.4). As described in (4.2.12), we obtain from this iterated cofibration a spectral sequence E_r^{**} just as required in B). We have

$$\text{(4.4.2)} \qquad E_1^{p,q} = [S^{p+q} Z_p, F] = \bigtimes_{Z_p^*} \pi_{p+q}(F) \qquad (q \geq 0, \ p \geq 1)$$

and otherwise $E_1^{p,q} = 0$. To see this, set $A_p = S^{p-1}Z_p$ in the definition of (3.2.11). By (3.2.13) we have $E_n^{n,0} = J_n(X, Y; u)$, and by (4.1.3) λ^* induces an isomorphism

$$\lambda^* : E_2^{p,q} \cong H^{p+q}(X_{p+q}, Y; u \bullet \pi_{p+q})$$

for $p + q \geq 2$, $p \geq 1$. This is because the differential $d_1^{p,q} : E_1^{p,q} \to E_1^{p+1,q-1}$ is induced by $\partial_p^{p+q} = E^{q+1}(\nabla fp)$, as follows from

(3.2.4) and the remark after (3.2.11). In accordance with (4.1.13), we have an isomorphism $\lambda^* : C^p(X, Y; u_{*}\pi_{p+q}) \cong [S^{p+q}Z_p, F] = E_1^{p,q}$ which enables us to regard $d_1^{p,q} = \langle\ \ , u\rangle^p_{p+q}$ as equivalent to the boundary operator δ^p_{p+q}. This proves claim B).

The naturality claim C) follows from the corresponding naturality for the exact couples, that is for the classification sequences (B) and ($\underline{B}$) in (2.4), see (1.2.14). It should be remembered in this connection that, given a cellular map $h : X' \to X$, we obtain maps $h(n)$ making

$$(4.4.3)\qquad \begin{array}{ccc} S^n Z'_{n+1} & \xrightarrow{h(n)} & S^n Z_{n+1} \\ {\scriptstyle f'_n}\downarrow & & \downarrow{\scriptstyle f_n} \\ X'_n & \xrightarrow{h|X'_n} & X_n \end{array}$$

homotopy commute such that for a suitable homotopy H, the map $\bar{H} : C_{f'_n} = X'_{n+1} \to C_{f_n} = X_{n+1}$ in (1.2.13) is homotopic to $h|X'_{n+1}$. The reader can see this directly or else refer to the general result 4.4. of Rutter [107]. ☐

We will now give some simple applications.

(4.4.4) <u>Corollary</u>: <u>Let</u> $n \geq 2$ <u>and</u> $H^p(X, Y; u_{*}\pi_{p+1}) = 0$ <u>for</u> $1 \leq p \leq n-2$. <u>Then</u> $J_n(X, Y; u) = H^n(X_n, Y; u_{*}\pi_n)$, <u>and if</u> $\pi_{n+1} = \dots = \pi_{n+k-1} = 0$ <u>for</u> $k \geq 1$ <u>then there is a homomorphism</u> $d^u_{k+1} : H^{n-1}(X, Y; u_{*}\pi_n) \longrightarrow H^{n+k}(X_{n+k}, Y; u_{*}\pi_{n+k})$ <u>for which</u> $J_{n+k}(X, Y; u) = \operatorname{Coker} d^u_{k+1}$.

We will show in (4.4.8) that d^u_{k+1} does not depend on all of u, but only on the restriction $u_{k+1} = u|X_{k+1}$.

<u>Proof of</u> (4.4.4): Consider the E_2-term of the spectral sequence (4.4.1) It follows from the assumptions and (4.4.3) that $E_2^{p,1} = 0$ for $p \leq n-2$, and so $E_n^{n,0} = E_2^{n,0}$, $n \geq 2$, and the first statement is

proved. The second statement can be proved in a similar way. The E_2-term looks like this

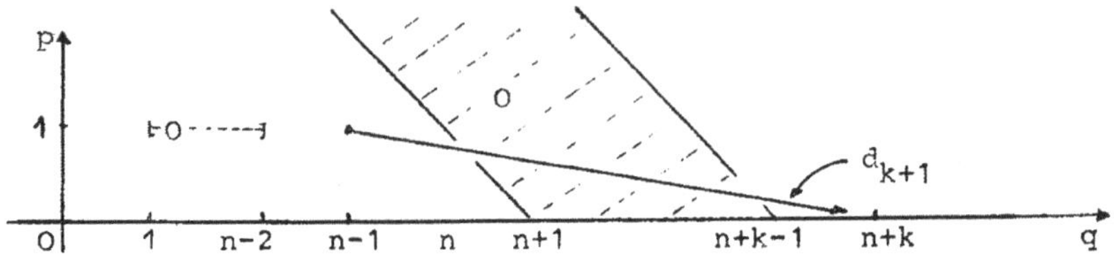

We have $E_{n+k}^{n+k,0} = E_{k+2}^{n+k,0}$ = Coker d_{k+1} with $d_{k+1} : E_{k+1}^{n-1,1} \to E_{k+1}^{n+k,0}$. Since $\pi_{n+1} = \dots = \pi_{n+k-1} = 0$, there is a surjective map $\pi : E_2^{n-1,1} \to E_{k+1}^{n-1,1}$ and an isomorphism $E_2^{n+k,0} = E_{k+1}^{n+k,0}$. The composition $d_{k+1}\pi$ gives us the homomorphism d_{k+1}^u. $\square$

From (4.4.4) we derive a statement about maps.

(4.4.5) Corollary: Let U be a CW-space and $\pi_i = \pi_i(U)$. Suppose $\pi_i = 0$ for $i < n$ and $n < i < k$, with $2 \le n \le k$. Then there exists a cohomology class

$$\nabla U \in H^k(K(\pi_n, n-1) \times U, U; \pi_k)$$

with the following property. When X is a CW-complex with $\dim X \le k$, then $\deg : [X, U] \to H^n(X, \pi_n)$ is surjective, and there is a bijection

$$\deg^{-1}(\xi) \approx H^k(X, \pi_k) \Big/ \{(\alpha, \xi)^*(\nabla U) \mid \alpha \in H^{n-1}(X, \pi_n)\} .$$

Here $(\alpha, \xi) : (X, *) \to (K(\pi_n, n-1) \times U, U)$ denotes also a map with degree (α, ξ).

In (6.3.2) we will offer another interpretation of the class ∇U, that is $\nabla U = L(\Delta k_n)$, where Δk_n is the difference element of the primary k-invariant $k_n : K(\pi_n, n) \to K(\pi_k, k+1)$ of U and L is the partial loop operation.

Proof of (4.4.5): For $u : X \to U$ and $\deg u = \xi$ we have $\deg^{-1}(\xi) =$

$J_k(X, *; u) \approx \text{Coker } d^u$ where $d^u = d^u_{k-n+1}$ as in (4.4.4). Now we call upon the naturality of the spectral sequence (4.4.1). Suppose that $f = (\alpha, u) : X \to X' = K(\pi_n, n-1) \times U$ and $u' : X' \to U$ is the projection. Then $u'f = u$ and $d^u(\alpha) = d^u f^*(i) = f^* d^{u'}(i)$ where i is the fundamental class of $K(\pi_n, n-1)$ or X respectively. We have for the inclusion $g : U \subset X'$ that $g^* d^{u'}(i) = d^{id} g^*(i) = d^{id}(0) = 0$, when id is the identity of U. Therefore $d^{u'}(i)$ determines the cohomology class ∇U. □

The next corollary is a classical illustration of the preceding result.

(4.4.6) <u>Corollary</u> (Pontrjagin, Steenrod [96], [119]): <u>Let</u> X <u>be a CW-complex with</u> dim $X = n + 1$. <u>Then</u> $\deg : [X, S^n] \to H^n(X, Z)$ <u>is surjective and</u>

$$\deg^{-1}(\xi) \approx \begin{cases} H^3(X,Z)/2\xi \cup H^1(X,Z) & , n = 2 \\ H^{n+1}(X,Z)/Sq^2 \mu_* H^{n-1}(X,Z) & , n \geq 3. \end{cases}$$

μ_* is the coefficient homomorphism for $Z \to Z_2$, and Sq^2 is the Steenrod square. If we take U to be complex projective space $\mathbb{C}P_n$, then $\pi_i(\mathbb{C}P_n) = 0$ for $2 < i < 2n + 1$ and $\pi_2(\mathbb{C}P_n) \cong \pi_{2n+1}(\mathbb{C}P_n) \cong Z$.

(4.4.7) <u>Corollary</u> (Spanier): <u>Let</u> X <u>be a CW-complex with</u> $\dim X = 2n + 1$. <u>Then</u> $\deg : [X, \mathbb{C}P_n] \to H^2(X, Z)$ <u>is surjective and</u> $\deg^{-1}(\xi) \approx H^{2n+1}(X,Z)/(n + 1)\xi^n \cup H^1(X,Z)$.

$\xi^n \in H^{2n}(X,Z)$ is the n-fold cup product of ξ. With regard to (4.4.6) and (4.4.7) the reader may refer to pages 460 and 452 of Spanier's book [116].

Proof of (4.4.6) and (4.4.7): These corollaries follow from

(1) $\nabla S^n = Sq^2 \mu_* \in H^{n+1}(K(Z,n-1); Z_2),$ $n \geq 3$

$\nabla S^2 = 2i_1 \cup \gamma \in H^3(S^1 \times S^2, S^2; Z),$ $n = 2$

(2) $\nabla CP_n = (n+1)i_1 \cup \gamma^n \in H^{2n+1}(S^1 \times CP_n, CP_n; Z), n \geq 1.$

in virtue of (4.4.5). Here $i_1 \in H^1(S^1,Z) = Z$ and $\gamma \in H^2(CP_n,Z) = Z$ are generators. To prove (1), we first take $\eta \in \pi_n(S^{n-1})$ to be a generator. Then C_η is the $(n+1)$-skeleton of $K(Z, n-1)$. A consideration of the geometric definition of the differential d_2^u for $u = pr : C_\eta \times S^n \to S^n$ shows that $d_2^u : H^{n-1}(C_\eta \times S^n; \pi_n(S^n)) = \pi_n(S^n) \to H^{n+1}(C_\eta \times S^n, \pi_{n+1}(S^n)) = \pi_{n+1}(S^n)$ takes a generator $i_n \in \pi_n(S^n)$ to the generator $S_\eta \in \pi_{n+1}(S^n)$. Therefore $\nabla(S^n) = d_2^u(i_n)$ is not trivial, and so must be $Sq^2\mu_*$. To prove (2), we let $u = pr : X = S^1 \times CP_n \to S^n$ and $\pi_i = \pi_i(CP_n)$ and consider the differential

$$d_{2n}^u : H^1(X, \pi_2) = \pi_2 \to H^{2n+1}(X, \pi_{2n+1}) = \pi_{2n+1}.$$

$i_1 \cup \gamma^n \in H^{2n+1}(X; Z)$ is a generator. We must show that d_{2n}^u takes a generator $i_2 \in \pi_2$ to an $(n+1)$-multiple of a generator $h_{n+1} \in \pi_{2n+1}$. To do this, we go back to the geometric definition of d_{2n}^u. Let $\pi : S^1 \times S^1 \twoheadrightarrow S^2$ be the quotient map, and let W and

$$W(h_n) : S^{2n+1} \xrightarrow{W} (S^1 \times X)_{2n+1} \xrightarrow{\pi \times 1} (S^2 \times CP_n)_{2n+1} = S^2 \times CP_{n-1} \cup CP_n$$

be the attaching map of the $(2n+2)$-cells in $S^1 \times X$ and in $S^2 \times CP_n$. Suppose $\bar{u} : S^2 \times CP_{n-1} \cup CP_n \to CP_n$ is an extension of $(i_2, 1) : S^2 \vee CP_n \to CP_n$. Then $u(\pi \times 1)$ is an extension of $i_2\pi : S^1 \times S^1 \to CP_n$ and $u : X \to CP_n$, so that

$$d_{2n}^u(i_2) = W^*(\bar{u}(\pi \times 1)) = W(h_n)^*(\bar{u}) = t \cdot h_{n+1} \in \pi_{2n+1} = Z$$

t is the degree of $\bar{\bar{u}}^* : H^{2n+2}(CP_{n+1};Z) \to H^{2n+2}(S^2 \times CP_n;Z)$

where $\bar{\bar{u}} : C_{W(h_n)} \to C_{h_{n+1}}$ is the extension of $\bar{u}$ over the mapping cone. (2) now follows from $\bar{\bar{u}}^*(\gamma^{n+1}) = (\bar{u}^*\gamma)^{n+1} = (i_2 + \gamma)^{n+1} = (n+1)i_2 \cup \gamma^n$. ▭

As we pointed out in (4.1.12) and (4.1.13), the differential $d_1 = d_1(u)$ in the spectral sequence (4.4.1) depends only on the restriction $u_1 = u|X_1$. More generally,

(4.4.8) <u>Theorem: The r-th differential</u> $d_r = d_r(u)$ <u>of the spectral sequence (4.4.1) depends only on the restriction</u> $u_r = u|X_r$. <u>That is, if</u> $u, u' \in \langle X, \tilde{X}\rangle$ <u>and</u> $r \geq 1$ <u>we have</u>

$$u|X_{r-1} = u'|X_{r-1} \implies E_r^{**}(u) = E_r^{**}(u'),$$

$$u|X_r = u'|X_r \implies d_r(u) = d_r(u').$$

This is why we can write $d_r = d_r(u_r)$, where $u_r \in \langle X_r, \tilde{X}_r\rangle$ is the restriction of u. The theorem gives a general explanation for the difference in the results for $n = 2$ and $n > 3$ in the classification theorem (4.4.6) of Pontryagin and Steenrod. The case $n \geq 3$ corresponds to a stable situation.

<u>Proof of (4.4.8):</u> We will work with maps instead of sections. Let $u : X \to U$ be given, and let $v = u|Y$ and $lu_r = u|X_r$ be the restrictions. It should now be remembered from (3.2.11) that the differential d_r is induced by $\beta\alpha^{-r+1}$. It follows from the definitions that the composition $\beta\alpha^{-r+1}$ is represented by the maps in

$$\begin{array}{ccccc} S^{q-1+p+r}Z_{p+r} & \xrightarrow{Wf_{p+r}} & \Sigma^q_Y X_{p+r-1} \cup X_{p+r} & \xrightarrow{\bar{x}} & U \\ & & \uparrow{\scriptstyle \alpha^{r-1}} & & \uparrow{\scriptstyle (x,u_{p+r})} \\ & & \Sigma^q_Y X_p \cup X_{p+r} & \xrightarrow{\mu \cup 1} & S^{q+p}Z_p \vee X_{p+r} \end{array}$$

α^{r-1} is the inclusion and Wf_{p+r} is defined as in (2.4.6). μ is the composition in diagram (2.4.11) with $f = f_p$. The differential $d_r = d_r(u)$ can now be described as follows. If $x \in [S^{q+p}Z_p, U] = E_1^{p,q}$, then $d_r(x)$ is defined if and only if $(\bar{\mu} \cup 1)^*(x, u_{p+r})$ can be extended to a map $\bar{x}$ as in the diagram. When it is defined, $d_r(x)$ is represented by $(Wf_{p+r})^*(x) \in E_1^{p+r,q-1}$. ($d_r(u)$ can be described in an analogous way when we take $u \in \langle X, \tilde{X} \rangle$ to be a section.)

We now see that this description of d_r allows us to replace the space Y by X_{p-1} in the diagram. Then

(2) $\quad \bar{\mu} \cup 1 : \Sigma^q_{X_{p-1}} X_p \cup X_{p+r} \to S^{q+p}Z_p \vee X_{p+r}$

is a homotopy equivalence. Let $V_1 = S^{q+p}Z_p \vee Y$. By induction over r, we obtain a space V_r with $X_{r-1} \subset V_r$ and a homotopy equivalence $\bar{\mu}$ as in

(3)
$$\begin{array}{ccc} \Sigma^q_{X_{p-1}} X_{p+r-1} \cup X_{p+r} & \xrightarrow{\bar{\mu} \cup 1} & V_r \cup_{X_{r-1}} X_{p+r} \\ \Big\uparrow {\scriptstyle Wf_{p+r}} & & \Big\uparrow \\ S^{q-1+p+r}Z_{p+r} & \overset{W}{- - - - \to} & V_r \cup_{X_{r-1}} X_r \end{array}$$

Since Wf_{p+r} is trivial on X_{p+r} (that is $\rho_*(Wf_{p+r}) = 0$), there exists a map W completing a homotopy commutative diagram as indicated. Furthermore, W is trivial on X_r. The reader may compare the proof of (3.5.13) (4). With $V_{r+1} = C_W$ as the mapping cone, we now obtain a homotopy equivalence

$$\bar{\mu} : \Sigma^q_{X_{p-1}} X_{p+r} \simeq V_{r+1} \cup_{X_r} X_{p+r} \qquad \text{(see (1.2.13))}$$

which extends $\bar{\mu} \cup 1$ in diagram (3) and also (2). The statement in (4.4.8) then follows from diagram (1).

When $p \geq r$, V_r can also be described using principal reduction (3.5.13). In this case there exist an ex-cofiber space B

over $X_{r-1} = D$ and a map $g : B \to X_{p-1}$ extending $X_{r-1} \subset X_{p-1}$, so that $X_{p+r-1} \simeq C_D g$. This yields homotopy equivalences

(4) $\Sigma^q_{X_{p-1}} X_{p+r-1} \simeq \Sigma^q_{X_{p-1}} C_D g \simeq S_D^{q+1} B \cup_D X_{p+r-1}$

of ex-cofiber spaces over X_{p+r-1}. Therefore $V_r = S_D^{q+1} B$. ☐

(4.4.9) Example: We will now show that $d_r(u)$ can indeed depend on u_r. Consider the homotopy set $[S^m \times S^n, U]$ $(n > m \geq 2)$ which we have already characterized in (3.3.15). In the derived spectral sequence, only two non-trivial differentials d_n and d_m appear:

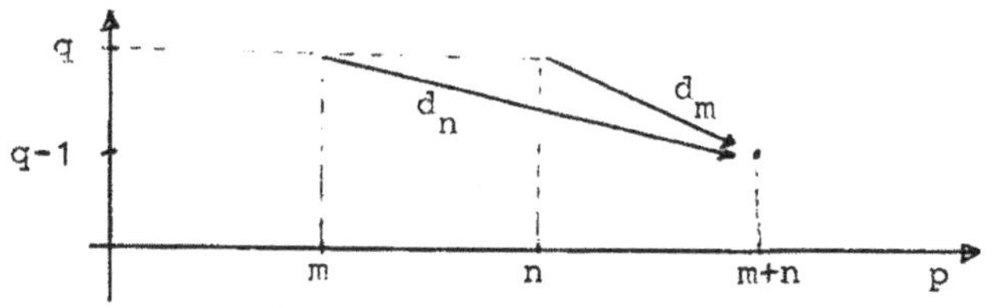

Given a map $u : S^m \times S^n \to U$, let $u^m = u | S^m$ and $u^n = u | S^n$. Then for $\alpha \in E_m^{n,q} = \pi_{n+q}(U)$ and $\beta \in E_n^{m,q} = \pi_{m+q}(U)$ we have

$$d_m(\alpha) = [u_m, \alpha] \in \pi_{n+q+m-1}(U) = E_m^{m+n,q-1}$$

$$d_n(\beta) = [u_n, \beta] + [u_m, \pi_{n+q}(U)] \in E_n^{m+n,q-1}$$

where $[\ ,\ ]$ denotes the Whitehead product. The reader may care to interpret for himself the other results of (3.3) in terms of the behavior of differentials in the spectral sequence.

We saw in (3.2.14) that the above spectral sequence can be used to describe higher-order differences. We now want to investigate higher-order obstructions as well. To start with, we recall that the secondary obstruction to extending a section from X_n to X_{n+2} is a cohomology class. This is so because, as follows from (4.3.12) for $n \geq 1$, $X_n \subset X_{n+2}$ is a principal cofibration relative X_1. Principal reduction of a CW-decomposition tells us that $X_n \subset X_{n+r}$ is always a principal cofibration relative X_{r-1} when $n \geq r - 1$. We will use this fact in the proof of the following generalization of the obstruction theorem (4.2.13).

Suppose we are given $u_{n-1} \in \langle X_{n-1}, \tilde{X}_{n-1} \rangle^V$. The restrictions α^{r-1} and $f_n^{\#}$ as in

$$\langle X_{n-r}, \tilde{X}_{n-r} \rangle^V \xleftarrow{\alpha^{r-1}} \langle X_{n-1}, \tilde{X}_{n-1} \rangle^V \xrightarrow{f_n^{\#}} [S^{n-1}Z_n, F]$$

yield the r-th order obstruction

$$\mathcal{H}_r = \mathcal{H}_r(u_{n-r}) = f_n^{\#}(\alpha^{r-1})^{-1}(u_{n-r}) \tag{4.4.10}$$

where u_{n-r} is the restriction of u_{n-1}. $\mathcal{H}_r$ is therefore not empty, and there is a filtration

$$\mathcal{H}_1 \in \mathcal{H}_2 \subset \ldots \subset \mathcal{H}_{n-1} \subset \mathcal{H}_n \subset [S^{n-1}Z_n, F]$$

with $\mathcal{H}_1 = f_n^{\#}(u_{n-1})$ and $\mathcal{H}_n = \operatorname{Im} f_n^{\#}$. In accordance with the obstruction theorem (4.2.13), we can regard $\mathcal{H}_2$ as a cohomology class and $\mathcal{H}_3, \ldots, \mathcal{H}_{n-1}$ (not $\mathcal{H}_n$) as unions of cohomology classes. That is, we can write

$$\mathcal{H}_2 \in \mathcal{H}_3 \subset \ldots \subset \mathcal{H}_{n-2} \subset \mathcal{H}_{n-1} \subset H^n(X, Y; u \cdot \pi_{n-1})$$

where $\mathcal{H}_2 = \underline{c}(u_{n-1})$. More generally we obtain the following structure theorem for higher-order obstructions.

(4.4.11) <u>Obstruction theorem:</u> If $1 \leq r \leq (n+1)/2$ <u>then</u> $\mathcal{H}_r = \mathcal{H}_1 + B_r$ <u>is a coset of a subgroup</u> $B_r \subset [S^{n-1}Z_n, F]$, <u>and the obstructions of higher order</u>

$$\mathcal{H}_r \in \mathcal{H}_{r+1} \subset \ldots \subset \mathcal{H}_{n-r+1} \subset [S^{n-1}Z_n, F]/B_r$$

<u>are unions of cosets of</u> B_r. <u>If</u> $r \geq 2$ <u>then also</u>

$$\mathcal{H}_{n-r+1} \subset H^n(X,Y;\, u_*\pi_{n-1})/\underline{B}_r \subset [S^{n-1}Z_n, F]/B_r$$

<u>where</u> $\underline{B}_r = B_r/B_2$ <u>is a subgroup of</u> $H^n(X,\, Y;\, u_*\pi_{n-1})$.

<u>Proof</u>: Just as in the inductive construction for principal reduction of $X_{n-r} \subset X_n$, there exist an ex-cofiber space B over $D = X_{r-2}$ and a classifying map $g : B \to X_{n-r}$ of $X_{n-r} \subset X_{n-1}$ such that

$$S^{n-1}Z_n \xrightarrow{\ f_n\ } X_{n-1} \simeq C_D g$$

is functional with respect to g, that is $f_n \in E_g(\xi)$ and $\xi \in [S^{n-1}Z_n,\, B \cup {}_D X_{r-1}]_2$, see (3.5.13). Let $u_{n-1} : X_{n-1} \to \tilde{X}_{n-1}$ represent $u_{n-1} \in \langle X_{n-1}, \tilde{X}_{n-1}\rangle$, and set $u_k = u_{n-1}| X_k$. Consider the diagram

$$\begin{array}{ccccc} \langle S_D B\ \rho^* D\rangle^{u_{r-2}} & \xrightarrow{\ y^+\ } & \langle X_{n-1}, \tilde{X}_{n-1}\rangle^{u_{r-1}} & \xrightarrow{\ \alpha^{r-1}\ } & \langle X_{n-r}, \tilde{X}_{n-r}\rangle^{u_{r-1}} \\ & \searrow{\scriptstyle \langle\ ,u_{r-1}\rangle^1_\xi} & \downarrow{\scriptstyle f_n^\#} & \searrow{\scriptstyle \alpha^{k-1}} & \downarrow{\scriptstyle \alpha^{k-r}} \\ & & [S^{n-1}Z_n, F] & & \langle X_{n-k}, \tilde{X}_{n-k}\rangle^{u_{r-1}} \end{array}$$

with exact row. α^{r-1} and α^{k-1} for $r < k < n - r + 1$ are the restrictions, and $\langle\ ,\ u_{r-1}\rangle^1_\xi$ is the homomorphism induced from $E\xi$, see (3.5.8). The exactness of the row in the diagram follows from (2.4.1). Thus if $y \in \langle X_{n-1}, \tilde{X}_{n-1}\rangle^{u_{r-1}}$ then

(1) $\quad \operatorname{Im} y^+ = (\alpha^{r-1})^{-1}(x)$ with $x = \alpha^{r-1}y$.

The maps in the diagram satisfy as well

(2) $\quad \mathcal{H}_k = f_n^\#(\alpha^{k-1})^{-1}(u_{n-k})$

where $u_{n-k} \in \langle X_{n-k}, \tilde{X}_{n-k}\rangle^{u_{r-1}}$ and $\mathcal{H}_k$ is defined as in (4.4.10). Since

(3) $\quad f_n^\#(y + \beta) - f_n^\#(y) = \langle \beta\ ,\ u_{r-1}\rangle^1_\xi$

by (3.5.9), it follows from (1) that

(4) $\mathcal{H}_r(x) = f_n^{\#}(y) + \mathrm{Im}\langle\ ,\ u_{r-1}\rangle_\xi^1$

Therefore $\mathcal{H}_r = \mathcal{H}_r(u_{n-r}) = \mathcal{H}_1 + B_r$, where $B_r = \mathrm{Im}\langle\ ,\ u_{r-1}\rangle_\xi^1$ is a subgroup of $[S^{n-1}Z_n, F]$, see (3.5.10). Furthermore,

$$\mathcal{H}_k = \bigcup\{\mathcal{H}_r(x)\ |x \in (\alpha^{k-r})^{-1}(u_{n-k})\}$$

and so the second statement of the obstruction theorem follows from (4). ☐

The questions about linearity of higher-order obstructions can now be partially answered. The best way to do this is by considering a kind of spectral sequence $\underline{E}_r^{p,q}$. As before, suppose we are given $u_{n-1} \in \langle X_{n-1}, \tilde{X}_{n-1}\rangle$ and that $X = X_n$. In accordance with (4.3.1), we can define a spectral sequence $E_r^{**} = E_r^{**}(X_{n-1}, Y; u_{n-1})$ in the first quadrant. We want to extend the sequence, that is to increase the number of index pairs (p, q) for which the term $E_r^{p,q}$ is not trivial. The following picture indicates how the non-trivial part of the extension $\underline{E}_r^{p,q}$ looks.

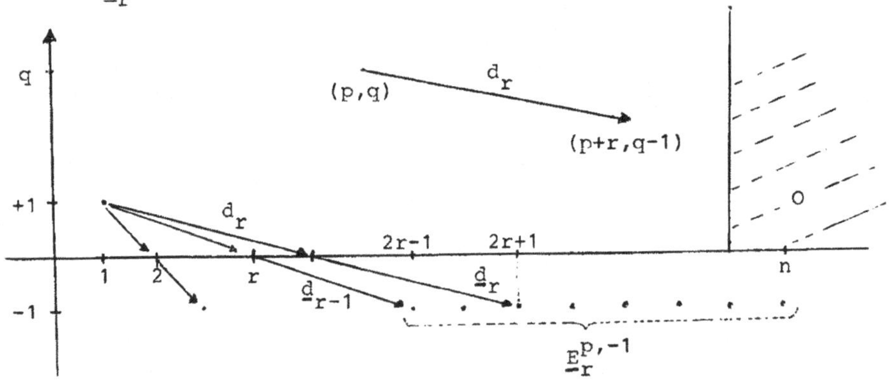

(4.4.12) <u>Definition</u>: $\underline{E}_r^{p,q}$ is defined inductively as follows.

$\underline{E}_r^{p,q} = E_r^{p,q}$ for $q \geq 1$,

$\underline{E}_r^{p,0} = \mathrm{Ker}\ \underline{d}_{r-1}/\mathrm{Im}\ d_{r-1} \subset E_r^{p,0}$ for $q = 0,\ p \geq r$,

$$\underline{E}_r^{p,-1} = [S^{p-1}Z_p, F]/\mathrm{Im}\,\underline{d}_{r-1} \qquad \text{for } q = -1,\ p \geq 2r-1.$$

The differential d_r is as in the spectral sequence $E_r^{p,q}$, and $\underline{d}_r : \underline{E}_r^{p-r,0} \longrightarrow \underline{E}_r^{p,-1}$ is defined by

$$(*) \quad \underline{d}_r\{\alpha\} = \mathcal{H}_r(u_{p-r} + \alpha) - \mathcal{H}_r(u_{p-r}) \quad \text{for } 2r+1 \leq p \leq n.$$

The element $\alpha \in [S^{p-r}Z_{p-r}, F] = E_1^{p-r,0}$ represents the class $\{\alpha\} \in E_r^{p-r,0}$. Since u_{p-r} has u_{n-1} as an extension $\mathcal{H}_r(u_{p-r}) = 0$ for $p < n$. $\underline{d}_r$ depends only on the restriction u_r, see (4.3.8).

When $r = 2$, the differential $\underline{d}_1$ in (4.4.12) is given by

$$(**) \quad \underline{d}_1\alpha = \mathcal{H}_1(u_{p-1} + \alpha) - \mathcal{H}_1(u_{p-1}), \qquad 3 \leq p \leq n$$

where $\alpha \in \underline{E}_1^{p,0} = E_1^{p,0}$. Equation (**) is equivalent with

$$\delta_p^p(\alpha) = c(u_{p-1} + \alpha) - c(u_{p-1}),$$

from $\Delta 1)$ of (4.2.9), see (4.2.8) a) and (4.1.13). The differential $\underline{d}_1$ on $\underline{E}_1^{p,0}$ is therefore nothing but the boundary operator δ_p^p, and so it follows from (4.4.1) that

$$\underline{E}_2^{p,q} = H^p(X^{p+q+1}, Y;\ u_*\pi_{p+q}) \quad \text{for } p+q \geq 2,\ q \geq -1.$$

The definition of the differential $\underline{d}_r$ in (*) of (4.4.12) is thus just a generalization of the classical formula $\Delta 1)$ in the obstruction theorem (4.2.9). With the subgroup B_r defined as in the obstruction theorem (4.4.11), we find that for $r > 1$

$$\underline{E}_r^{n,-1} = [S^{n-1}Z_n, F]/B_r.$$

The above formula (*) for $\underline{d}_r$ does not give us a differential, mapping into $\underline{E}_r^{2r-1,-1}$ or $\underline{E}_r^{2r,-1}$, although these groups are defined. The differential $\underline{d}_r$ goes to $E_r^{p,-1}$ only when $p \geq 2r+1$, and this is why the groups $E_{r+1}^{p,-1}$ are defined for $p > 2r + 1$. Indeed, the form formula (*) for $\underline{d}_r$ allows us in light of (4.4.11) to define a function

(4.4.13) $\Gamma : \underline{E}_r^{r,0} \to \underline{E}_r^{2r,-1}$ with $\Gamma\{\alpha\} = \mathcal{H}_r(u_r + \alpha) - \mathcal{H}_r(u_r)$

which describes the size of $\mathcal{H}_{r+1}$. This function is not a homomorphism in general, but it can be shown that Γ is always quadratic. That is, the function

$$(\alpha, \beta) \longmapsto \Gamma(\alpha + \beta) - \Gamma(\alpha) - \Gamma(\beta) = \Delta(\alpha, \beta)$$

is bilinear. We will give an example of this in (4.4.15). It is thus not very likely that we can obtain a further extension of $\underline{E}_r^{p,q}$ as a 'normal' spectral sequence. The algebraic structure of a complete extension of the spectral sequence $E_r^{p,q}$ is an open problem. A special case of the function Γ is the function

(4.4.14) $\Gamma : H^2(X, Y; u_* \pi_2) \to H^4(X, Y; u_* \pi_3)$,

$$\Gamma\{\alpha\} = \underline{c}(u_2 + \alpha) - \underline{c}(u_2)$$

for $r = 2$, where the obstruction cohomology class $\underline{c}$ is defined as in (4.2.13).

(4.4.15) Example (see (4.4.9): Let $X = S^r \times S^r$ with $n = 2r$ and let $u_{n-1} = (u_1, u_2) \in \pi_r \times \pi_r$, writing $\pi_r = \pi_r(U)$. Then

$$\Gamma : \underline{E}_r^{r,0} = \pi_r \times \pi_r \longrightarrow \underline{E}_r^{2r,-1} = \pi_{2r-1}$$

is given by the Whitehead product $\Gamma(\alpha, \beta) = [u_1 + \alpha, u_2 + \beta] - [u_1, u_2]$.

Remark: Kervaire and Milnor investigated in [62] the map Γ in (4.4.13) in the course of classifying maps $[X, S^r]$, where X is a 2r-dimensional $(r - 1)$-connected manifold. They use Γ to define the Arf invariant of Kervaire invariant of X . Hirzebruch and Hopf calculate in [50] an example of the function Γ in (4.4.14).

(4.4.16) Note (higher-order obstructions in the stable range): We call a relative CW-complex (X, Y) stable when it has cells $e \subset X - Y$ only in dimensions $N < \dim e < 2N - 1$ for some N, so that $(X, Y) = (X_{2N-1}, X_{N-1})$. Given an element $u \in \langle X, \tilde{X} \rangle$, we see from (4.3.12) that we can derive a complete spectral sequence $\underline{E}_r^{**}$, that is, in this case the differentials d_r for $r \geq 1$ are defined at every point (p, q). Thus the higher-order obstructions in a stable (X, Y) can be characterized in terms of the differentials of this spectral sequence. These differentials are the higher-order additive cohomology operations used by McClendon in 5.2 of [80] to describe higher-order obstructions.

In the stable case, the spectral sequence $\underline{E}_r^{**}$ can be used to prove the following classification theorem. We will denote by α_p : $\langle X, \tilde{X} \rangle^v \to \langle X_p, \tilde{X}_p \rangle^v$ the restriction.

(4.4.17) Classification theorem: Let (X, Y) be stable, that is $(X, Y) = (X_{2N-1}, X_{N-1})$. If $u \in \langle X, \tilde{X} \rangle^v$, then $\langle X, \tilde{X} \rangle^v$ has a group structure with u as neutral element. The filtration

$$\langle X, \tilde{X} \rangle^v = K_{N-1} \supset K_N \supset \dots \supset K_{2N-1} = \{u\}$$

with $K_p \subset \alpha_p^{-1}(u_p)$ and $u_p = \alpha_p(u)$ is a filtration of subgroups with

$$K_{p-1}/K_p = J_p(X, Y; u) \cap \operatorname{Im} \alpha_p = \underline{E}_p^{p,0}$$

The group $J_p(X, Y; u)$ is defined as in (4.4.4). This classification theorem is equivalent to McClendon's classification in (5.1) of [80], however he uses a Postnikov decomposition to prove his result. An example of (4.4.17) is the classification theorem (4.2.14). The group structure of $\langle X, \tilde{X} \rangle^{v}$ in (4.4.17) is also defined by (2.1.20), as follows from principal reduction (2.3.8).

Note: Let X be a CW-complex with $\dim X \leq 2n - 2$, and let Y be (n - 1)-connected. Then the suspension map on [X, Y] is a bijection and $[X, Y] = \varinjlim [S^n X, S^n Y] = \sigma_o^X(Y)$ is a stable homotopy group, see (3.1.11). This group can be studied with the aid of the Adams spectral sequence [1] . The spectral sequence $\underline{E}_r^{**}$ amounts to a generalization of the Adams sequence, see the papers of J.F. McClendon [83], J.P. Meyer [84] and C.A. Robinson [105].

CHAPTER 5 : POSTNIKOV HOMOTOPY DECOMPOSITION

(5.1) Difference fibrations.

Given two sections $u_o, u_1 : X \to \tilde{X}$ of a fibration $\tilde{X} \to X$, we can define the difference fibration $P(u_o, u_1) \to X$. The characteristic class of this fibration is just the primary cohomology difference of u_o and u_1. We prove various properties of the difference fibration that we will need for the classification of fibrations with fiber an Eilenberg-MacLane space in (5.2).

(5.1.1) Definition: Let $p : \tilde{X} \to X$ be a fibration with fiber F and suppose $u_o, u_1 : X \to \tilde{X}$ are sections of p. The *difference fibration* $P(u_o, u_1) \to X$ makes the diagram

$$\begin{array}{ccc} P(u_o, u_1) & \longrightarrow & X^I \\ \downarrow & & \downarrow \bar{p} \\ X & \xrightarrow{(u_o, u_1, c)} & X \times X \times X^I \end{array}$$

cartesian. $\bar{p}$ is the fibration defined in (2.2.4) with $p(\tau) = (\tau(0), \tau(1), p\tau)$, and c is the map to constant paths. The fiber of the difference fibration is ΩF. We have

$$P(u_o,u_1) = \{(x,\tau) \in X \times X^I \mid \tau(0)=u_o(x),\ \tau(1)=u_1(x),\ \mathrm{Im}\,\tau \subset p^{-1}(x)\}$$

If $u_o : X \to \tilde{X}$ is a section, we define the fibration $\Omega(u_o) = P(u_o, u_o) \to X$, which also has ΩF as fiber.

(5.1.2) Note: Suppose $(A; \sigma, \rho)$ is an ex-fiber space over D and that $f : X \to A$ and $d = \rho f$. Then the relative fibration $P_D f \to X$ is a special case of the difference fibration, because the fibration $d^*A \to X$ has two sections f and $d^*\sigma$ for which $P_D f =$

$P(f, d^*\sigma)$ see (0.0.5). On the other hand, the difference fibration $P(u_o, u_1) \to X$ can be regarded as a special case of the relative principal fibration. Since $A = (X; u_1, p)$ is an ex-fiber space over X and $P(u_o, u_1) = P_X(u_o)$. In particular, $\Omega(u_1) = \Omega_X(A)$ is a relative loop space. We introduce the notion of difference fibration because it is symmetric in u_o and u_1, and because it is more directly related to section homotopies.

(5.1.3) The difference fibration $P(u_o, u_1) \to X$ has a section H if and only if there exists a section homotopy $H : u_o \simeq u_1$.
A section homotopy $H : u_o \simeq u_1$ is a map $H : I \times X \to \tilde{X}$ whose adjoint $\bar{H} : X \to \tilde{X}^I$ has precisely the property $\bar{p}\bar{H} = (u_o, u_1, c)$ Thus $\bar{H} : X \to P(u_o, u_1)$ is a section, since the diagram in (5.1.1) is cartesian. We will also call the section $\bar{H}$ and the section homotopy H adjoint.

(5.1.4) Naturality: A map $f : Y \to X$ induces a canonical homeomorphism

$$f^*P(u_o, u_1) = P(f^*u_o, f^*u_1)$$

of fibrations. $f^*P(u_o, u_1) \to Y$ is the induced fibration, and $P(f^*u_o, f^*u_1)$ is the difference fibration of the induced sections $f^*u_o, f^*u_1 : Y \to f^*\tilde{X}$.

Proof: There are inclusions $f^*P(u_o, u_1) \subset Y \times X \times \tilde{X}^I$, $P(f^*u_o, f^*u_1) \subset Y \times Y^I \times \tilde{X}^I$. The homeomorphism above is defined by $(y, f(y), \sigma) \to (y, c(y), \sigma)$. □

(5.1.5) Additive property: Let $u_o, u_1, u_2 : X \to \tilde{X}$ be sections of $\tilde{X} \to X$. We define $P(u_o, u_1, u_2)$ as the space making the square

$$\begin{array}{ccc} P(u_o, u_1, u_2) & \longrightarrow & P(u_o, u_1) \\ \downarrow & \overset{\mu}{\searrow} \; P(u_o, u_2) \; \searrow & \downarrow \\ P(u_1, u_2) & \longrightarrow & X \end{array}$$

cartesian. The addition of paths determines a map μ extending the diagram. That is, $\mu(x, \sigma, x, \sigma') = (x, \sigma + \sigma')$ where $(x, \sigma) \in P(u_o, u_1)$ and $(x, \sigma') \in P(u_1, u_2)$. If $f : Y \to X$ is a map, μ induces the function

$$+ \; : \; [Y, P(u_o, u_1)]_f \times [Y, P(u_1, u_2)]_f \to [Y, P(u_o, u_2)]_f$$

defined by $H + G = \mu_*(H, G)$. This function adds homotopies, and has properties similar to those for addition of path classes in a fundamental groupoid, see 10.1 [24]. In particular, $+$ is a group multiplication for $\Omega(u_o)$ and in this case μ is as in (2.2.12). Furthermore

$$[Y, P(u_o, u_1)]_f \approx [Y, \Omega(u_o)]_f : H_1 \to -H_o + H_1$$

is the inverse of the bijection (2.2.11), where $H_o \in [Y, P(u_o, u_1)]_f$.

(5.1.6) Example: In the trivial fibration $X \times X \to X$ with fiber X, the diagonal map d and the inclusion $i : X \times * \subset X \times X$ are sections whose difference fibration $P(d, i) \approx P(X)$ is just the path space $P(X)$ over X. ▭

We now show that the primary obstruction for the existence of a section in a difference fibration is just a primary difference. Let C_f be the mapping cone of $f : A \to X$, and let $\tilde{C}_f \to C_f$ be a fibration with fiber F. Suppose $u_o, u_1 : C_f \to \tilde{C}_f$ are sections and that $H : u_o|X \simeq u_1|X$ is a section homotopy. Then the primary difference $d(u_o, H, u_1) \in [SA, F]$ of (1.2.16) is defined. If $P(u_o, u_1) \to C_f$

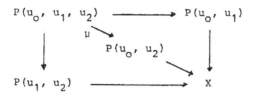

is the difference fibration of u_o and u_1, the section homotopy H provides us with a section $\overline{H} = H_{u_o u_1} : X \to i^*P(u_o, u_1)$ where $i : X \subset C_f$ is the inclusion. The obstruction $f^{\#}(H_{u_o u_1}) \in [A, \Omega F]$ to this section is defined as in (1.2.11). In view of the bijection $\vartheta : [SA, F] = [A, \Omega F]$ we have the

(5.1.7) <u>Theorem</u> $f^{\#}(H_{u_o u_1}) = \vartheta d(u_o, H, u_1)$

<u>Proof:</u> The fibration $\widetilde{C}_f \to C_f$ and the sections u_o and u_1 can be pulled back over the quotient map $\pi_f : CA \to C_f$. Due to the naturality of the construction in (6.1.8), we see that it suffices to prove the theorem for $f = 1_A$. But in this case $\widetilde{CA} \to CA$ is equivalent oto a trivial fibration. Therefore we only have to show (5.1.7) for the trivial fibration $CA \times F \to CA$. □

We now want to compare with each other the local groups defined by $\widetilde{X} \to X$ and $P(u_o, u_1) \to X$. Suppose then that $p : \widetilde{X} \to X$ is a fibration with fiber F and that $\pi_1(F) = 0$. The local group $\pi_n(F)_p$ in X is defined in (4.1.22). Since $\pi_1(\Omega F)$ operates trivially on $\pi_n(\Omega F)$, the difference fibration $\overline{p} : P(u_o, u_1) \to X$ of the sections u_o and u_1 also defines a local group $\pi_n(\Omega F)_{\overline{p}}$.

(5.1.8) <u>Theorem: The isomorphism</u> $\vartheta : \pi_n(\Omega F) = \pi_{n+1}(F)$ <u>is an equivariant isomorphism</u>

$$\vartheta : \pi_n(\Omega F)_{\overline{p}} = \pi_{n+1}(F)_p$$

<u>with respect to the operation</u> of $\pi_1(X)$.

We assume F is simply connected so that ΩF is path-connected.

<u>Proof:</u> If $\alpha : S^1 \to X$ then $\alpha^*(\pi_n(F)_p) = \pi_n(F)_{\alpha^* p}$
$\alpha^* P(u_o, u_1) = P(\alpha^* u_o, \alpha^* u_1)$ by (4.1.23). It therefore suffices to prove the theorem for $X = S^1$. Suppose $\widetilde{S}^1 \to S^1$ is a fibration with

fiber F. Since $\pi_1(F) = 0$, the sections $u_o, u_1 : S^1 \to \tilde{S}^1$ are section homotopic. We therefore may take $u_o = u_1 = u$. Since the fiber F is well-pointed, we can construct the relative lifting diagram

$$\begin{array}{ccc} F \times \{0\} \cup \{*\} \times I & \xrightarrow{(i,\, u)} & \tilde{S}^1 \\ \downarrow & \overset{g}{\nearrow} & \downarrow \\ F \times I & \xrightarrow{pr} & S^1 : (x,\, t) \to t. \end{array}$$

where $u : I \to \tilde{S}^1$ is defined by the section u. From g we can construct the lifting diagram

$$\begin{array}{ccccc} (\Omega F) \times \{0\} \cup \{*\} \times I & \xrightarrow{(i,u)} & \Omega(u) & \longrightarrow & \tilde{S}^{1\,I} \\ \downarrow & \overset{g}{\nearrow} & \downarrow & \text{pull} & \downarrow \\ (\Omega F) \times I & \xrightarrow{pr} & S^1 & \xrightarrow{(u,u,\ )} & \tilde{S}^1 \times \tilde{S}^1 \times S^{1\,I} \end{array}$$

Here $\bar{u}(t) = (t, cu(t))$, and i denotes as before the inclusion of the fiber. The lifting g is defined by

$$\bar{g}(\sigma, t) = (t, \tau_{\sigma,t}) \quad \text{with} \quad \tau_{\sigma,t}(t') = g(\sigma(t'), t).$$

The element $\alpha = id \in \pi_1(S^1)$ operates on $\xi \in \pi_{n+1}(F)$ by $\xi^{-\alpha} = (g_1)_*(\xi)$, see (4.1.25). Similarly α operates on $\eta \in \pi_n(\Omega F)$ by $\eta^{-\alpha} = (\bar{g}_1)_*(\eta)$, where $g_1 : F \to F$ and $\bar{g}_1 : \Omega F \to \Omega F$ are the restrictions to $F \times \{1\}$ and $(\Omega F) \times \{1\}$ of, respectively, g and $\bar{g}$. Since $\bar{g}_1 = \Omega g_1$, (5.1.8) follows. ▭

We can now show that the characteristic class $\tilde{c}(P(u_o, u_1))$ of the difference fibration $P(u_o, u_1) \to X$ is just the primary cohomology difference $\tilde{\Delta}(u_o, u_1)$. More generally

(5.1.9) <u>Theorem</u>: <u>Suppose</u> (X, L) <u>is a CW-pair and</u> $p : X \to X$ <u>is a fibration with fiber</u> F <u>and that</u> q <u>is the smallest integer such that</u>

$\pi_q(F) \neq 0$, $q > 1$. Suppose further that $u_o, u_1 : X \to \tilde{X}$ are sections with $u_o|_L = u_1|_L = u$. Then u defines a section $\bar{u} : L \to i^*P(u_o, u_1)$ where $i : L \subset X$ such that $\vartheta_* \tilde{c}(\bar{u}) = \tilde{\Delta}(u_o, u_1)$ see (4.3.7), where ϑ is the coefficient isomorphism of (5.1.8) and

$$\vartheta_* : H^q(X, L; \pi_{q-1}(\Omega F)_{\bar{p}}) \cong H^q(X, L; \pi_q(F)_p).$$

Proof: The theorem is a consequence of (5.1.7) in virtue of (5.1.8) and (4.2.8). □

From (5.1.6) we derive the following application.

(5.1.10) Corollary: Let X be a CW-space and q the smallest integer such that $\pi_q(X) \neq 0$, $q > 1$. The characteristic class $\bar{c}(PX)$ of the path fibration $PX \to X$ and the fundamental class $\Delta(X)$ are then related to each other by $\vartheta_* \bar{c}(PX) = \Delta(X)$ where $\vartheta_* : H^q(X, \pi_{q-1}(\Omega X)) = H^q(X; \pi_q(X))$.

(5.2) Classification of fibrations with fiber an Eilenberg-Mac Lane space

We will give a classification of fibrations $\tilde{X} \to X$ with fiber $K(G, n)$, where X is CW-space and G is abelian. The case of non-abelian G for $n = 1$ is treated by Hill in [45]. An essential tool in the construction of the Postnikov decomposition of a fibration is the approximation theorem (5.2.8) proved in this section. The classifying space for $K(G, n)$-fibrations is described in an appendix.

(5.2.1) Definition: A fibration $p : E \to X$ together with a homotopy equivalence $i : K(G, n) \to p^{-1}(*)$ is a $K(G, n)$-fibration. There is thus an induced isomorphism $i_* : G = \pi_n(p^{-1}(*))$, that determines in the manner of (4.1.22) a group operation of $\pi_1(X)$ on G that we will denote by G_p. Two $K(G, n)$-fibrations (p, i) and (p', i') are equivalent, denoted $p \sim p'$, when there exists a homotopy equivalence $h : E \to E'$ over X such that $hi \simeq i'$, see (3.3.21).

We assume that G is an abelian group from now on, and that X is a path-connected CW-space. With the characteristic class $\bar{c}(p)$ as defined in (4.3.15) we have the

(5.2.2) Classification theorem:

Existence : If $\tilde{G}$ is an operation of $\pi_1(X)$ on G and $\xi \in H^{n+1}(X, \tilde{G})$ is any cohomology class, there exists a $K(G, n)$-fibration $(n > 1)$ $p : E_\xi(\tilde{G}, n) \to X$ with $G_p = \tilde{G}$ and $c(p) = \xi$.

Uniqueness: If $p : E \to X$ and $p' : E' \to X$ are $K(G, n)$-fibrations, then $p \sim p' \iff G_p = G_{p'}$ and $\bar{c}(p) = \bar{c}(p')$.

(5.2.3) Corollary: There is a bijection

$$\mathcal{F}(K(G, n), X) \approx \bigcup_{\tilde{G}} H^{n+1}(X, \tilde{G})$$

defined by $p \to \bar{c}(p)$, where $\mathcal{F}(K(G, n), X)$ is the set of equivalence classes of $K(G, n)$-fibrations over X. $\tilde{G}$ runs through all operations of $\pi_1(X)$ on G.

Example: If $n = 1$ and $X = K(\pi, 1)$, then (5.2.1) gives us a classification of group extensions as follows. A $K(G, 1)$-fibration $E \to K(\pi, 1)$ has the short exact sequence (with $H = \pi_1(E)$)

$$0 \longrightarrow G \xrightarrow{i} H \xrightarrow{p} \pi \longrightarrow 0$$

at the right end of its long homotopy sequence. (i, p) and (i', p') are said to be equivalent when there is an isomorphism $h : H \cong H'$ such that $hi = i'$ and $p'h = p$. The set of equivalence classes of such short exact sequences for (G, π) is then in 1 - 1 correspondence

(5.2.4)

with $\bigcup_{\tilde{G}} H^2(K(\pi, 1), \tilde{G})$ by (5.2.3), see [71]. In proving the classification theorem for $n = 1$, we will mainly rely on (4.3.6).

Definition: If $\tilde{G}$ is an operation of π on G, we denote by $L(\tilde{G}, n)$ a $K(G, n)$-fibration $p : L(\tilde{G}, n) \to K = K(\pi, 1)$ with a section u and such that $G_p = \tilde{G}$. $L(\tilde{G}, n)$ thus denotes an ex-fiber space over K.

Let $\pi = \pi_1(X)$ and suppose $w : X \to K(\pi, 1)$ is a map inducing the identity on the fundamental groups. It follows from (4.3.7) and the above classification theorem that there is an equivalence

$$w^*L(\tilde{G}, n) \sim E_o(\tilde{G}, n)$$

of $K(G, n)$-fibrations over X. The next theorem also follows from (4.3.7).

(5.2.4) Theorem: Suppose $i : A \subset X$ is a closed cofibration. There exist group isomorphisms making the diagram

$$\begin{array}{ccccccc} [A, \Omega_K L(\tilde{G},n)]_{wi} & \xrightarrow{\delta} & [X, L(\tilde{G}, n)]_K^A & \xrightarrow{j} & [X, L(\tilde{G},n)]_w & \xrightarrow{i^*} & [A, L(\tilde{G},n)]_{wi} \\ \wr\| & & \wr\| & & \wr\| & & \wr\| \\ H^{n-1}(A, \tilde{G}) & \xrightarrow{\delta} & H^n(X,A; \tilde{G}) & \xrightarrow{j} & H^n(X, \tilde{G}) & \xrightarrow{i^*} & H^n(A, \tilde{G}) \end{array}$$

commutative. The rows are exact. The upper row comes from (2.2.15), the lower one from (4.1.7). The isomorphisms are defined using the primary cohomology difference by $u_1 \mapsto \tilde{\Delta}(u_1, u)$.

If the coefficients are simple, that is $L(\tilde{G}, n) = K \times K(G, n)$, then we recover from this theorem the isomorphsim $[X, K(G, n)] = H^n(X, G)$ of (0.5.5). The fibrations $L(\tilde{G}, n) \to K$ play a role in the local coefficient case much like that played by Eilenberg-Mac Lane spaces $K(G, n)$ in the case of simple coefficients. In the proof of the classification theorem we construct a fibration $L(\tilde{G}, n)$ with the aid of a bundle in Steenrod's sense. To this end we need the

(5.2.5) Lemma: There exists an Eilenberg-Mac Lane space $K(G, n)$ with the property that $\mathrm{Aut}\,(G)$ acts as a transformation group on it, namely by basepoint-preserving homeomorphisms such that

$$\begin{array}{ccc} \pi_n(K(G, n)) & = & G \\ \Big\downarrow (\gamma)_* & & \Big\downarrow \gamma \\ \pi_n(K(G, n)) & = & G \end{array}$$

commutes for every $\gamma \in \mathrm{Aut}\,(G)$. The realization of the semisimplicial $K(G, n)$ has this property.

(5.2.6) Construction of $L(\tilde{G}, n)$: The operation $\tilde{G}$ of π on G induces the adjoint homomorphism $\chi : \pi \to \mathrm{Aut}\,(G)$. As described in 13.9 of Steenrod [120], χ determines an $\mathrm{Aut}\,(G)$-principal bundle $E(\tilde{G}) \to K = K(\pi, 1)$. As in (5.2.5), we take the associated bundle with fiber $K(G, n)$, which is the desired bundle $L(\tilde{G}, n) \to K$. Explicitly,

$$L(\tilde{G}, n) = (\hat{K} \times K(G, n))/\sim$$

where $\hat{K}$ is the universal covering and $\sim$ is the equivalence relation $(x, y) \sim (x \cdot \alpha, y \cdot \chi(\alpha))$ for $\alpha \in \pi$. If $q : \hat{K} \to K$ is the projection, then $p : L(\tilde{G}, n) \to K$ is given by $p[x, y] = qx$, and a section $u : K \to L(\tilde{G}, n)$ is such that $u(x) = [x', *]$ with $x' \in q^{-1}(x)$. Since p is locally trivial it is a fibration. Using (5.2.5) and proceeding as in the proof of (5.1.8) we can show that $G_p = \tilde{G}$.

(5.2.7) Construction of $E_\xi(G, n)$: If $\xi \in H^{n+1}(X, \tilde{G})$, then by (5.2.4) there exist two sections $u_o, u_1 : X \to E_o(\tilde{G}, n + 1)$ with $\tilde{\Delta}(u_o, u_1) = \xi$. The difference fibration

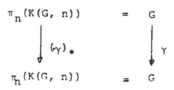

$$\bar{p} \; : \; E_\xi(\tilde{G}, n) = P(u_o, u_1) \to X$$

defined in (5.1.1) has the loop space $\Omega K(G, n+1) \overset{\lambda}{\simeq} K(G, n)$ as fiber, see (0.5.6). λ induces the adjunction isomorphism

$$\pi_n(\Omega K(G, n+1)) \overset{\vartheta}{=} \pi_n(K(G, n+1)) = G.$$

Therefore $G_{\bar{p}} = \tilde{G}$ by (5.1.8), and $\tilde{c}(\bar{p}) = \tilde{\Delta}(u_o, u_1) = \xi$ by (5.1.9). In view of (5.1.2), $E_\xi(\tilde{G}, n)$ can also be obtained as a relative principal fibration in the following way. Consider the commutative diagram (see (2.2.5)).

$$\begin{array}{ccc}
E_\xi(\tilde{G}, n) & \longrightarrow & P_K L(\tilde{G}, n+1) \\
\downarrow & \text{pull} & \downarrow \\
X & \xrightarrow{\quad u_o \quad} & L(\tilde{G}, n+1) \\
& {\scriptstyle w} \searrow \quad \swarrow \nearrow {\scriptstyle u} & \\
& K &
\end{array}$$

where $E_\xi(\tilde{G}, n) = P(u_o, w^* u) = P_K(u_o)$ if $\tilde{\Delta}(u_o, w^* u) = \xi$.

<u>Remark</u>: For the case that $\pi_1(X)$ acts trivially on G, Lamotke gives in 4.8 p. 236 of [67] a semisimplicial classification of $K(G, n)$-fibrations. The result in its present form has also been obtained by Siegel [113] and Robinson [104] using other methods. Siegel employs in his proof a Serre spectral sequence in local coefficients, and a bundle construction instead of the difference fibration. The following approximation theorem was also used by McClendon [80] and Siegel (see 1.6 of [113]). It directly implies the uniqueness statement of (5.2.1), and is an essential tool in Postnikov decomposition procedure described in (5.3).

(5.2.8) <u>Approximation theorem</u>: <u>Let</u> $p : E \to X$ <u>be a fibration with fiber</u> F <u>such that</u> X <u>and</u> F <u>are path-connected CW-spaces. Suppose that</u> $\pi_i(F) = 0$ <u>for</u> $i < n$ <u>and that</u> $\pi_1(F)$ <u>is abelian if</u> $n = 1$.

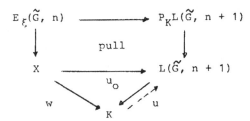

Then there exists a map h *making*

$$\begin{array}{ccc} F & \xrightarrow{\Delta F} & K(\pi_n(F),\, n) \\ \big\downarrow & & \big\downarrow \\ E & \overset{h}{\dashrightarrow} & E_{\bar{c}(p)}(\pi_n(F)_p,\, n) \\ & {\scriptstyle p}\searrow \quad \swarrow{\scriptstyle p'} & \\ & X & \end{array}$$

commute. The fibration p' *arises from the characteristic class* $\bar{c}(p) \in H^{n+1}(X,\, \pi_n(F)_p)$ *as in (5.2.7) and* ΔF *is a map representing the fundamental class* $\Delta F \in H^n(F,\, \pi_n(F))$, *see (4.3.11).*

The next fact, proved for bundles by G.W. Whitehead, will be used in the proof of the approximation theorem, see Steenrod [120] 37.16 p. 189.

(5.2.9) *Lemma: With the assumptions and notation of (5.2.8), we have that* $\delta\Delta F = p^*\bar{c}(p)$ *for the homomorphisms*

$$H^n(F,\, \pi_n(F)) \xrightarrow[\delta]{} H^{n+1}(E,\, F;\, p^*\pi_n(F)_p) \xleftarrow[p^*]{} H^{n+1}(X,\, *;\, \pi_n(F)_p).$$

The map $\tau : \alpha \mapsto p^{*^{-1}}\delta(\alpha)$ from the subgroup $\delta^{-1}\,\mathrm{Im}\, p^*$ of $H^i(F;\, \pi_n(F))$ into the quotient group $H^{i+1}(X,\, *,\, \pi_n(F)_p)/\mathrm{Ker}\, p^*$ is called the *transgression of the fibration* p. If $i = n$ then p^* in the lemma is injective, so in this case the characteristic class $\bar{c}(p)$ can be defined with the aid of the transgression as $\bar{c}(p) = \tau(\Delta F)$. (The claim that p^* is injective follows from (5.2.4) and (1.4.14) and the fact that the fiber of $p : E/F \twoheadrightarrow X$ is $SF \wedge \Omega X$ and so is $(n + 1)$-connected, see (6.4.9).)

Proof of (5.2.9): Consider the diagram

$$\begin{array}{ccccc} F \times F & \subset & E^2 & \longrightarrow & E \\ {\scriptstyle pr}\big\downarrow & & \big\downarrow{\scriptstyle p'} & \text{pull} & \big\downarrow{\scriptstyle p} \\ F & \subset & E & \xrightarrow[p]{} & X \end{array}$$

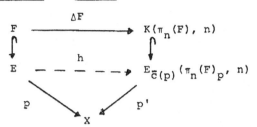

We now apply (4.3.7). The projection pr has two sections $i : F \to F \times F$, $x \to (x, *)$ and $d : F \to F \times F$, $x \to (x, x)$ for which $\tilde{\Delta}(d, i) = \Delta(F)$ by definition of the fundamental class in (4.3.11). Therefore $\delta\Delta(F) = \delta\tilde{\Delta}(d, i) = \tilde{c}(i) - \tilde{c}(d)$. Since d can be extended as a section of p', $\tilde{c}(d) = 0$. It follows from naturality that $\tilde{c}(i) = p^*\tilde{c}(0)$, where $\tilde{c}(0) = \bar{c}(p)$ by definition of the characteristic class of p. ▭

<u>Proof of the approximation theorem (5.2.8)</u>: Let $G = \pi_n(F)$. Consider the diagram

$$\begin{array}{ccccccc} F \times \Omega K \subset p^*E' & & \longrightarrow & & E' = P(u_0, u_1) = E_{\bar{c}(p)}(G_p, n) & & \\ \mathrm{pr}\downarrow \qquad \downarrow & & \overset{h}{\nearrow} & & \downarrow p' & & \\ F \underset{i}{\subset} E & & \xrightarrow{\quad p \quad} & & X & \overset{u_0}{\underset{u_1}{\rightrightarrows}} & E_0(G_p, n+1) \end{array}$$

arising from the fibration p. $P(u_0, u_1)$ is the difference fibration of (5.2.7) for sections u_0, u_1 such that

(1) $\bar{c}(p) = \tilde{\Delta}(u_0, u_1)$.

F is the fiber of p and i is the inclusion. Denoting by $K = K(G, n)$ the fiber of $q : E_0(G_p, n + 1) \to X$ we have $\Omega K \subset E' = P(u_0, u_1)$ as the fiber of p'. Since $u_0(*) = u_1(*) = *$, the sections p^*u_0 and p^*u_1 are equal on F, namely

(2) $p^*u_0|F = p^*u_1|F = u : F \to F \times \Omega K, \quad u(x) = (x, *)$.

Therefore $\tilde{\Delta}(p^*u_0, p^*u_1) \in H^{n+1}(E, F; p^*G_p)$ is the primary cohomology difference. Now look at the diagram

$$\begin{array}{ccccc} H^n(F, G) & \xrightarrow{\delta} & H^{n+1}(E, F; p^*G_p) & \longrightarrow & H^{n+1}(E, p^*G_p) \\ \Delta(F) \in & & \uparrow p^* & & \\ & & H^{n+1}(X, *; G_p) & & \end{array}$$

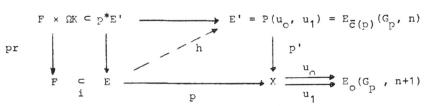

By (5.2.9) and (1),

(3) $\delta\Delta(F) = p^*\bar{c}(p) = p^*\tilde{\Delta}(u_o, u_1) = \tilde{\Delta}(p^*u_o, p^*u_1)$.

It follows from (5.1.9) for that section u of (2) that

(4) $\tilde{\Delta}(p^*u_o, p^*u_1) = \tilde{c}(u)$

Suppose now that $u' : F \to F \times \Omega K$ is a section such that

(5) $\Delta(u', u) = \Delta(F) \in H^n(K, G)$

where we have identified $\pi_n(\Omega K) \overset{\vartheta}{=} \pi_{n+1}(K) = G$.

There $u' = (1, \lambda(\Delta F)) : F \to F \times \Omega K$ with λ the homotopy equivalence of (0.5.6). By (4.3.7) and (5) then,

(6) $\delta\Delta(F) = \delta\tilde{\Delta}(u', u) = \tilde{c}(u) - \tilde{c}(u')$

It follows that

(7) $\tilde{c}(u') = \tilde{c}(u) - \delta\Delta(F)$ see (6),

$= \tilde{c}(u) - \tilde{\Delta}(p^*u_o, p^*u_1)$ see (3),

$= \tilde{c}(u) - \tilde{c}(u)$ see (4),

$= 0.$

Therefore u' can be extended to a section $h : E \to p^*E'$ by reason of (4.3.7). This section determines a map h in the commutative diagram of (5.2.8). This completes the proof of the approximation theorem. It should be noted that we used (1.4.21). □

(5.2) <u>Appendix</u>: <u>The classifying space for fibrations with fiber an Eilenberg-Mac Lane space</u>.

Let G be an abelian group. We denote by $\hat{G}$ the operation of the automorphism group Aut (G) on G, $G \times \text{Aut}(G) \to G$, $(g, \phi) \mapsto \phi^{-1}(g)$.

As in (5.2.6), the Eilenberg-Mac Lane space $K = K(\mathrm{Aut}\,(G), 1)$ determines a fibration $\rho : L(\hat{G}, n+1) \to K$ with section σ. The relative path fibration

$$p : P_K(L(\hat{G}, n+1)) \to L(\hat{G}, n+1)$$

is then a $K(G, n)$-fibration which classifies all $K(G, n)$-fibrations, in the sense of Stasheff [117].

(5.2.10) *Theorem: For a path-connected CW-space* X

$$[X, L(\hat{G}, n+1)] \to \mathfrak{F}(K(G, n), X)$$

defined by $f \to f^*p$ *is a bijection.*

A proof can easily be derived from (5.2.2) and (5.2.7). This we leave to the reader, who may refer to [113], [105] .

(5.3) Postnikov decompositions

We use the approximation theorem (5.2.8) to construct the Postnikov decomposition of an arbitrary fibration. We will show that Postnikov decompositions are natural for maps and unique up to homotopy.

(5.3.1) *Theorem* (Postnikov decomposition): *Let* $p : E \to B$ *be a fibration with fiber* F *and let* E, F, B *be path-connected CW-spaces. Then there exist fibrations* q_n *and maps* h_n *making the diagram*

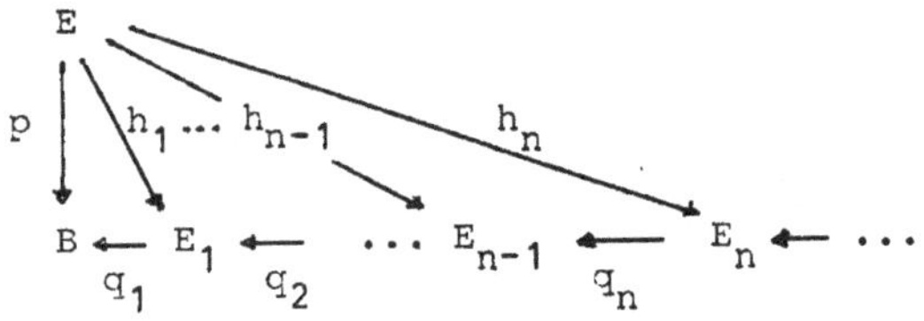

commute, and such that for $n > 1$

(i) q_n *is a fibration with fiber* $K(\pi_n(F), n)$,

(ii) h_n <u>is $(n + 1)$-connected.</u>

A system $\{E_n, q_n, h_n\}_{n \geq 1}$ as described in the theorem is called a <u>Postnikov decomposition</u> of $p : E \to B$. (More generally, any map $f : X \to Y$ has a Postnikov decomposition, namely that of the fibration $W_f \to Y$, see (0.1.10)). We call $p_n = q_n q_{n-1} \cdots q_1 : E_n \to B$ the n-th approximation for $E \to B$. We have

$$p_{n*} : \pi_i(E_n) \cong \pi_i(B) \quad \text{for} \quad i \geq n + 2,$$

$$h_{n*} : \pi_i(E_n) \cong \pi_i(E) \quad \text{for} \quad i \leq n.$$

The characteristic class

(5.3.2) $\quad k_n = \bar{c}(q_n) \in H^{n+1}(E_{n-1}, \widetilde{\pi_n(F)}), \quad n \geq 2$

of the fibration q_n is called the n-th <u>Postnikov invariant</u>, or n-th <u>k-invariant</u>, of the fibration p. By (5.2.2), the fibration q_n is determined by k_n up to equivalence. The local group $\widetilde{\pi_n(F)}$ is given by $h^*_{n-1}\widetilde{\pi_n(F)} = \pi_n(F)^p$ in accordance with (4.1.15), since $h_{n-1*} : \pi_1(E) \to \pi_1(E_{n-1})$ is an isomorphism for $n \geq 2$. If $\pi_1(F)$ is abelian, there is also a k-invariant for $n = 1$. We will show in the proof that the k-invariants k_n are just the characteristic classes of $h_{n-1} : E \to E_{n-1}$ made into a fibration.

<u>Proof of (5.3.1)</u>: We construct the fibrations q_n inductively. We will obtain a fibration $q_1 : E_1 \to B$ only when $\pi_1(F)$ is abelian. For the construction of q_1 when $\pi_1(F)$ is not abelian, the reader should see the paper of Hill [46]. Let $E_o = B$, $q_o = 1_B$, $h_o = p$, and suppose (E_i, q_i, h_i) has been constructed for $i \leq n - 1$, $n \geq 1$. We first make h_{n-1} into a fibration q making

(1)

$$
\begin{array}{ccc}
 & & P_{n-1} = P_{h_{n-1}} \\
 & & \uparrow \downarrow \\
E & \longrightarrow & W_{n-1} = W_{h_{n-1}} \\
\searrow h_{n-1} & \nearrow q & \\
 & E_{n-1} &
\end{array}
$$

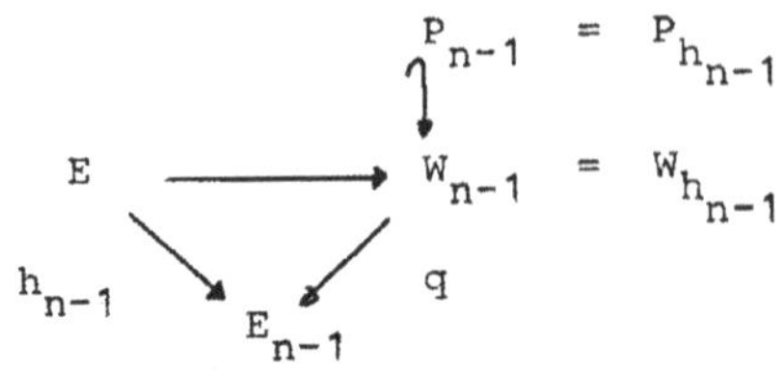

commute, see (0.1). The homotopy-theoretic fiber P_{n-1} is $(n-1)$-connected because h_{n-1} is n-connected. There is an isomorphism

(2) $\pi_j(P_{n-1}) = \pi_j(F)$ for $j \geq n$.

This is because, if we regard $E \xrightarrow{h_{n-1}} E_{n-1} \xrightarrow{P_{n-1}} B$ as a triple of spaces, then the exact homotopy sequence of this triple is

$$
\ldots \to \pi_j(P_{n-1}) \to \pi_j(F) \to \pi_j(F_{n-1}) \to \pi_{j-1}(P_{n-1}) \to \ldots
$$

$F_{n-1} = p_{n-1}^{-1}(*)$ is the fiber of $E_{n-1} \to B$ for which by assumption $\pi_j(F_{n-1}) = 0$ for $j \geq n$. It follows from (2) and the approximation theorem (5.2.8) that there is a $(n+1)$-connected map h making

$$
\begin{array}{ccc}
h_n : E & \overset{\alpha}{\longleftarrow} W_{n-1} & \xrightarrow{h} E_{\bar{C}(q)}(\widetilde{\pi_n}(F), n) = E_n \\
\searrow h_{n-1} & \downarrow q & \nearrow q_n \\
 & E^{n-1} &
\end{array}
$$

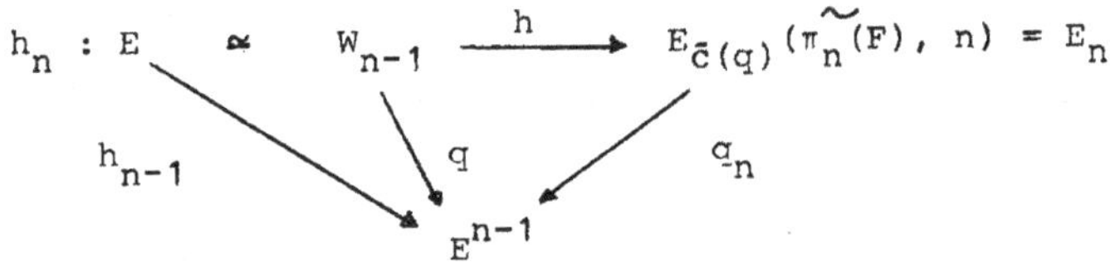

commute. The construction of E_n, q_n, h_n is done and the proof is complete. ▭

If $\pi_n(F) = 0$, then $q_n : E_n \to E_{n-1}$ is the identity. If $0 < n_1 < n_2 < \ldots < n_k < \ldots$ are the integers for which $\pi_{n_j}(F) \neq 0$ (that is $\pi_i(F) = 0$ for $i \neq n_j$), then the Postnikov decomposition consists only of the fibrations q_{n_i}. We will use the complete index set n to avoid having to write an extra index in n_i.

If F_n is the fiber of $p_n : E_n \to B$, we can restrict the maps in (5.3.1) to obtain the commutative diagram

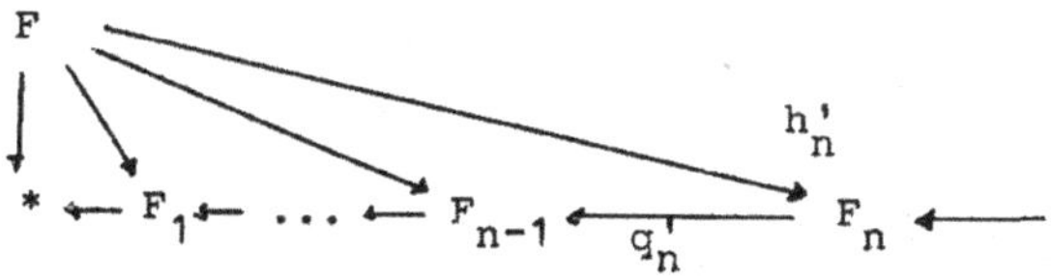

where $F_1 = K(\pi_1(F), 1)$. By (5.3.1),

(5.3.3) <u>Corollary</u>:

(i) q_n' <u>is a fibration with fiber</u> $K(\pi_n(F), 1)$

(ii) h_n' <u>is</u> $(n + 1)$-<u>connected</u>.

Thus $\{F_n\}$ is a Postnikov decomposition of the CW-space Γ. F_n is called the <u>n-th Postnikov section</u> of F. The Postnikov decomposition of a space can also be obtained by the Cartan-Serre-Whitehead technique of killing homotopy groups, see § 17 of [38], see also [3] and [134].

<u>Remark</u>: The Postnikov decomposition of a fibration was described for the oriented-case by Hermann [44, 45]. Moore [87] presented the decomposition in the semisimplicial category. Mc Clendon [80], J. Siegel [113] and Robinson [104] also obtained the decomposition above. Hill [46] treated the case when the fundamental group is non-abelian. Our proof diverges from those of the papers just mentioned in that we use only primary cohomology obstructions, see (4.3.10) B).

Next we want to show that the Postnikov decomposition is uniquely determined up to homotopy equivalence and is natural for maps. Suppose we have a commutative diagram

$$(5.3.4) \qquad \begin{array}{ccccc} E & \xrightarrow{\;\tilde{f}\;} & E' & \xrightarrow{\;\tilde{g}\;} & E'' \\ {\scriptstyle p}\downarrow & & {\scriptstyle p'}\downarrow & & \downarrow{\scriptstyle p''} \\ B & \xrightarrow{\;f\;} & B' & \xrightarrow{\;g\;} & B'' \end{array}$$

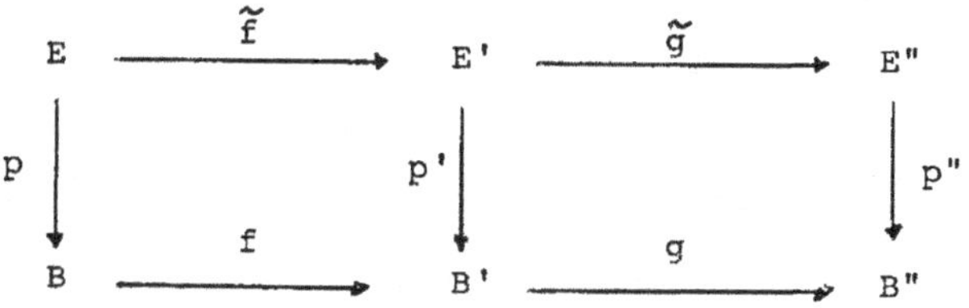

with fibrations p, p', p'' as in (5.3.1). Using the Postnikov decompositions of p and p' we obtain the commutative diagram

(5.3.5)

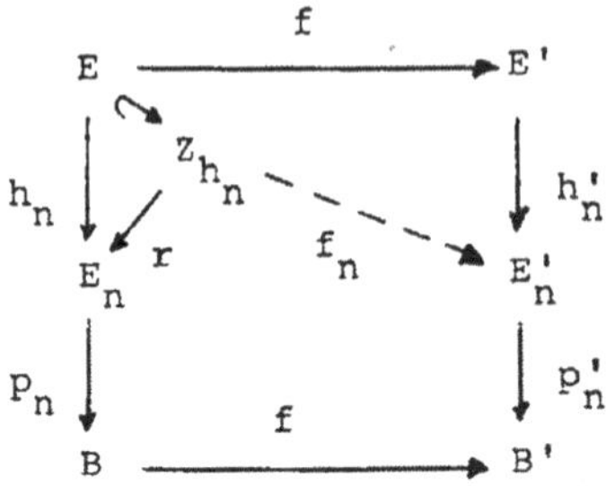

where Z_{h_n} is the mapping cylinder of h_n, and the retraction r is a homotopy equivalence.

(5.3.5) Theorem (Naturality of the Postnikov decomposition): Up to homotopy under E and over B' there exists exactly one map f_n extending the diagram commutatively.

For the maps $(f, \tilde{f})$, $(g, \tilde{g})$ and $(gf, \tilde{g}\tilde{f})$ in diagram (5.3.4) we thus obtain

$$f_n : Z_{h_n} \to E'_n, \quad g_n : Z_{h'_n} \to E''_n, \quad (gf)_n : Z_{h_n} \to E''_n.$$

We define

(5.3.6) $\qquad g_n \square f_n : Z_{h_n} \to E''_n$

by

$$(g_n \square f_n)(x) = g_n(f_n(x)) \qquad \text{for } x \in E_n$$

$$(g_n \square f_n)(x) = \begin{cases} g_n\, f_n(x, 2t) & \text{for } (x,t) \in E \times I, \ 0 \leq t \leq 1/2 \\ g_n\, (\tilde{f}(x), 2t-1) & \text{for } (x,t) \in E \times I, 1/2 \leq t \leq 1 \end{cases}$$

The picture

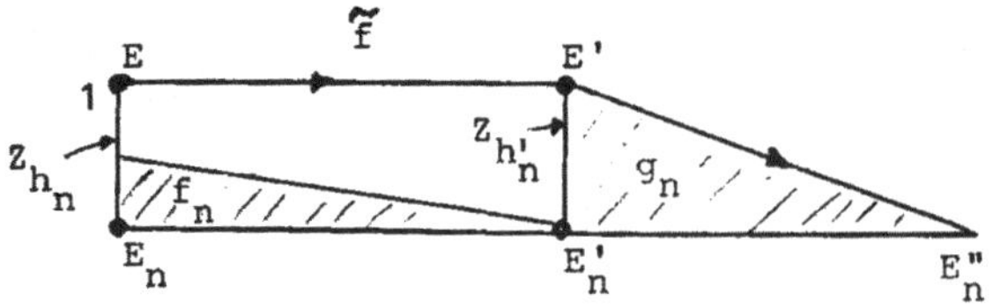

illustrates this map. Since each of the maps $(gf)_n$ and $g_n \square f_n$ makes

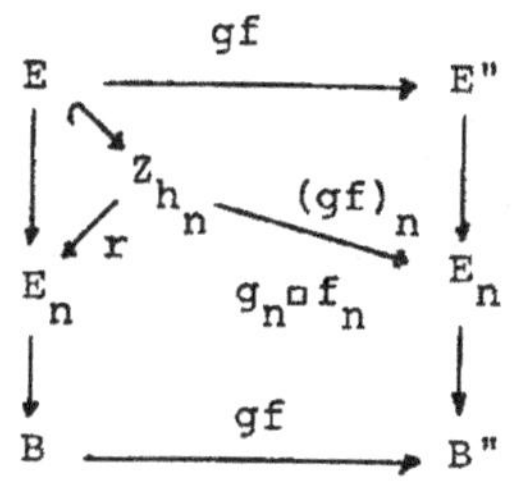

commute, it follows from uniqueness that

(5.3.7) Corollary: *There is a homotopy* $(gf)_n \simeq g_n \square f_n$ *under* E *and over* B".

In the semisimplicial category the Postnikov decomposition is strictly functorial, but it only obeys (5.3.7) in the topological category. The following uniqueness of the Postnikov decomposition follows from (5.3.7).

(5.3.8) Corollary: *Suppose* $\{E_n, h_n, q_n\}$ *and* $\{E'_n, h'_n, q'_n\}$ *are Postnikov decompositions of the fibration* $p : E \to B$ *that satisfy (i) and (ii) in (5.3.1). Then there is a homotopy equivalence* ε_n *making*

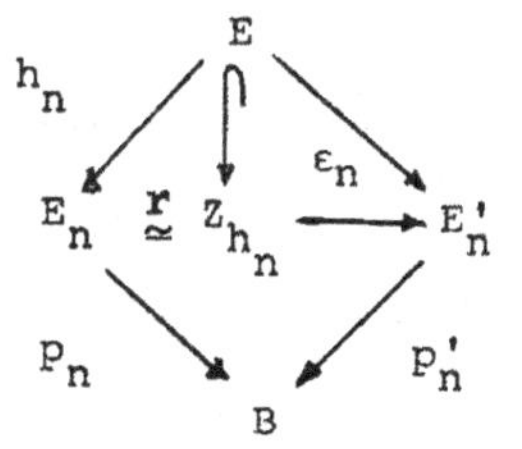

commute. ε_n *is determined as a map up to homotopy under* E *and over* B.

Proof of (5.3.5): We have the fibration

$$q : \tilde{Z} = (rp_n f)^* E'_n \to Z_{h_n}$$

over the mapping cylinder Z_{h_n}, where $q = (rp_n f)^* p'_n$ is induced by the maps in the diagram to (5.3.5). We consider

(5.3.9)

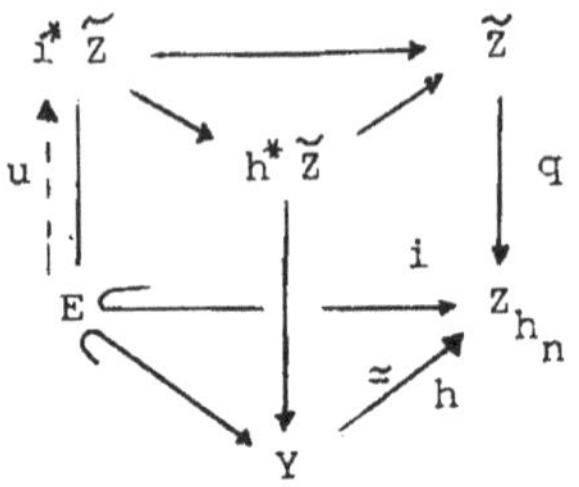

h is a CW-model under E as defined in (1.4.6). The map fh_n' defines a section u of the fibration i^*q. The fiber F_n' of q is such that $\pi_i(F_n') = 0$ for $i > n$. Since $Y - E$ has cells only in dimensions $> (n + 1)$, the section u can be extended over Y, and any two such section extensions are section homotopic, see (4.2.9). The section homotopy set $\langle Y, i^*\tilde{Z}\rangle^u = \{\bar{u}\}$ therefore contains exactly one element. Since h is a homotopy equivalence relative E, it induces a bijection (see (1.1.6))

$$\langle Y, i^*\tilde{Z}\rangle^u = \langle Z_{h_n}, \tilde{Z}\rangle^u.$$

Thus $\langle Z_{h_n}, \tilde{Z}\rangle^u$ contains exactly one element, which fact is equivalent to (5.3.5). $\square$

(5.3) Appendix: Principal approximation of $K(G, n)$-fibrations.

We will use the Postnikov decomposition to show that nilpotent and complete fibrations $E \to B$ are Postnikov spaces over B. This yields the proof of (1.5.12) that we deferred earlier.

(5.3.9) Theorem: Let $p : E \to B$ be a $K(G, k)$-fibration, $k \geq 1$. Let $\Phi = G^p$ denote the operation of $\pi_1(E)$ on G, and suppose

$$\ldots \subset \Gamma^n \subset \Gamma^{n-1} \subset \ldots \subset \Gamma \subset G = \Gamma^0$$

are Φ-invariant subgroups of G such that Γ^n is normal in Γ^{n-1}, $G_n = \Gamma^{n-1}/\Gamma^n$ is abelian and the operation of $\pi_1(E)$ on G_n induced *by* Φ *is trivial. Then there exist fibrations* q_n *and maps* h_n *for which*

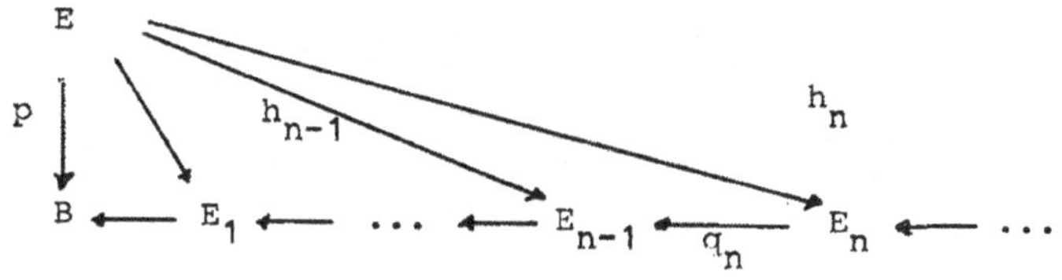

commutes and

(i) q_n *is a principal* $K(G_n, k)$*-fibration*

(ii) $p_n = q_1 \cdots q_n$ *is a* $K(G/\Gamma^n, k)$*-fibration, and* h_n *induces on the homotopy of the fibers of* p *and* p_n *the quotient map* $G \to G/\Gamma^n$.

Such a filtration Γ^n is the central series $\Gamma^n_\Phi G$ of (1.5.10). Recalling the definitions in (1.5.11), we can now prove the

(5.3.10) *Corollary: Suppose* $p : E \to B$ *is a* $K(G, k)$*-fibration with* $k > 1$. *If* p *is nilpotent, then it has a finite principal refinement. That is, there exist abelian groups* G_i *and principal* $K(G_i, k)$*-fibrations* q_i *such that*

$$p = q_n \cdots q_1 : E = E_n \xrightarrow[q_n]{} E_{n-1} \to \cdots \to E_1 \to B$$

If p *is complete,* E *is a Postnikov space over* B.

Proof of the corollary: In (5.3.9) we set $\Gamma^n = \Gamma^n_\Phi G$. If p is a nilpotent fibration with nil $(p) = n$, then the map h_n in (5.3.9) is a homotopy equivalence over B. If p is a complete fibration, then $\varprojlim h_n : E \to \varprojlim E_n$ is a weak homotopy equivalence, as follows from (1.3.33) and the remark after it, and so p is a Postnikov space over B. $\square$

The prece ding theorem is proved by induction using the naturality of $K(G, n)$-fibrations as expressed in the

(5.3.11) Lemma: Let X be a path-connected CW-space and $\bar{f} : \tilde{G} \to \tilde{H}$ a homomorphism of local groups in X. Then to every $\xi \in H^{k+1}(X, \tilde{G})$ there is a map

$$f : E_\xi(\tilde{G}, k) \longrightarrow E_{\bar{f}_* \xi}(\tilde{H}, k)$$

over X that on the fibers $K(G, n) \to K(H, n)$ is given by $\bar{f}$.

Proof of (5.3.9): We prove the theorem only for abelian G. We construct h_1 and q_1 as follows. Suppose $\bar{h}_1 : G \to G/\Gamma$ is the quotient map and $q_1 : E_1 = E_\xi(G/\Gamma, k) \to B$ is induced by $\xi = (\bar{h}_1)_* \bar{c}(p)$. The map h_1 exists by the preceding lemma. By regarding h_1 as a $K(\Gamma, k)$-fibration, we obtain q_2 and h_2 similarly, where $h_2 : \Gamma \to \Gamma/\Gamma^2$ is the quotient map. Iterating this procedure completes the proof. If G is not abelian and $n = 1$, the methods used to prove 2.9 on p. 65 of [49] yield a proof here. ☐

(5.3.11) Proof of (1.5.12): It follows directly from (5.3.10) and the Postnikov decomposition (5.3.2) that (1.5.12) holds, in view of the next lemma.

(5.3.12) Lemma: If $\{E_n, q_n, h_n\}$ is a Postnikov decomposition of the fibration $p : E \to B$, then $h = \varprojlim h_n : E \to \varprojlim E_n$ is a weak homotopy equivalence.

This is proved in analogy with (1.3.33).

(5.4) Cohomology obstructions of higher order

These are obstructions to the factorability of certain diagrams. Two sets of criteria for factorability, expressed in terms of CW-decomposition and then of Postnikov decomposition, will be shown to be equivalent. We will see that higher-order cohomology obstructions are natural for maps. A particular case of naturality yields the formulas of Olum [95], which in this way become especially easy to understand.

Suppose

(A)

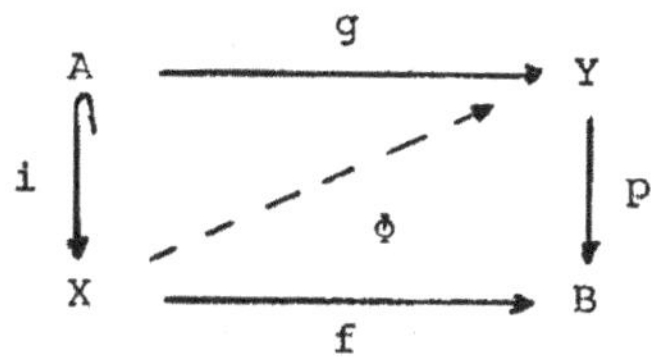

is a commutative diagram of path-connected CW-spaces, with i a closed cofibration and p a fibration with path-connected fiber F. Our initial considerations in (1.1) were centered around such a diagram. We now give a complete list of algebraic obstructions to the existence of a factoring map ϕ.

The first obstruction depends on whether the fundamental groups factor correspondingly, that is whether there exists a homomorphism θ extending the diagram

(B)

$$
\begin{array}{ccc}
\pi_1(A) & \xrightarrow{\ g_*\ } & \pi_1(Y) \\
{\scriptstyle i_*}\big\downarrow & {\theta}\nearrow & \big\downarrow{\scriptstyle p_*} \\
\pi_1(X) & \xrightarrow[\ f_*\ ]{} & \pi_1(B)
\end{array}
$$

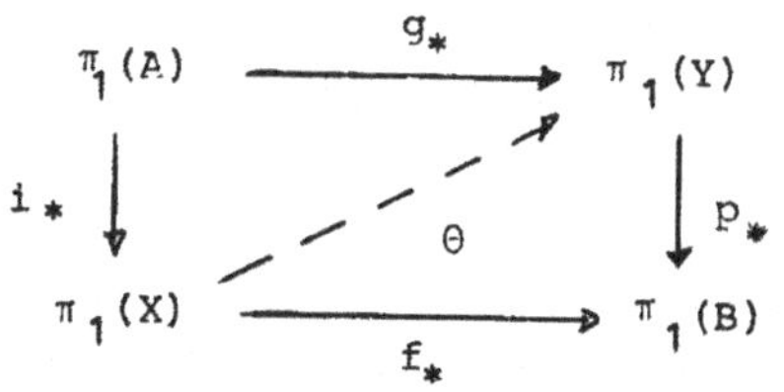

commutatively. We call a factoring map ϕ in (A) __compatible with__ θ when $\phi_* = \theta$. As in (1.1.2) and (1.1.3), we consider now commutative diagrams

(5.4.1)

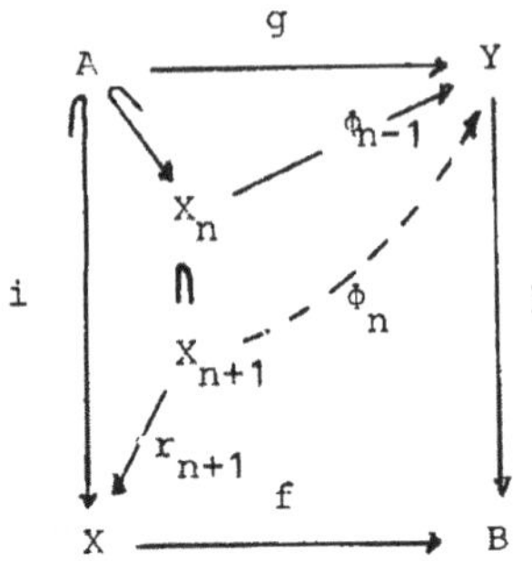

(C)

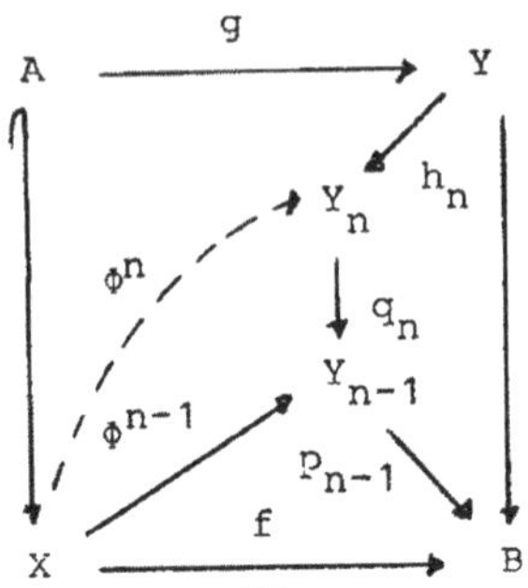

(D)

X_n is the n-skeleton of a CW-model $\overline{X} \xrightarrow{r} X$ of (X, A) (which we can take to be strictly pointed), see (1.4.5)), and r_{n+1} is the restriction of r. Suppose that $\{Y_n, q_n, h_n\}$ is a Postnikov decomposition of $Y \to B$ as in (5.3.1). For $n \geqslant 1$ we call ϕ_n in (C) and ϕ^n in (D) <u>compatible with</u> Θ when respectively

$$\Theta = \phi_{n*}(r_{n+1\,*})^{-1} \quad : \quad \pi_1(X) \cong \pi_1(X_{n+1}) \to \pi_1(Y),$$

and $$\Theta = (h_{n*})^{-1}\phi^n_{\;*} \quad : \quad \pi_1(X) \to \pi_1(Y_n) \cong \pi_1(Y).$$

For these two cases we can define the sets

$$[X_{n+1}, Y]^A_{B,\Theta}, \qquad [X, Y_n]^A_{B,\Theta}$$

of homotopy classes of maps ϕ_n respectively ϕ^n under A and over B that are compatible with Θ. Let $j : [X_{n+1}, Y]^A_B \to [X_n, Y]^A_B$ be the restriction. There is a canonical bijection

$$\text{(5.4.1)} \qquad \lambda : j[X_{n+1}, Y]^A_{B,\Theta} \approx [X, Y_n]^A_{B,\Theta} \quad .$$

<u>Proof</u>: With $j\,\phi_n = \phi_n|X_{n-1} = \phi_{n-1}$ in diagram (C), the map $h_n\,\phi_{n-1}$ can be extended over X, and such an extension represents $\lambda(j\phi_n)$. Conversely, a map $\phi^n|X_{n+1}$ in diagram (D) can be lifted to ϕ_n with $h_n\,\phi_n = \phi^n|X_{n+1}$. We have used (1.4.13) and (1.4.14) in these claims. ☐

(5.4.2) Definition: If $p : Y \to B$ is a fibration with path-connected fiber F, then for $n \geq 1$ the group operation $\pi_n(F)^p$ of $\pi_1(Y)$ on $\pi_n(F)$ is defined as in (4.1.15) Consider the exact homotopy sequence

$$\pi_2(Y) \xrightarrow{p_*} \pi_2(B) \xrightarrow{\delta} \pi_1(F) \xrightarrow{i_*} \pi_1(Y) \xrightarrow{p_*} \pi_1(B) \longrightarrow 0$$

The abelian subgroup $\underline{\pi}_1(F) = \delta\pi_2 B = \operatorname{Ker} i_*$ of $\pi_1(F)$ is invariant under the group operation of $\pi_1(Y)$ and lies in the center of $\pi_1(F)$. We therefore can define an operation of $\pi_1(B)$ on $\underline{\pi}_1(F)$ in the manner of (4.1.22). This local group in B we will denote by $\underline{\pi}_1(F)_p$. In short it is defined by the condition $p^*(\underline{\pi}_1(F)_p) = \underline{\pi}_1(F)^p \subset \pi_1(F)^p$.

A further necessary condition for the existence of factoring maps Φ_n and Φ^n for $n \geq 1$ is that

(E) $\quad f_* \pi_2(X) \subset p_* \pi_2(Y) \subset \pi_2(B)$

because $\pi_2(X_2) \to \pi_2(X)$ is surjective. Sufficient conditions are expressed in

(5.4.3) Theorem: Suppose (f, g) are as in diagram (A) and that we are given a homomorphism Θ extending the diagram (B) of fundamental groups, and suppose (E) holds. Then we can construct

$$\mathcal{O}^1_\Theta(f, g) \in H^2(X, A; f^* \underline{\pi}_1(F)_p)$$

and $\quad \mathcal{O}^n_\Theta(f, g) \subset H^{n+1}(X, A; \Theta^* \pi_n(F)^p)$ for $n \geq 2$,

such that for $n \geq 1$ the following statements are equivalent.

1) $0 \in \mathcal{O}^n_\Theta(f, g)$.

2) $\mathcal{O}^{n+1}_\Theta(f, g)$ is not empty.

3) In diagram (C) there exists an extension Φ_n compatible with Θ.

4) In diagram (D) there exists an extension Φ^n compatible with Θ.

If statement 3) or 4) holds, then we call diagram (A) <u>(Θ, n)-factorable</u>. This property is thus independent of the choice of a CW-model $\bar{X} \to X$ in diagram (C). The subset $\mathcal{O}^n_\Theta(f, g)$ is called the <u>n-th order cohomolgoy obstruction</u> to factoring diagram (A).

(5.4.4) <u>Construction of $\mathcal{O}^1_\Theta(f, g)$</u> : In diagram (A) we form the induced fibrations $p' : f^* Y = \tilde{X} \to X$ and $i^*\tilde{X} = \tilde{A} \to A$. We regard the map g as a section $A \to \tilde{A}$. We now have

(1)

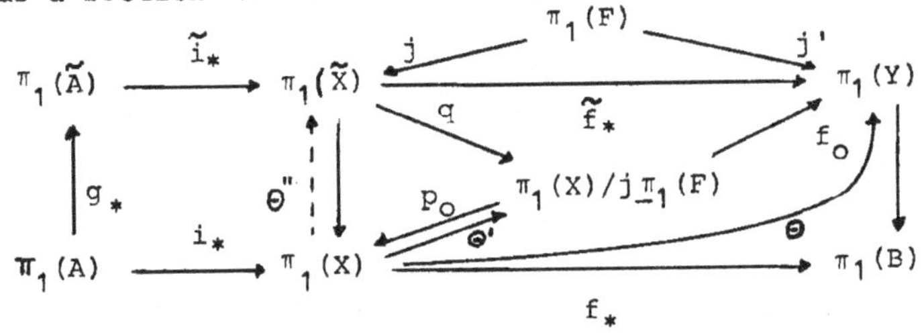

where $\tilde{i}$ and $\tilde{f}$ are induced from i and f, and j, j' are induced from the inclusion of the fiber F. Condition (E) $f_* \pi_2 X \subset p_* \pi_2 Y$ is equivalent to j being injective. It now follows from (4.3.1) that diagram (A) is $(\Theta, 1)$-factorable if and only if there exists a splitting Θ'' of p'_* such that

$$(2) \qquad \Theta'' i_* = \tilde{i}_* g_* \quad \text{and} \quad \tilde{f}_* \Theta'' = \Theta .$$

We will derive $\mathcal{O}^1_\Theta(f, g)$ as the obstruction to the existence of such a splitting. $\tilde{f}_*$ and p'_* factor over the quotient map q of the quotient $\pi = \pi_1(\tilde{X})/j\pi_1(F) = \pi_1(\tilde{X})/j \operatorname{Ker} j'$, giving us the homomorphisms f_o and p_o in (1). The homomorphism Θ defines a canonical splitting Θ' of p_o such that $f_o\Theta' = \Theta$.
(Definition of Θ': Let $\alpha \in \pi_1(X)$. Since $\pi_o(F) = 0$ there is an $\alpha' \in \pi_1(\tilde{X})$ with $p'_*(\alpha') = \alpha$. Therefore $-\Theta(\alpha) + \tilde{f}_*(\alpha') = j'\alpha''$ for some $\alpha'' \in \pi_1(F)$. The class $\Theta'\alpha$ is represented by $\alpha' - j\alpha''$.)

From (1) we extract the diagrams

$$(3)\qquad \begin{array}{ccc} \pi_1(A) & \xrightarrow{\tilde{i}_* g_*} & \pi_1(\tilde{X}) \\ {\scriptstyle i_*}\downarrow & \overset{\Theta''}{\nearrow} & \downarrow{\scriptstyle q} \\ \pi_1(X) & \xrightarrow[\Theta']{} & \pi \end{array} \qquad\qquad \begin{array}{ccc} K(\pi_1 A,\ 1) & \xrightarrow{\underline{\tilde{i}_* g_*}} & K(\pi_1\tilde{X},1) \\ {\scriptstyle \underline{i}_*}\downarrow & \overset{\underline{\Theta}''}{\nearrow} & \downarrow{\scriptstyle \underline{q}} \\ K(\pi_1 X,\ 1) & \xrightarrow[\underline{\Theta}']{} & K(\pi,\ 1). \end{array}$$

The right-hand one is the Eilenberg-Mac Lane guise of the one to the left. We can assume that $\underline{i}_*$ is a closed cofibration and $\underline{q}$ is a fibration.

A Θ'' that splits p_*' and satisfies (2) exists exactly when (3) factors in the indicated way. Such a factoring map Θ'' exists exactly when an extension $\underline{\Theta}''$ exists. The sole obstruction to the existence of $\underline{\Theta}''$ is, by (4.3.6), the element

$$(4)\qquad \tilde{c}(u) \in H^2(K(\pi_1 X,\ 1),\ K(\pi_1 A,\ 1);\quad f^*\ \underline{\pi}_1(F)_p) = H^2.$$

u is the section $\underline{\tilde{i}_* g_*}$ defined on the base space $K(\pi_1 A,\ 1)$ of the induced fibration $(\underline{\Theta}')^*\underline{q}$.

The map $h : (X,\ A) \to (K(\pi_1 X,\ 1),\ K(\pi_1 A,\ 1))$, which is an isomorphism on the fundamental groups, induces an injection $h^* : H^2 \to H^2(X,\ A;\ f^*\underline{\pi}_1(F)_p)$ in cohomology, in view of (1.4.14) and (5.2.4). We now set

$$(5)\qquad \mathcal{O}^1_\Theta(f,\ g) \;=\; h^*\tilde{c}(u).$$

It is clear that this element makes the statement of theorem (5.4.3) for $n = 1$ valid.

(5.4.5) <u>Construction of</u> $\mathcal{O}^n_\Theta(f,g)$ <u>for</u> $n > 2$: We may assume that $r : \bar{X} \to X$ is the identity, thus simplifying the presentation. Consider the diagrams

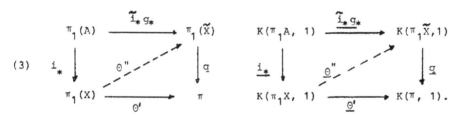

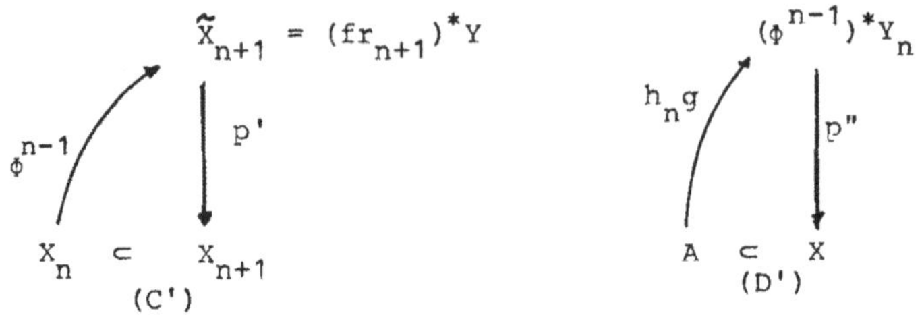

induced by the diagrams (C) and (D). We can regard Φ_{n-1} and $h_n g$ as partial sections of the induced fibrations p' and p" respectively. The local groups defined by (C') and (D') are such that

(1) $\quad \Phi_{n-1} \bullet \pi_n(F) = \Theta^*\pi_n(F)^p = \pi_n(F)_{p''}$

Therefore the cohomology classes

(2) $\quad c'(j\Phi_{n-1}) = \underline{c}(\Phi_{n-1}), \quad c''(\phi^{n-1}) = \tilde{c}(h_n g) \in H^{n-1}(X, A, \Theta^*\pi_n(F)^p)$

are defined, see (4.2.13) and (4.3.7). We have $c'(j\Phi_{n-1}) = 0$ exactly when there exists in (C) an extension Φ_n satisfying $\Phi_n | X_{n-1} = \Phi_{n-1} | X_{n-1} = j\Phi_{n-1}$. We have $c''(\phi^{n-1}) = 0$ exactly when there exists in (D) an extension ϕ^n satisfying $q_n \phi^n = \phi^{n-1}$. It is easy to see from the definitions that

(3) $\quad$ if $\lambda(j\,\Phi_{n-1}) = \phi^{n-1}$ then $c'(j\,\Phi_{n-1}) = c''(\phi^{n-1})$,

where λ is the bijection of (5.4.1). We now define

$$\mathcal{O}^n_\Theta(f, g) = \left\{ c'(j\Phi_{n-1}) \mid \Phi_{n-1} \in [X_n, Y]^A_{B,\Theta} \right\}$$
$$= \left\{ c''(\phi^{n-1}) \mid \phi^{n-1} \in [X, Y_{n-1}]^A_{B,\Theta} \right\}$$

It is clear from these definitions that $\mathcal{O}^n_\Theta(f, g)$ makes the statement of (5.4.3) for $n \geq 2$ valid.

Regarding $\mathcal{O}^n_\Theta(f, g)$ as a higher-order obstruction to the existence of sections of $f^*Y \to X$, we see that it has similar properties to those expressed in (4.4.11).

<u>Remark</u>: In [95] Olum also defines the obstructions $\mathcal{O}^n_\Theta(f, g)$ for $n \geqslant 2$. As we have seen, they can be derived from the classical obstruction theorem (4.2.13) for sections. Our $\mathcal{O}^1_\Theta(f, g)$ construction is somewhat less direct than that for the higher dimensions, and does not seem to occur in the literature. The resulting cohomology criterion for when $\mathcal{O}^2_\Theta(f, g)$ is not empty, that Olum suppresses as "elementary", is in fact less easy to derive than the higher-dimensional ones. Only with this criterion in hand is it possible to set up a complete list of algebraic obstructions to the existence of a factoring map Φ in diagram (A). Eckmann and Hilton also show in [27] that cohomology obstructions for Postnikov and CW-decompositions are the same, i.e. statement (3) in (5.4.5).

We will now discuss naturality of the cohomology obstructions $\mathcal{O}^n_\Theta(f, g)$. Consider the commutative diagram

$$\begin{array}{ccccccc}
A' & \xrightarrow{\;w'\;} & A & \xrightarrow{\;g\;} & Y & \xrightarrow{\;v\;} & Y' \\
\downarrow{\scriptstyle i'} & & \downarrow{\scriptstyle i} & \overset{\Phi}{\nearrow} & \downarrow{\scriptstyle p} & & \downarrow{\scriptstyle p'} \\
X' & \xrightarrow{\;w\;} & X & \xrightarrow{\;f\;} & B & \xrightarrow{\;v'\;} & B'
\end{array}$$

of path-connected CW-spaces, with cofibrations i, i' and fibrations p, p' as in diagram (A). If there is an extension Φ for (f, g), then $v\Phi$ is an extension for $(v'f, vg)$ and Φw is an extension for (fw, gw'). Thus we expect the corresponding cohomology obstructions to stand in some relation to each other. Let Θ be as in (B). The restriction of v to the fibers $v : F \to F'$ induces the coefficient homomorphism $v_* : \Theta^*\pi_n(F)^p \to (v_*\Theta)^*\pi_n(F')^{p'}$ for $n \geqslant 2$ and v_* :

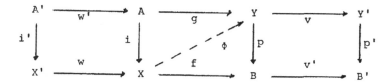

$f^* \underline{\pi}_1(F)_p \to (v'f)^* \underline{\pi}_1(F')_{p'}$ for $n = 1$ between local groups over X. The pair of maps $w = (w, w')$ induces a cohomology homomorphism w^* for which $(n \geq 1)$

(5.4.6) <u>Theorem:</u> $w^* \mathcal{O}^n_\Theta(f, g) \subset \mathcal{O}^n_{\Theta w_*}(fw, gw')$,

$$(v_\#)_* \mathcal{O}^n_\Theta(f, g) \subset \mathcal{O}^n_{v_* \Theta}(v'f, vg).$$

When $n = 1$ we assume condition (E) holds, so that $\mathcal{O}^1_\Theta(f, g)$ is defined. This theorem expresses naturality. It is an easy consequence of the naturality of the corresponding cohomology obstructions of (4.2.13).

(5.4.7) <u>Definition:</u> Let $\tilde{G}$ be a local group in X. We associate to a cohomology class $c \in H^{n+1}(X, A; \tilde{G})$ the homomorphism

$$c_\# : \pi_{n+1}(X, A) \to G$$

defined by $c_\#(\eta) = \eta^* c$, where $\eta^* : H^{n+1}(X, A; \tilde{G}) \to H^{n+1}(E^{n+1}, S^n; G) = G$ is induced by η.

As a particular case we regard $c \in H^{n+1}(Z_p, Y; \tilde{G})$, where Z_p is the mapping cylinder of the fibration $p : Y \to B$ and $\tilde{G}$ is a local group in $B \simeq Z_p$. Then $\pi_n(F) = \pi_{n+1}(Z_p, Y) \xrightarrow{c_\#} G$ is in fact a coefficient homomorphism $c_\# : \pi_n(F)^p \to p^*\tilde{G}$ of local groups in Y.

The pair of maps (f, g) in diagram (A) defines a map $f^o : F_i \to F$ between the fibers of i and p that induces

$$(f, g)_\# : \pi_{n+1}(X, A) = \pi_n(F_i) \xrightarrow{f^o_*} \pi_n(F).$$

In addition, (f, g) defines a map $(X, A) \to (Z_p, Y)$ of pairs and induces

$$(f, g)^* : H^{n+1}(Z_p, Y; \tilde{G}) \longrightarrow H^{n+1}(X, A; f^*\tilde{G}).$$

With this notation, we can derive the formulas of Olum

(5.4.8) *Theorem: With* (f, g) *as in diagram (A) let* Θ *be given as in (B), and suppose* $n \geq 1$ *and* $\mathcal{O}^n_\Theta(f, g)$ *is not empty. Let* $\tilde{G}$ *be a local group over* B. *Then*

1) *For any* $c \in H^{n+1}(Z_p, Y; \tilde{G})$ *we have*

$$(f, g)^*c = (c_\#)_*(\mathcal{O}^n_\Theta(f, g))$$

where $c_\# : \Theta^*\pi_n(F)^p \longrightarrow \Theta^*p^*\tilde{G} = f^*\tilde{G} \qquad (n \geq 2)$
is the coefficient homomorphism associated to c.

2) *For any* $\eta \in \pi_{n+1}(X, A)$ *we have*

$$(f, g)_\# \eta = \eta^*(\mathcal{O}^n_\Theta(f, g))$$

and in fact $\eta^*\mathcal{O}^n_\Theta(f, g) = c_\#(\eta)$ *for any* $c \in \mathcal{O}^n_\Theta(f, g)$.

Proof: The formulas follow easily from naturality of the obstructions $\mathcal{O}^n_\Theta(f, g)$. Consider the diagram

$$\begin{array}{ccccccc}
S^n & \xrightarrow{w'} & A & \xrightarrow{g} & Y & \xrightarrow{v} & E_{jc}(\tilde{G},n) \\
\downarrow & & \downarrow & & \downarrow p & & \downarrow p' \\
E^{n+1} & \xrightarrow{w} & X & \xrightarrow{f} & B & \xrightarrow{1} & B
\end{array}$$

where $(w, w') \in \eta$, and the $K(G, n)$-fibration p' is derived from $j(c) \in H^{n+1}(B, \tilde{G})$ as in (5.2.7). There is a section v of p' and $\tilde{c}(v) = c$ and

$$v_\# = c_\# : \Theta^*\pi_n(F)^p \longrightarrow (v_*\Theta)^* p'^*\tilde{G} = f^*\tilde{G}$$

It follows from (5.4.6) that

$$\eta^*(\mathcal{O}^n_\Theta(f, g)) \subset \mathcal{O}^n_{\Theta w_*}(fw, gw') = (f, g)_\#(\eta)$$

$$(c_\#)_*(\mathcal{O}^n_\Theta(f, g)) \subset \mathcal{O}^n_{v_*\Theta}(f, vg) = (f, g)^*(c)$$

Since we assumed that $\mathcal{O}^n_\Theta(f, g)$ is not empty, (5.4.8) follows. (The map v is not uniquely defined by c. However, c determines up to homotopy the restriction of v to the fibers.) ▭

CHAPTER 6 : ITERATED PRINCIPAL FIBRATIONS

Until Chapter 3 we always endeavored to develop both approaches to obstruction theory - through principal fibrations and through principal cofibrations - along parallel lines where this was possible. This procedure was interrupted for chapters 4 and 5, in which we took up the important cases of CW-complexes and Postnikov towers as treated by classical obstruction theory. In this chapter we recur to our parallel development by deriving the dual statements to chapter 3. As before, we will attempt to maintain duality in the definitions as well as in the methods of proof.

(6.1) The partial loop operation.

We here discuss various properties of a partial loop construction that generalizes the usual loop functor Ω. The important fact will be proved that the partial loop of a cup product map is again a cup product map. In the appendix we construct in analogy with the Steenrod algebra, the algebra of partially stable cohomology operations. It turns out that this algebra is just the Massey-Peterson algebra. The dual presentation is reflected in the notation (6.1.i) for the item corresponding to (3.1.i), $i \geq 1$.

Let $\sigma = (0, 1) : Y \subset B \times Y$ be the inclusion with $\sigma(y) = (*, y)$. We say that $\xi \in [B \times Y, A]$ is trivial on Y when $\sigma^*\xi = 0$. We write $[B \times Y, A]_2$ for the set of elements trivial on Y. It can be seen from the cofiber sequence of σ that $[B \times Y, A]_2 = [B \rtimes Y, A]$, where A is an H-group and $B \rtimes Y = B \times Y/\{*\} \times Y$. Let PB be the path space. The commutative diagram

$$\begin{array}{ccccc}
PB \times Y & \xrightarrow{p_o} & B \times Y & \longrightarrow & C(p_o) \\
\uparrow\cup & & \uparrow \sigma & & \uparrow \lambda_o \\
\Omega B \times Y & \xrightarrow{\rho} & Y & \longrightarrow & C(\rho)
\end{array}$$

in which $p_o(\sigma, y) = (\sigma(0), y)$ and ρ is the projection, gives us a map λ_o between mapping cones. If A is an H-group, we obtain the homomorphisms

$$\begin{array}{ccccc}
 & & [C(p_o), A] & \xrightarrow[\cong]{\partial} & [B \times Y, A]_2 \\
 & & \downarrow \lambda_o^* & & \\
[\Omega B \times Y, \Omega A]_2 & \xrightarrow[\cong]{j} & [C(\rho), A] & &
\end{array}$$

where the isomorphisms ∂ and j arise in the corresponding cofiber sequences. We call the homomorphism

(6.1.1) $\quad L : [B \times Y, A]_2 \longrightarrow [\Omega B \times Y, \Omega A]_2$ with $L = j^{-1} \lambda_o^* \partial^{-1}$

the partial loop. With $\xi : B \times Y \to A$ we obtain a representative of $L\xi$ as follows. If $\xi(Y) = *$ we set

$$(L\xi)(\sigma, y)(t) = \xi(\sigma(t), y) \quad \text{for } t \in I, \ \sigma \in \Omega B, \ y \in Y.$$

If $\xi(Y) \neq *$, there exists a null homotopy $\bar{\xi} : Y \to PA$ lifting $\xi\sigma$ Let $\rho : \Omega B \times Y \to Y$ be the projection. Then

$$H_\xi : \Omega B \times Y \subset (B \times Y)^I \xrightarrow{\xi^{id}} A^I$$

is a self homotopy of $\xi\sigma\rho$ and so, as in (3.1.2), we see that

(6.1.2) $\quad L\xi = d(\bar{\xi}\rho, H_\xi, \bar{\xi}\rho) \in [\Omega B \times Y, \Omega A]_2$

is the difference (1.3.5).

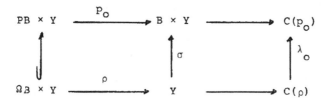

The partial loop construction appears often in the literature, for instance p. 501 [55], p. 205 [79], [106]. It generalizes the usual loop operation Ω, specifically the diagram

$$(6.1.3)\qquad \begin{array}{ccc} [B \times Y, A]_2 & \xrightarrow{L} & [\Omega B \times Y, \Omega A]_2 \\ \uparrow \rho_1^* & & \uparrow \rho_1^* \\ [B, A] & \xrightarrow{\Omega} & [\Omega B, \Omega A] \end{array}$$

commutes. ρ_1 is the projection. Thus $L = \Omega$ when $Y = *$. The following property of L generalizes a familiar one of Ω.

(6.1.4) Theorem: If B is b-connected, and $\pi_j(A) = 0$ for $j \geq m_A$, then

$$L : [B \times Y, A]_2 \to [\Omega B \times Y, \Omega A]_2 \text{ is } \begin{cases} \text{isomorphic for } m_A \leq 2b+2 \\ \text{monomorphic for } m_A \leq 2b+3. \end{cases}$$

We will prove this theorem in the more general form (6.4.7). By iterating we obtain the homomorphism

$$L^n : [B \times Y, A]_2 \longrightarrow [\Omega B \times Y, \Omega^n A]_2$$

natural in B, Y and A. That is, if $u : Y \to Y'$, $v : B \to B'$ and $w : A' \to A$ then we have

$$(6.1.5)\qquad \begin{cases} (\Omega^n v \times u)^* \circ L^n = L^n \circ (v \times u)^* \\ (\Omega^n w)^* L^n = L^n w^* \end{cases}$$

There is another kind of naturality. Suppose A and B are H-groups. If $\xi \in [D \times Y, A]_2$ and $k : Y \to X$, we have the diagram

$$\begin{array}{ccc} [D \times Y, B]_2 & \xleftarrow{(\xi, k)^*} & [A \times X, B]_2 \\ \downarrow L & & \downarrow L \\ [\Omega D \times Y, \Omega B]_2 & \xleftarrow{(L\xi, k)^*} & [\Omega A \times X, \Omega B]_2 \end{array}$$

It is commutative, because dually to (3.1.6) we have

(6.1.6) Theorem: $L(\eta) \circ (L\xi, k) = L(\eta \circ (\xi, k))$ for $\eta \in [A \times X, B]_2$.

Suppose still that A is an H-group. Then A^D and $A^{\wedge D}$ are also H-groups if D is compact hausdorff, see (0.1.8). If ξ is trivial on Y, then there exists up to homotopy exactly one map $\xi^{\hat{\wedge} D}$ extending the diagram

$$\begin{array}{ccc} A^D & \xleftarrow{\xi^D} & B^D \times Y^D \\ i \Big\uparrow & & \Big\uparrow i \times c \\ A^{\wedge D} & \xleftarrow{\xi^{\hat{\wedge} D}} & B^{\wedge D} \times Y . \end{array}$$

This yields the homomorphism

(6.1.7) $$[B \times Y, A]_2 \longrightarrow [B^{\wedge D} \times Y, A]_2 : \quad \xi \to \xi^{\hat{\wedge} D}$$

It is easy to see that

(6.1.8) $$(\xi^{\hat{\wedge} D})^{\hat{\wedge} D'} = \xi^{\hat{\wedge}(D \wedge D')}$$

(6.1.9) $$L^n \xi = \xi^{\hat{\wedge} S^n}$$

Dually to (3.1.10), we can represent $L^n \xi$ by

(6.1.10) $$L^n \xi = (W\xi)^*(i \times \bar{\xi}) : \Omega^n B \times Y \to B^{S^n} \times_B P_\xi =$$
$$= \underline{\Omega}^n_Y (B \times Y) \times_{B \times Y} P_\xi \xrightarrow{W_\xi} \Omega^n A$$

where $\bar{\xi}: Y \to P_\xi$ lifts $Y \subset B \times Y$, and $W\xi$ is defined as in (2.5.6). When $n = 1$, this is just the representation of $L\xi$ in (6.1.2).

We will now show that the partial loop of a cup product map is also a cup product map. In a sense, this situation is dual to (3.1.1) for the Whitehead product map.

Let G and π be abelian groups. The Eilenberg-Zilber theorem gives us a well-defined element $a \cup b \in H^{m+n}(X, \pi \otimes G)$, the cup product of

$a \in H^m(X, \pi)$ and $b \in H^n(X; G)$. We also have the cross product $x \times y = p_1^*(x) \cup p_2^*(y) \in H^{m+n}(X \times Y, \pi \otimes G)$ of elements $x \in H^m(X; \pi)$ and $y \in H^n(Y; G)$, where p_1 and p_2 are the projections of $X \times Y$ onto X and Y, see for instance [116]. The inclusion $i : X \vee Y \subset X \times Y$ induces $i^*(x \times y) = 0$, therefore $x \times y \in \operatorname{Im} s^*$ where $s : X \times Y \to X \wedge Y$ is the identification map. We see from the exact sequence in (0.3) that s^* is injective, so the fundamental classes $i_m \in H^m(K(\pi, m), \pi)$ and $i_n \in H^n(K(G, n), G)$ give us a well-defined element

$$(\sigma^*)^{-1}(i_m \times i_n) \in H^{m+n}(K(\pi, m) \wedge K(G, n); \pi \otimes G) .$$

This element yields the cup product map

$$(6.1.11) \quad \cup : \quad K(\pi, m) \times K(G, n) \xrightarrow{s} K(\pi, m) \wedge K(G, n) \xrightarrow{\hat{\cup}} K(\pi \otimes G, m+n)$$

where $\hat{\cup} \in \gamma^{-1}(s^{*-1}(i_m \times i_n))$, see (0.5.5). This map induces the cup product, i. e. $\gamma^{-1}(a \cup b) = \cup_*(\gamma^{-1}(a), \gamma^{-1}(b))$. For the homotopy equivalence λ of (0.5.6) we then have

(6.1.12) <u>Theorem: The partial loop</u> $L\cup$ <u>of the cup product map is also a cup product map. That is</u>

$$\begin{array}{ccc} \Omega K(\pi, m) \times K(G, n) & \xrightarrow{L\cup} & \Omega K(\pi \otimes G, m+n) \\ \downarrow \lambda \times 1 & & \downarrow \lambda \\ K(\pi, m-1) \times K(g, n) & \xrightarrow{\cup} & K(\pi \otimes G, m+n-1) \end{array}$$

<u>is homotopy commutative</u>.

This was proved by Rutter in 2.2, 2.3 of [106].

<u>Proof:</u> Write $K_m = K(\pi, m)$, and let

$$S^* : \tilde{H}^{m-1}(X, \pi) \overset{\delta}{\cong} H^m(CX, X; \pi) \overset{\pi_0^*}{\cong} \tilde{H}^m(SX; \pi)$$

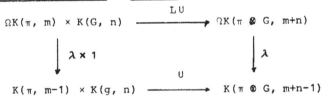

be the suspension homomorphism. The diagram with coefficients in π

$$\begin{array}{ccc}
\tilde{H}^{n-1}(X) & \xrightarrow{\delta} & H^n(CX, X) \\
{\scriptstyle a \mapsto i_1 \times a}\downarrow & \searrow^{s^*} & \uparrow \pi_0^* \\
H^n(S^1 \times X) & \xleftarrow{s^*} & H^n(SX)
\end{array}$$

is commutative, where $i_1 \in H^1(S^1, Z)$ is the fundamental class of S^1 and $n \geq 1$. Therefore for $a \in H^{n-1}(X)$ we have

(1) $$s^*s^*(a) = i_1 \times a \quad .$$

This follows from 6 p. 250 of [116], for example. If $\hat{U} : S^1 \wedge K_{n-1} \to K_n$ is the cup product map, then by (1)

(2) $$\gamma(\hat{U}) = s^*(i_{n-1}) \in H^n(S^1 \wedge K_{n-1}).$$

Let $R : S\Omega K_n \to K_n$ be the evaluation map. Then

(3) $$\begin{array}{ccc}
S^1 \times \Omega K_n & \xrightarrow{s} & S^1 \wedge \Omega K_n \\
\downarrow 1\times\lambda & & \downarrow R \\
S^1 \times K_{n-1} & \xrightarrow{U} & K_n
\end{array}$$

is homotopy commutative, since R induces s^*. Now we are in a position to prove (6.1.11). Given $U_{m,n} = U : K_m \times K_n \to K_{m+n}$ we can form the partial loop $L(U_{m,n})$ for which

(4) $$L(U_{m,n})(\sigma, x)(t) = U_{m,n}(\sigma(t), x), \quad \text{see (6.1.2)}.$$

This leads to the diagram

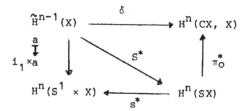

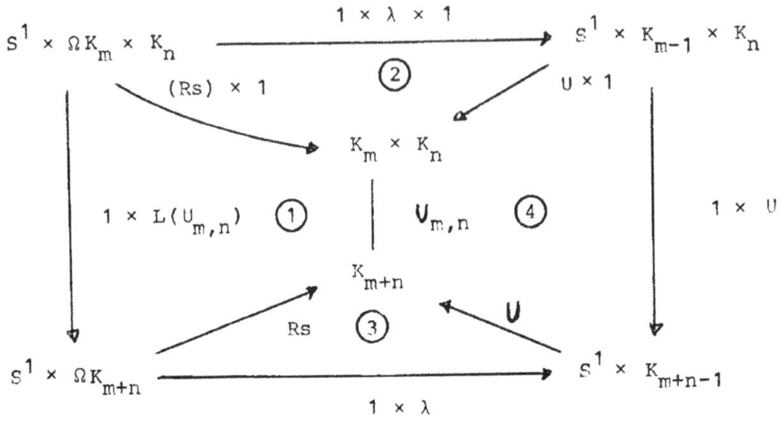

① commutes because of (4), ② and ③ are homotopy commutative by (3), and ④ is homotopy commutative because the $\cup$-product is associative. Therefore there is a homotopy

(5) $\quad \cup(1 \times \lambda)(1 \times L(\cup_{m,n})) \simeq \cup(1 \times \cup)(1 \times \lambda \times 1).$

This fact is equivalent to

(6) $\quad i_1 \times \gamma(\lambda \circ L(\cup_{m,n})) = i_1 \times (\gamma(\lambda) \times i_n)$

in $H^{m+n}(S^1 \times \Omega K \times K_n)$. Since $i_1 \times \; : a \mapsto i_1 \times a$ is injective by (1), it follows that

(7) $\quad \gamma(\lambda \circ L(\cup_{m,n})) = \gamma(\lambda) \times i_n$

which is equivalent to the statement of (6.1.12). ▭

(6.1) Appendix: The algebra of stable and partially stable cohomology operations, the Steenrod algebra and the Massey-Peterson algebra.

Definition: An Ω-spectrum $E = \{E_n, e_n\}$ is a sequence of CW-spaces E_n and homotopy equivalences $e_n : E_n \to \Omega E_{n+1}$, $n \geq 0$, where for all n sufficiently large E_n is path-connected.

If π is an abelian group, the spectrum $H\pi = \{K(\pi, n), \lambda\}$ of Eilenberg-MacLane spaces is an Ω-spectrum. An Ω-spectrum E yields a cohomology theory by setting $H^n(X, E) = [X, E_n]$ for any CW-space X. Such cohomology theories can be studied more extensively in Gray's book [38], for instance.

Given Ω-spectra Y, A let $\sigma_Y^*(A)$ denote the graded abelian group defined by

$$(6.1.13) \qquad \sigma_Y^q(A) = \varprojlim [Y_n, A_{n+q}]$$

where the inverse limit is taken over the loop maps

$$[Y_n, A_{n+q}] \xrightarrow{\Omega} [\Omega Y_n, \Omega A_{n+q}] \overset{\varepsilon}{=} [Y_{n-1}, A_{n-1+q}]$$

The isomorphisms $\varepsilon = (e_{n-1})^*(e^*_{n-1+q})^{-1}$ arise from the homotopy equivalences e_n of the spectra. The elements of $\sigma_Y^q(A)$ are <u>stable cohomology operations</u> of degree q for the cohomology theories $H^*(\ \ , Y)$ and $H^*(\ \ , A)$, see in this connection 27.4 on p. 295 of Gray [38].

In a similar manner, we define the graded abelian group for a space X by setting

$$(6.1.14) \qquad \sigma_Y^q(A; X) = \varprojlim [Y_n \times X, A_{n+q}]_2$$

where the inverse limit is taken over the partial looping maps

$$[Y_n \times X, A_{n+q}]_2 \xrightarrow{L} [\Omega Y_n \times X, \Omega A_{n+q}]_2 \overset{\varepsilon}{=} [Y_{n-1} \times X, A_{n-1+q}]_2.$$

As before, ε arises from the homotopy equivalences e_n. We will call the elements of $\sigma_Y^q(A, X)$ <u>partially stable cohomology operations.</u> Of course we have the inclusions $\sigma_Y^*(A) \subset \sigma_Y^*(A; X)$ and $\sigma_Y^*(A) = \sigma_Y^*(A; *)$.

Due to (6.1.6), partially stable cohomology operations can be composed. That is, for Ω-spectra A, B, D there is a bilinear pairing

$$(6.1.15) \qquad \circ : \sigma_B^q(A;\, X) \times \sigma_A^p(D,\, X) \to \sigma_B^{q+p}(D;\, X)$$

defined in analogy with (3.1.15). This pairing defines an associative multiplication on $\sigma_A^*(A;\, X)$. Thus $\sigma_A^*(A;\, X)$ is a graded algebra with subalgebras $\sigma_A^*(A)$ and $\sigma_A^0(A)$. We now consider some examples of such algebras.

Let p be a prime and HZ_p the Eilenberg-Mac Lane spectrum of Z_p. Then the algebra of stable cohomology operations

$$(6.1.16) \qquad \sigma_{HZ_p}^*(HZ_p) = A_p$$

for singular cohomology $H^*(\;\;, Z_p)$ is just the <u>Steenrod algebra</u> A_p, whose generators and relations are known completely, see for instance [121] or § 27, 28 in [38]. From the Steenrod algebra we can construct the <u>Massey-Peterson algebra</u> of a space X $A_p \odot H^*(X,\, Z_p)$ as follows, see [76].

(6.1.17) <u>Definition</u>: As Z_p- vector space, the <u>semi-tensor algebra</u> $A_p \odot H^*(X,\, Z_p)$ is just the tensor product $A_p \otimes H^*(X,\, Z_p)$. The multiplication is defined by the aid of the coproduct $\psi : A_p \to A_p \otimes A_p$ to be

$$(a \otimes d)(b \otimes e) = \sum_i (a_i'b) \otimes (a_i''(e) \cup d)$$

where $\psi(a) = \sum_i a_i' \otimes a_i''$. $a_i'b$ is multiplication in A_p, and $a_i''(e)$ is the operation of A_p from the left on $H^*(X,\, Z_p)$, and $\cup$ is the cup product in $H^*(X,\, Z_p)$. The coproduct $\psi(a)$ is characterized by the equation

$$a(x \cup y) = \sum_i a_i'(x) \cup a_i''(y)$$

for all $x, y \in H^*(X,\, Z_p)$.

(The tensor product $V \otimes W$ of graded vector spaces $V = \{V_q\}$ and $W = \{W_q\}$ is defined by $(V \otimes W)_q = \oplus_i V_i \otimes W_{q-i}$. Since A_p operates from the left on $H^*(X, Z_p)$ by composition, we put A_p on the left in the tensor product of (6.1.17).)

(6.1.18) <u>Theorem:</u> <u>There is a canonical algebra isomorphism</u>

$$\alpha : \sigma^*_{HZ_p}(HZ_p, X) \cong A_p \otimes H^*(X, Z_p).$$

The reader should compare this theorem with (3.1.18) and (3.1.25). To prove it we will need the following lemma. Let L^∞ be the canonical map on the inverse limit.

(6.1.19) <u>Lemma:</u> <u>For an Ω-spectrum</u> A,

$$L^\infty : \sigma^A_q(H\pi; X) \to [A_N \times X, K(\pi, N+q)]_2$$

<u>is an isomorphism for</u> N <u>sufficiently large.</u>

<u>Proof:</u> Since A_N is connected when N is large enough, there is a constant c such that A_N is $(N - c)$-connected. It follows from (6.1.4) that the statement of the lemma is valid for $N > q + 2c - 3$. ▭

<u>Proof of (6.1.18):</u> When N is large enough, the preceding lemma gives us the isomorphism (coefficients in Z_p, $K_n = K(Z_p, n)$)

$$\begin{aligned}\alpha : \sigma^q = \sigma^q_{HZ_p}(HZ_p, X) &= H^{N+q}(K_N \rtimes X) \\ &= \bigoplus_{q \geq j \geq 0} \tilde{H}^{N+j}(K_N) \otimes H^{q-j}(X) \\ &= \bigoplus_{q \geq j \geq 0} A^j_p \otimes H^{q-j}(X) \\ &= (A_p \otimes H^*(X))^q\end{aligned}$$

Since α is defined by the cup product, it follows from (6.1.12) that this isomorphism is independent of N. We will now show that α is an algebra isomorphism. Suppose $\xi \in \sigma^p$ and $\eta \in \sigma^q$ are represented by maps $\xi' : K_M \times X \to K_{M+p}$ and $\eta' : K_N \times X \to K_{N+q}$. Then $\xi \circ \eta$ is represented by the composition

$$\xi''(\eta'', 1) : K_{M+N} \times X \to K_{M+N+q} \times X \to K_{M+N+p+q}$$

where $\eta'' = (\varepsilon L)^M \eta'$ and $\xi'' = (\varepsilon L)^{N+q} \xi'$. It suffices to consider the case $\alpha(\xi) = a \otimes d$ and $\alpha(\eta) = b \otimes e$. Using the cup product maps of (6.1.11) and simplifying the notation, we have

$$\xi''(\eta'', 1) = (a(b \cup e)) \cup d = (\sum_i a_i'(b) \cup a_i''(e)) \cup d$$

and therefore

$$\alpha(\xi \cdot \eta) = \sum_i (a_i' b) \otimes (a_i''(e) \cup d). \quad \square$$

(6.2) Iterated principal fibrations and the derived spectral sequence for the classification of maps and retractions.

We will use the classification sequences $(A', \underline{A}')$ and $(B', \underline{B}')$ of (2.5) to construct a spectral sequence for iterated principal fibrations. This procedure also leads to a solution of the classification problem in (1.1). We describe the first differential of this spectral sequence in terms of the classifying maps and the partial loop operation. In an appendix we introduce a stable chain complex related to this spectral sequence. Our derivation of it throws light on McClendon's axiomatic development of "twisted cohomology operations of higher order". Certain classification theorems of James-Thomas and Nomura can be interpreted in terms of the spectral sequence.

Let A be an H-group. Given a map $\xi : B \times X \to A$ trivial on X, the partial loops $L^n\xi : \Omega^n B \times X \to \Omega^n A$ are defined. Given a space U, or a closed cofibration $i : X \subset \overline{X}$ with cofiber $j : X \to \overline{X}/X = F$ the $L^n\xi$ induce for $n \geq 0$ the products

$$[\ ,\ ]^n_\xi : [U, \Omega^n B] \times [U, X] \to [U, \Omega^n A], \quad [\beta, u]^n_\xi = (L^n\xi)^*(\beta, u)$$

(6.2.1)

$$\langle\ ,\ \rangle^n_\xi : [F, \Omega^n B] \times \langle \overline{X}, X\rangle \to [F, \Omega^n A], \quad \langle \beta, u\rangle^n_\xi = j^{*-1}[j^*\beta, u]^n_\xi$$

$j^* : [F, \Omega^n A] \to [X, \Omega A]$ is injective when a retraction $u \in \langle \overline{X}, X\rangle$ exists. The exactness of the cofiber sequence implies that $\langle \beta, u\rangle^n_\xi$ is well-defined. For fixed u, the maps $[\ , u]^n_\xi$ and $\langle\ , u\rangle^n_\xi$ are homomorphisms of abelian groups when $n \geq 1$. As before, we call $[\ ,\ ]^n_\xi$ the product, and $\langle\ ,\ \rangle^n_\xi$ the <u>twisted</u> product, induced by ξ. In (6.3) we will investigate the properties of these products more thoroughly.

Now suppose we are given a double path fibration

$$\begin{array}{ccccc} & & P_f & & \\ & & \downarrow & & \\ & & X = P_g & \xrightarrow{\ f\ } & A \\ & & \downarrow & & \\ Y & \leftarrow & X_o & \xrightarrow{\ g\ } & B \end{array}$$

where A is a co-H-group. We define an element Δf as follows. The projections $\rho_1 : \Omega B \times P_g \to \Omega B$ and $\rho_2 : \Omega B \times P_g \to P_g$ define $\rho_2 + \rho_1 : \Omega B \times P_g \to P_g$ which is the operation μ of (1.3.7) up to the order of B and P_g. The element

(6.2.2) $\quad \Delta f = -f_*(\rho_2) + f_*(\rho_2 + \rho_1) \in [\Omega B \times P_g, A]_2$

is then trivial on P_g. From Δf we can construct partial loopings $L^n(\Delta f) \in [\Omega^{n+1}B \times P_g, \Omega^n A]_2$, which induce operations as in (6.2.1) with $\xi = \Delta f$.

Let $u : U \to X = P_g$ be a map which can be lifted to P_f. More generally, let $P_f \subset \overline{P_f}$ be a closed cofibration with cofiber F, and suppose $u : \overline{P}_g \to P_g$ is a retraction of the induced cofibration that can be lifted to P_f. Then for $n \geq 1$ there are homomorphisms

(6.2.3)

$$\Delta^n(u,f) : [U, \Omega^{n+1}B] \xrightarrow{\gamma} [U, \underline{\Omega}^n_Y X]_u \xrightarrow{\underline{\Omega}^n f} [U, \Omega^n A]$$

(6.2.3)

$$\bar{\Delta}^n(u,f) : [F, \Omega^{n+1}B] \xrightarrow{\gamma} \langle \sigma_* \bar{X}, \underline{\Omega}^n_Y X \rangle_u \xrightarrow{\underline{\Omega}^n f} [F, \Omega^n A]$$

Here $\underline{\Omega}^n f = \underline{\Omega}^n(u,f)$ are homomorphisms as in (2.5.3) from the classification sequences $(\underline{A}')$, $(\underline{B}')$ for P_f. The homomorphisms $\gamma = (-1)^{n+1}u^+$ come from the classification sequences for P_g. For the operation defined in (6.2.1) we have

(6.2.4) <u>Theorem:</u> $\Delta^n(u, f)(\beta) = [\beta, u]^n_{\Delta f}$,

$$\bar{\Delta}^n(u, f)(\beta) = \langle \beta, u \rangle^n_{\Delta f}.$$

Although the proof of (6.2.4) is fairly complicated, it is strictly dual to the proof of (3.2.4). Thus we may entrust it to the reader, who should use (6.1.10) in place of (3.1.10), as well as the statement dual to (2.4.11). If $Y = X_o$ then γ in (6.2.3) is an isomorphism, as follows from (1.3.8) and (1.3.23). Thus we can use γ to replace the relevant groups in the classification sequences $(A', \underline{A}')$ and $(B', \underline{B}')$. The homomorphism $\underline{\Omega}^n f$ then becomes $\Delta^n(u, f)$ and $\bar{\Delta}^n(u, f)$ respectively. This results in two new classification sequences, which we will not write out. However, we use them to derive a classification theorem corresponding to (2.5.4).

(6.2.5) <u>Corollary:</u> <u>Let</u> u <u>be a lifting of</u> $v : U \to X_o$ <u>or of the retraction</u> $v : \bar{X}_o \to X_o$. <u>Then there are bijections</u>

$$[U, P_f]_v \approx \bigcup_{\beta} [U, \Omega A]/\mathrm{Im}\,[\ ,\ u + \beta]^1_{\Delta f}$$

$$\langle \bar{P}_f, P_f \rangle_v \approx \bigcup_{\beta} [F, \Omega A]/\mathrm{Im}\,\langle\ ,\ u + \beta \rangle^1_{\Delta f}$$

<u>The first union is taken over all</u> $\beta \in [U, \Omega B]$ <u>with</u> $f_*(u + \beta) = 0$, <u>The second is taken over all</u> $\beta \in [F, \Omega B]$ <u>with</u> $f_*(u + \beta) = 0$.

The bijections are defined by $\alpha \mapsto u_\beta + \alpha$, where u_β is a lifting of $u + \beta$ to P_f. In (6.5.7) we will give a condition on a double path fibration under which the homomorphism $\langle\ ,\ u + \beta\rangle^1_{\Delta f}$ is no longer dependent on β.

We now generalize the classification in the corollary to iterated mapping path spaces with the aid of a spectral sequence derived from an exact couple along dual lines to (3.2.11). Let $X = \lim X_i \to \dots \to X_1 \to X_o = Y$ be an iterated principal fibration with classifying maps $f_i : X_{i-1} \to A$. We assume that A_i is an H-group for $i \geq 1$. Let u be a map $F \to X$, or a retraction $\bar{X} \to X$ of a closed cofibration $X \subset \bar{X}$ with cofiber F. Denote by $u_i : \bar{X}_i \to X_i$ the retractions induced by u and let $v = u_o$. The maps

$$q_n : [F, X_n]_v \to [F, X_{n-1}]_v,\ q_n : \langle \bar{X}_n, X_n\rangle_v \to \langle \bar{X}_{n-1}, X_{n-1}\rangle_v$$

induced by $q : X_n \to X_{n-1}$ give us abelian groups

(6.2.6) $$Q_n(X, Y; u) = q_n^{-1}(u_{n-1}) = [F, \Omega A_n]/\mathrm{Im}\ \underline{\Omega}\ (u_{n-1}, f_n)$$

in which u_n is the zero element. The group structure is defined by the operation of $[F, \Omega A_n]$, see (2.5.2). By (2.5.3), the abelian group $Q_n(X, Y; u)$ depends only on the homotopy class $u \in [F, X]$, respectively $u \in \langle \bar{X}, X\rangle$. This group can be calculated with the aid of the following spectral sequence, thus solving the classification problem of (1.1). We define bigraded groups $E_1^{**} = \{E_1^{p,q}\}$ and $D^{**} = \{D^{p,q}\}$ for $p, q \in Z$. The groups E_1^{**} are abelian. Let

$$E_1^{p,q} = [F, \Omega^{q+1} A_p] \qquad \text{for } q \geq 0 \text{ and } p \geq 1,$$

$$D^{p,q} = \langle \sigma_* \bar{X}_p, \underline{\Omega}^q_Y X_p\rangle_{u_p} \qquad \text{for } q \geq 1 \text{ and } p \geq 1.$$

For $q = 0$ and $p \geq 1$ let

$$D^{p,0} = \bigoplus_{n=1}^{p} Q_n(u) \quad \text{with} \quad Q_n(u) = Q_n(X, Y; u) \quad \text{and let}$$

$$E_1^{p,q} = D^{p,q} = 0 \quad \text{for} \quad q < 0 \quad \text{or} \quad p \leq 0.$$

From the exact classification sequence $(B', \underline{B}')$ in (2.5) we obtain for each $p, q \in Z$ a long exact sequence of groups and group homomorphism

$$\ldots \longrightarrow E_1^{p,q} \xrightarrow{\gamma^{p,q}} D^{p,q} \xrightarrow{\alpha^{p,q}} D^{p-1,q} \xrightarrow{\beta^{p-1,q}} E_1^{p,q-1} \xrightarrow{\gamma} \ldots$$

where for $q \geq 1$ and $p \geq 1$

$$\gamma^{p,q} = (-1)^{q+1} u_p^+ \qquad \text{as in (6.2.3),}$$

$$\alpha^{p,q} = q_* \qquad \text{is induced by } q : X_p \to X_{p-1},$$

$$\beta^{p-1,q} = \underline{\Omega}^q f_q \qquad \text{as in (6.2.3).}$$

If $q = 0$ and $p \geq 1$

$$\gamma^{p,0} = i_p \gamma_p \quad : \quad E_1^{p,q} \longrightarrow Q_p(u) \subset \bigoplus_{i \leq p} Q_i(u)$$

$$\alpha^{p,0} = pr \; : \; \bigoplus_{i \leq p} Q_i(u) \longrightarrow \bigoplus_{i \leq p-1} Q_i(u).$$

The map γ_p is the surjective homomorphism which induces the group structure of $Q_p(u)$ in (6.2.6). i_p is the inclusion and pr the projection. For all other index pairs, the homomorphisms in the sequence are zero. This exact sequence can also be written as an exact couple

(6.2.7)

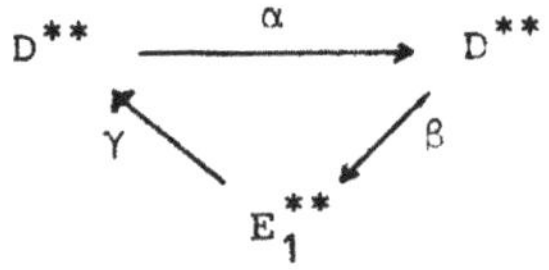

where α, β, γ have bidegree $(-1, 0)$, $(+1, -1)$ and $(0, 0)$ respectively.

<u>Note:</u> By definition (6.2.3), the first differential $d_1 = \beta\gamma$:

$E_1^{p,q} \to E_1^{p+1,q}$ for $p, q \geq 1$ is just the homomorphism $\bar{\Delta}_q(u_p, f_{p+1})$. By (6.2.4), this differential on E_1^{**} is induced by the q-th partial looping of the element $\Delta f_{p+1} \in [\Omega A_p \times X_p, A_{p-1}]^2$. Accordingly, the homology $E_2^{**} = H(E_1^{**}, d_1)$ can be computed with the aid of the primary operations in (6.2.1). In the appendix to this section we use the elements Δf_{p+1} to obtain a chain complex related to the cochain complex (E_1^{**}, d_1).

The exact couple above yields in the manner of (3.2.8) a spectral sequence $\{E_r^{**}, d_r\}_{r \geq 1}$ in which d_r has bidegree $(r, -1)$. As in (3.2.12) and (3.2.13), we can show that

(6.2.8) Theorem: The spectral sequence $\{E_r^{**}, d_r;\ r \geq 1\}$ depends only on the homotopy class $u \in \langle \bar{X}, X \rangle$, so we can write $E_r^{**} = E_r^{**}(X, Y; u)$. Moreover

$$Q_p(X, Y; u) = E_p^{p,0}(X, Y, u) \quad \text{for} \quad p \geq 1.$$

There is a dual to (3.2.14) as well.

When the iterated principal fibration $\to X_i \to X_{i-1} \to \cdots$ is the Postnikov decomposition of an orientable fibration $X \to B$, the reader may compare the above spectral sequence with the spectral sequence in (4.4.1).

(6.2.9) Remark: In the case of maps, the following differentials have been explicitly treated in the literature

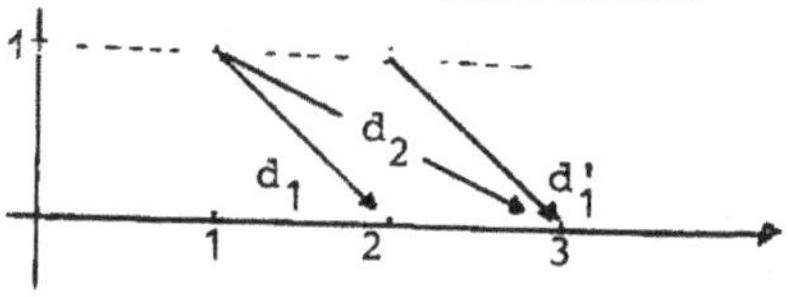

James and E. Thomas in 4.3 of [57] write the differential d_1 : $[F, \Omega^2 A_1] \to [F, \Omega A_2]$ as $\Delta(\Theta, \eta)$ (where $f_2 = \Theta$, $u_1 = \eta$). Our statement (6.2.5) on the lifting set $[U, P_f]_v$ is identical with

their classification theorem. Nomura considers in [91] the set $[U, X]$, where $X = X_3$, $Y = X_o = *$. His non-stable secondary operation $\Phi_\Theta(\rho, v)$ is exactly the differential $d_2 : E_2^{1,1} \to E_2^{3,0}$ (where $\Theta = f_1$, $\rho = f_2$, $v = u_1$). Thus theorem A in [91] can be understood as the statement that $d_2 = 0$ when $\ker d_1 = 0$.

(6.2) Appendix: Iterated principal fibrations and the associated chain complex.

As before, let $X = \lim X_i \to \dots \to X_1 \to X_o = Y$ be an iterated principal fibration with classifying maps $f_i : X_{i-1} \to A_i$ where the A_i are H-groups. The first differential d_1 of the spectral sequence just defined is induced by elements $\Delta f_{p+1} \in [\Omega A_p \times X_p, A_{p+1}]_2$. That $d_1 d_1 = 0$ follows also from

(6.2.10) Theorem: The composition

$$(\Delta f_{p+1})(L(\Delta f_p)(1 \times q), pr) : \Omega^2 A_{p-1} \times X_p \to \Omega A_p \times X_p \to A_{p+1}$$

is null-homotopic, where $q : X_p \to X_{p-1}$.

This theorem is proved exactly as was (3.2.15). Important instances of iterated principal fibrations are those where the A_i Eilenberg-Mac Lane spaces. Somewhat more generally,

Definition: Suppose that to every A_p there is given an Ω-spectrum B_p such that $B_{p,o} = A_p$. Then we define $df_{p+1} \in \sigma_{B_p}^{-1}(B_{p+1};X)$ to be the image of Δf_{p+1} under the homomorphism

$$[\Omega A_p \times X_p, A_{p+1}]_2 \stackrel{\varepsilon}{=} [B_{p,1} \times X_p, B_{p+1,o}] \xrightarrow{L^\infty} \sigma_{B_p}^{-1}(B_{p+1}, X_p)$$

$$\xrightarrow{q} \sigma_{B_p}^{-1}(B_{p+1}, X)$$

where q is induced by the fibration $X \to X_p$.

The composition defined in (6.1.5) is then seen by the preceding theorem to be such that

(6.2.11) <u>Corollary</u>: $(df_{p+1}) \cdot (df_p) = 0, \quad p \geq 1.$

Suppose now that p is a prime and that HZ_p is the Eilenberg-Mac Lane spectrum. Consider the following condition.

(*) Let Z_n be a finite graded set, in which we denote the degree of $e \in Z_n$ by $|e|$. Suppose $A_n = \underset{e \in Z_n}{\times} K(\mathbb{Z}_p, |e| + 1)$ and further that $|e| > n$ for $e \in Z_n$.

In this situation we can regard df_{n+1} as an element

$$df_{n+1} \in \underset{e \in Z_{n+1}}{\times} \ \bigoplus_{d \in Z_n} \sigma_{H\mathbb{Z}_p}^{|e|-|d|+1} (H\mathbb{Z}_p; X)$$

where $|e| - |d| + 1 > 2$. Now let $\sigma^* X = \sigma^*_{H\mathbb{Z}_p} (H\mathbb{Z}_p; X) = A_p \otimes H^*(X, \mathbb{Z}_p)$ be the Massey-Peterson algebra defined in (6.1.18). The elements df_{n+1} induce homomorphisms

$$(*)' \longrightarrow \bigoplus_{e \in Z_{n+1}} \sigma^* X \xrightarrow{df_{n+1}} \bigoplus_{e \in Z_n} \sigma^* X \xrightarrow{df_n} \bigoplus_{e \in Z_{n-1}} \sigma^* X \longrightarrow \cdots$$

of free $(\sigma^* X)$-modules, via linear extension.

(6.2.12) <u>Corollary</u>: <u>If (*) is satisfied, (*)' is a chain complex of $(\sigma^* X)$-modules, so in particular the composition</u> $(df_n)(df_{n+1}) = 0$ <u>is zero</u>

<u>Remark</u>: The chain complex (*)' is precisely a chain complex C as on p. 189 of [79], with which McClendon somewhat abruptly begins the axiomatic development of "twisted cohomology operations of higher order". We have here shown how the chain complex C arises naturally. McClendon's treatment generalizes those of Maunder [77] and Adams [2].

(6.3) The first differential of the spectral sequence. Cup products and relative cup products.

In the last section we derived a spectral sequence whose first differential on the E_1-term was defined in terms of operators $\Delta_n(u, f)$ and $\bar{\Delta}_n(u, f)$. We will show that these operators are additive in f, and we will give a composition formula. These results are dual to the corresponding ones in (3.3). As examples of these operators we will consider the cup product and the relative cup product. It turns out that the relative cup product plays a role in the classification of retractions, similar to that played by the cup product in the classification of maps. We will apply these results to classify retractions onto a complex projective space.

Let A be an H-group and B and H-space. From the projections ρ_i: $B \times B \to B$ we form $\rho_2 + \rho_1 : B \times B \to B$, which is just the multiplication on B up to the order of addition. Given a map $f : B \to A$, we define the difference element

$$(6.3.1) \qquad \Delta f = -f_*(\rho_2) + f_*(\rho_2 + \rho_1) \in [B \times B, A]_2$$

as in (6.2.2). It is trivial on the second factor B, so partial looping gives us elements $L^n \Delta f$ inducing the operations

$$\Delta^n(u, f) = [\ , u]^n_{\Delta f} : [U, \Omega^n B] \longrightarrow [U, \Omega^n A]$$

$$\bar{\Delta}^n(u, f) = \langle\ , u\rangle^n_{\Delta f} : [F, \Omega^n B] \longrightarrow [F, \Omega^n A]$$

where $u \in [U, B]$ or $u \in \langle \bar{B}, B\rangle$ denotes a retraction of a closed cofibration $B \subset \bar{B}$ with cofiber F. We write $\Delta = \Delta^1$. Now let $P_f \subset \bar{P}_f$ be a closed cofibration with cofiber F and let $B \subset \bar{B}$ denote the induced cofibration. A partial extension of the classification in (6.2.3) is

(6.3.2) <u>Classification theorem: There are bijections</u>

$$[U, P_f] \approx \bigcup_u [U, \Omega A]/\mathrm{Im}\, \Delta(u, f),$$

$$\langle \bar{P}_f, P_f \rangle \approx \bigcup_u [F, \Omega A]/\mathrm{Im}\, \bar{\Delta}(u, f),$$

where the first union is taken over all $u \in [U, B]$ with $f_*(u) = 0$ and the second over all $u \in \langle \bar{B}, B \rangle$ with $f_\#(u) = 0$.

Again, the bijections are defined by $\alpha \to u_0 + \alpha$, where u_0 is a lifting of u to P_f. If $B = \Omega B'$ is a loop space, this result reiterates (6.2.5) for the special case in which $X_0 = *$.

(6.3.3) Note: Suppose B' is an H-space and $B = \Omega B'$. If a retraction $u_0 : \bar{P}_f \to P_f$ exists, the second bijection of (6.3.2) can be expressed as

$$\langle \bar{P}_f, P_f \rangle \approx \bigcup_\beta [F, \Omega A]/\mathrm{Im}\, \bar{\Delta}(u + \beta, f)$$

where $u : \bar{B} \to B$ is the retraction induced by u_0. The union is taken over all $\beta \in [F, B]$ with $\langle \beta, u \rangle^{\circ}_{\Delta f} = 0$. This can be proved exactly as was (3.3.3).

Using the properties of the partial loop operation, we can immediately deduce the following theorems.

(6.3.4) Composition theorem: Let $P \xrightarrow{g} B \longrightarrow A$ be such that A, B are H-groups and P is either a mapping path space or an H-space. If $u \in [U, P]$ then

$$\Delta^n(u, fg) = \Delta^n(gu, f) \circ \Delta^n(u, g).$$

Let $P \subset \bar{P}$ be a closed cofibration. If $u \in \langle \bar{P}, P \rangle$ is a retraction, then

$$\bar{\Delta}^n(u, fg) = \bar{\Delta}^n(g_* u), f) \circ \bar{\Delta}^n(u, g)$$

where $g_* u \in \langle g_* \bar{P}, B \rangle$ denotes the induced retraction.

Proof: As with (3.3.5), the statements follow by partially looping $\Delta(f\ g) = (\Delta f)(\Delta g, g\rho_2)$. □

(6.3.5) Additivity theorem: Let A be an abelian H-group, and P an H-space or a mapping path space. Let $f, g \in [P, A]$ and either $u \in [U, P]$ or $u \in \langle \overline{P}, P\rangle$ with $P \subset P$, then

$$\bar{\Lambda}^n(u, f + g) = \bar{\Lambda}^n(u, f) + \bar{\Lambda}^n(u, g).$$

Proof: As with (3.5.7), the statement follows from $\Delta(f + g) = \Delta f + \Delta g$ which holds because A is an abelian H-group. □

In (3.3.12) we used a lemma of Ganea to express the products and $[\ ,\ ]^n_\xi$, $\langle\ ,\ \rangle^n_\xi$ in terms of Whitehead products. The dual products (6.2.1) cannot be similarly expressed in terms of a universal operation, because the dual of Ganea's lemma (3.1.20) does not hold. However, we do have

(6.3.6) Lemma: Let A be an H-group. Then

$$\Theta : [B, A] \times [X, A] \times [B \wedge X, A] \to [B \times X, A]$$

defined by $\Theta = p_1^* + p_2^* + s^*$ is a bijection, and if $f \in [X \wedge Y, A]$ then $\Omega(s^* f) = 0$.

p_1 and p_2 are the projections of $B \times X$ onto B and X and $s : B \times X \to B \wedge X$ is the identification map. This lemma is a consequence of the Puppe sequence for the cofibration $B \vee X \subset B \times X$, just as was the exact sequence in (0.3). The lemma shows that the operation $[\ ,\ ]^n_\xi$ of (6.2.1) for $\xi = p_1^*\ \xi_B + s^*\hat{\xi}$ decomposes as

$$(6.3.7) \qquad [\rho, u]^n_\xi = (\Omega^n\ \xi_B)_* \beta + [\rho, u]^n_{s^*\hat{\xi}}$$

where $\Omega[\beta\ ,\ u]^n_{s^*\hat{\xi}} = 0$. However $L^n(s^*\xi)$ cannot be expressed in terms of an operator dual to the Whitehead product. In any case though, (6.3.7) implies the dual statement to (3.3.13)

(6.3.8) <u>Theorem</u>: <u>Let</u> $\xi_B : B \subset B \times X \xrightarrow{\xi} A$. <u>Then</u>

1) $[\beta, 0]^n_\xi = (\Omega^n \xi_B)\beta$

2) <u>If</u> U <u>is a co-H-space, then</u> $[\beta, u]^n_\xi = (\Omega^n \xi_B)\beta$ <u>for all</u> u.

3) $\Omega[\beta, u]^n_\xi = \Omega((\Omega^n \xi_B)\beta)$.

Although the cup product cannot be regarded as a strict dual to the Whitehead product, many properties of the Whitehead product and the cup product make it possible to think of them as dual to each other. In rational homotopy theory it becomes clear in just what sense these two products are dual to each other. There is unfortunately no room to go into this in more detail in this book, so we refer the interested reader to [20].

In analogy with (3.3.8), we make the

(6.3.9) <u>Definition</u>: Let $i : A \subset X$ be a cofibration with cofiber $j : X \to X/A = F$. If π is an abelian group, j induces an isomorphism $\chi : \tilde{H}^n(F;\ \pi) = H^n(X,\ A;\ \pi)$. The relative cup product

$$\cup : H^n(X,\ A;\ \pi) \otimes H^m(X,\ G) \longrightarrow H^{n+m}(X,\ A;\ \pi \otimes G)$$

thus induces an operation

$$\bar{\cup} : H^n(F;\ \pi) \otimes H^m(X,\ G) \longrightarrow H^{n+m}(F,\ \pi \otimes G)$$

by $\alpha\ \bar{\cup}\ \beta = \chi^{-1}(\chi(\alpha) \cup \beta)$ which we will also call the <u>relative cup product</u>.

Dually to (3.3.9) we have

$$j^*(\alpha \overline{\cup} \beta) = (j^*\alpha) \cup \beta \quad \text{in} \quad H^{n+m}(X, \pi \otimes G). \tag{6.3.10}$$

If $A \subset X$ has a retraction, j^* is injective. In this case the relative cup product can also be defined by $\alpha \overline{\cup} \beta = j^{*-1}(j^*(\alpha) \cup \beta)$ as in the similar case (6.2.1) for $\langle\ ,\ \rangle^n_\xi$. Thus the relative cup product plays a role in the classification of retractions, similar to that played by the cup product in the classification of maps. We now give some examples.

In what follows we identify $H^n(X, \pi)$ with $[X, K(\pi, n)]$, and $\Omega K(\pi, n+1)$ with $K(\pi, n)$. This means that we eliminate the isomorphism γ from (0.5.5), and the homotopy equivalence λ from (0.5.6).

The cup product map $\cup : B = K(\pi, r) \times K(G, s) \to A = K(\pi \otimes G, r+s)$ is defined on a loop space B. Therefore the element $\Lambda(\cup)$ is defined as in (6.3.1) and we obtain the operation

$$\Delta^n(u, \cup) : H^{r-n}(U, \pi) \oplus H^{s-n}(U, g) \to H^{r+s-n}(U, \pi \otimes G)$$

for $u = (u_1, u_2) \in H^r(U, \pi) \times H^s(U, G)$. Somewhat more generally, let $B \subset \overline{B}$ be a closed cofibration with cofiber F and $u = (u_1, u_2) \in \langle \overline{B}, B \rangle$ be a retraction. Then

$$\overline{\Delta}^n(u, \cup) : H^{r-n}(F, \pi) \oplus H^{j-n}(F, \pi) \to H^{r+s-n}(F, \pi \otimes G)$$

is also defined.

(6.3.11) <u>Theorem</u>: <u>For</u> $n \geq 1$,

$$\Delta_n(u, \cup)(\alpha, \beta) = \alpha \cup u_2 + u_1 \cup \beta,$$

$$\Delta_n(u, \cup)(\alpha, \beta) = \alpha \overline{\cup} u_2 + u_1 \overline{\cup} \beta.$$

<u>Proof</u>: Let B_1, B_2 be copies of B. The bilinearity of the cup product implies that $\Delta\cup : B_1 \times B_2 \to A$ is equal to

$$\begin{aligned}\Delta U &= -U_*(\rho_2) + U_*(\rho_2+\rho_1), \qquad \rho_i = (\rho_i', \rho_2''), \\ &= -\rho_2' \cup \rho_2'' + (\rho_2'+\rho_1') \cup (\rho_2''+\rho_1'') \\ &= \rho_1' \cup \rho_2'' + \rho_2' \cup \rho_1'' + \rho_1' \cup \rho_1''\end{aligned}$$

Since $\Omega(\rho_1' \cup \rho_1'') = 0$, the statement follows from (6.1.12). □

A consequence of the classification theorem (6.3.2) is

(6.3.12) Corollary: Let P be the mapping path space of $\cup$: $K(\pi, r) \times K(G, s) \to K(\pi \otimes G, r+s)$. There is a bijection

$$[U, P] = \bigcup_{u} \frac{H^{r+s-1}(U, \pi \otimes G)}{H^{r-1}(U, \pi) \cup u_1 + u_2 \cup H^{s-1}(U, g)}$$

where the disjoint union is taken over all $u = (u_1, u_2) \in H^r(U, \pi) \times H^s(U, G)$ with $u_1 \cup u_2 = 0$.

(6.3.3) implies a more complicated statement for retractions.

(6.3.13) Corollary: Let P be the mapping path space of $\cup$ as in the preceding corollary, let $P \subset \overline{P}$ be a closed cofibration with cofiber $j : \overline{P} \to F$ and $u : \overline{P} \to P$ be a retraction. There is a bijection

$$\langle \overline{P}, P \rangle \approx \bigcup_{\alpha,\beta} \frac{H^{r+s-1}(F, \pi \otimes G)}{H^{r-1}(F, \pi) \mathbin{\overline{\cup}} (u_1 + j^*\alpha) + (u_2 + j^*\beta) \mathbin{\overline{\cup}} H^{j-1}(F, G)}$$

The projection q of the path fibration defines the elements $(u_1, u_2) = qu : \overline{P} \to P \to K(\pi, r) \times K(G, s)$. The disjoint union is taken over all $(\alpha, \beta) \in H^r(F, \pi) \times H^s(F, G)$ with $\alpha \mathbin{\overline{\cup}} u_2 + u_1 \mathbin{\overline{\cup}} \beta + \alpha \cup \beta = 0$.

Let $A \subset X$ be a closed cofibration and $f : A \to P$ be a map which can be extended over X. Then the preceding corollary gives a classification of the homotopy set $[X, P]^f$ of extensions of f, for the reason that $[X, P]^f = \langle f_*X, P \rangle$ can be interpreted as the retraction homotopy set of the induced cofibration $P \subset f_* X$.

Remark: Similar classification theorems for $[U, P_f]$ and $\langle \bar{P}_f, P_f \rangle$ can be formulated for maps $f : B \to A$, where A and B are products of Eilenberg-Mac Lane spaces. Such an f can always be decomposed into known cohomology operations in such a way that Δf and $L^n \Delta f$ are computable. See in this connection the treatment by Rutter in [106].

We can apply the methods of (6.3.13) to the following more familiar spaces.

(6.3.14) *Theorem*: *Let* X *be a CW-complex containing the complex projective space* $\mathbb{C}P_n$ *as a subcomplex. Suppose that* $\dim (X - \mathbb{C}P_n) \leq 2n + 1$ *and that there is a retraction* $u : X \to \mathbb{C}P_n$. *Then there is a bijection*

$$\langle X, \mathbb{C}P_n \rangle \approx \coprod_{\alpha \in H^2(X, \mathbb{C}P_n)} \frac{H^{2n+1}(X, \mathbb{C}P_n)}{(n+1)(j\alpha + \deg u)^n \cup H^1(X, \mathbb{C}P_n)}$$

where $j : H^2(X, \mathbb{C}P_n) \to H^2(X)$, *and* $\deg(u) \in H^2(X)$ *is the degree of the map* u. *All cohomology groups have coefficients in* $\mathbb{Z}$.

For the trivial cofibration $\mathbb{C}P_n \subset X = \mathbb{C}P_n \vee Y$ we have $\langle X, \mathbb{C}P_n \rangle = [Y, \mathbb{C}P_n]$, and the theorem expresses the result (4.4.7) of Spanier. When $n = 1$ we have $\mathbb{C}P_1 = S^2$, and the theorem generalizes Pontryagin's result (4.4.6) to the case of retractions.

Proof: Let $i \in H^2(\mathbb{C}P_\infty)$ be a generator. Then for $f = i^{n+1}$ we have a map h as in the commutative diagram

$$\begin{array}{ccccc} \mathbb{C}P_n & \xrightarrow{h} & P_f & & \\ & \searrow & \swarrow & & \\ & & B = \mathbb{C}P_\infty & \xrightarrow{f} & K(\mathbb{Z}, 2n+2) = A \end{array}$$

and which is $(2n+2)$-connected. h induces a bijection

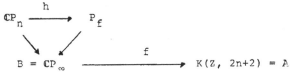

(1) $h_* : \langle X, \mathbb{C}P_n \rangle \approx \langle h_*X, P_f \rangle$

between retraction homotopy sets, in view of (1.4.13) applied to trivial fibrations. For $f = i^{n+1}$ it follows from bilinearity of the cup product that

(2) $$\Delta f = -f_*\rho_2 + f_*(\rho_2 + \rho_1) : B \times B \to A$$
$$= -\rho_2^{n+1} + (\rho_2 + \rho_1)^{n+1} = \sum_{k=0}^{n} \binom{n+1}{k} \rho_2^k \cup \rho_1^{n+1-k}$$

where $\rho_2^k \cup \rho_1^{n+1-k} = (-1)^* \rho_1^{n+1-k} \cup \rho_2^k = (-1)^* \cup (i^{n+1-k} \times i^k)$.

We conclude from (6.2.12) and (6.1.5) that

$$L(\rho_2^k \cup \rho_1^{n+1-k}) = (-1)^*(L\cup)\ (\Omega(i^{n+1-k}) \times i^k).$$

Since $\Omega i^r = 0$ for $r \geqslant 2$, it follows from (2) that

(3) $$L\Delta f = (n+1)\ \rho_2^n \cup \rho_1 \quad : \quad (\Omega B) \times B \to \Omega A.$$

Now (6.3.3), (6.3.10) and definition (6.2.1) imply the statement of the theorem. It should be remembered that $X \to h_*X$ induces an isomorphism $H^k(h_*X) \to H^k(X)$ for $k \leqslant 2n+1$. $\square$

(6.4) <u>The functional loop operation</u>

We now define a functional operation that we call the functional loop operation. A special case of it is the partial loop operation of (6.1). Thus the functional loop operation is a far-reaching generalization of the loop functor Ω. Surprisingly enough, there is an isomorphism theorem in the general case just as for Ω. We will show this using a result of Mather and Ganea. This isomorphism theorem will be an essential tool in the proof that a Postnikov decomposition has a principal reduction, see (2.3.4).

Let Y be a space over D, that is suppose we are given a map $d: Y \to D$. Let B be an ex-fiber space over D with projection ρ and section σ. A map $g : Y \to B$ over D and its relative mapping path space $P_D g$ fit into the commutative diagram

(6.4.1)

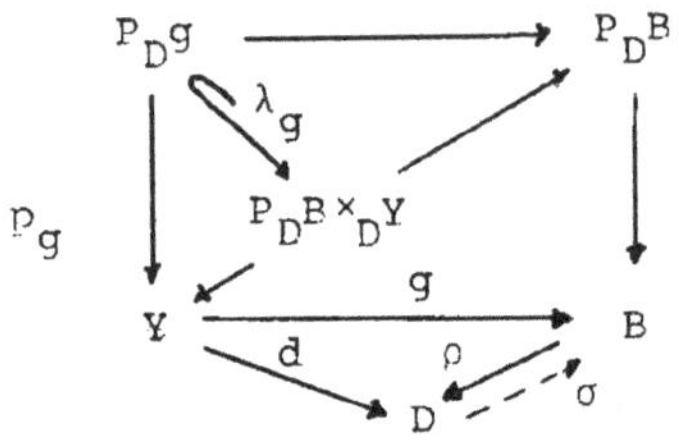

in which the square is cartesian. λ_g is the inclusion and $\times_D$ denotes a pullback, see (0.0). λ_g induces a diagram of cohomology groups that is dual to diagram (3.4.2) and whose rows are exact,

(6.4.2)

$$\begin{array}{ccccccccc} & & 0 \longrightarrow & H^n(B \times_D Y, P_D B \times_D Y) & \xrightarrow{j} & H^n(B \times_D Y) & \xrightarrow{\sigma^*} & H^n(Y) \to 0 \\ & & & \downarrow \lambda_g^* & & \downarrow (g,1)^* & & \\ H^{n-1}(Y) & \xrightarrow{p_g^*} & H^{n-1}(P_D g) \xrightarrow{\delta} & H^n(Y, P_D g) & \xrightarrow{j} & H^n(Y) & \to & \dots \end{array}$$

All the cohomology groups have local coefficients induced by the local group $\hat{G}$ in D. In our usual manner we write $H^*(B, A) = H^*(Z_f, A)$ for a map $f : A \to B$. The section $\sigma : D \to B$ induces a section $\sigma : Y \to B \times_D Y$. ρ^* splits σ^* in the upper row of the diagram, which is exact because of (2.4.4). This can be seen from the cohomology sequence (4.1.7) of a pair.

The last diagram has a somewhat more general form. Let D with $D \to K$ be a space over K, and let $A = \Omega_K A'$ be an ex-fiber space over K that is also a relative loop space of the indicated form. We then have the diagram

$$
\begin{array}{ccccccccc}
 & & 0 & \to [Z',A]_K^{Y'} & \xrightarrow{j} & [B \times_D Y, A]_K & \xrightarrow{\sigma^*} & [Y,A]_K & \to 0 \\
(6.4.3) & & & \downarrow \lambda_g^* & & \downarrow (g,1)_* & & & \\
[Y,A]_K & \xrightarrow{p_g^*} & [P_D g, \Omega_K A]_K & \xrightarrow{\delta} [Z,A]_K^Y & \xrightarrow{j} & [Y,A]_K & \to & \cdots &
\end{array}
$$

of group homomorphisms. The rows are exact, as portions of the exact sequence (2.2.15). Z and Z' are the mapping cylinders of p_g : $P_D g \to Y$ and $(p_o \times 1) : P_D B \times {}_D Y \to B \times {}_D Y = Y'$ respectively.

If in (6.4.3) we replace A by the ex-fiber space $L(\hat{G}, n)$ over K of (5.2.4), we obtain the commutative diagram (6.4.2). If $K = *$, then (6.4.3) is the exact dual of diagram (3.4.2). We say that a map $\xi : B \times {}_D Y \to A$ over K is <u>trivial on</u> Y when $\sigma^*\xi = 0 \in [Y,A]_K$. By

$$[B \times {}_D Y, A]_{K2} = \operatorname{Ker} \sigma^*, \; H^n(B \times {}_D Y; \hat{G})_2 = \operatorname{Ker} \sigma^* = H^n(B \times_D Y, Y; \hat{G})$$

we denote the group of elements trivial on Y. The functional operation for the preceeding diagrams is the <u>functional loop operation</u> $L_g = \delta^{-1} \lambda_g^* j^{-1}$ with

$$
(6.4.4) \quad
\begin{array}{l}
L_g : H^n(B \times {}_D Y; \hat{G})_2 \cap \operatorname{Ker}(g,1)^* \longrightarrow H^{n-1}(P_D g; G)/\operatorname{Im} p_g^* \\
L_g : [B \times_D Y, A]_{K2} \cap \operatorname{Ker}(g,1)^* \longrightarrow [P_D g, \Omega_K A]_K / \operatorname{Im} p_g^*
\end{array}
$$

The partial loop operation L of (6.1) is a special case of the functional loop operation. If $g = \sigma d : Y \to D \to B$, then $P_D g = \Omega_D B \times {}_D Y$ and we have dually to (3.4.5)

$$
(6.4.5) \quad
\begin{array}{l}
L : H^n(B \times_D Y, Y; \hat{G}) \longrightarrow H^{n-1}(\Omega_D B \times_D Y, Y; \hat{G}) \\
L : [B \times_D Y, A]_{K2} \longrightarrow [\Omega_D B \times_D Y, \Omega_K A]_{K2}
\end{array}
$$

where $L(\xi)$ is uniquely determined by the conditions $L(\xi) \in L_{\sigma d}(\xi)$, $\sigma^* L(\xi) = 0$. If $D = K = *$ then L is just the partial loop operation L of (6.1.1). The next theorem is in a sense dual to the suspension theorem (3.4.8).

(6.4.6) Theorem: Let B be an ex-fiber space over D and let

(B, D) be b-connected. For any map $g : Y \to B$ and any local co-

efficients $\hat{G}$ in D, the preceding cohomology operations L_g and

L are isomorphisms for $n \leqslant 2b + 1$ and monomorphisms for $n \leqslant 2b + 2$.

There is a corresponding theorem for the operations L_g and L when

$A = \Omega_K A'$ and $\pi_j(A, K) = 0$ for $j \geqslant n + 1 = m_A$, that implies

(6.1.4) for the case $K = D = *$.

Using the notation of the last theorem, we now exhibit an exact

sequence in a stable range.

(6.4.7) Corollary: When $n \leqslant 2b + 1$, the maps L and $(1 \times p_g)^*$ in

the following commutative diagram are isomorphisms and its row is exact.

$$\cdots \quad H^{n-1}(Y;\hat{G}) \xrightarrow{\ \ p_g^*\ \ } H^{n-1}(P_D g;\hat{G}) \xrightarrow{\ \overline{L_g}\ } H^n(B \times_D Y, Y;\hat{G}) \xrightarrow{\ (1,g)^*\ } H^n(Y;\hat{G}) \ \text{---}$$

$$-\rho_2^* + \mu^* \Big\downarrow \qquad \mu \quad H^{n-1}(\Omega_D B \times_D Y, Y;\hat{G})$$

$$\cong \Big\downarrow L$$

$$\cong \Big\downarrow (1 \times p_g)^*$$

$$H^{n-1}(\Omega_D B \times_D P_D g, \hat{G}) \xleftarrow{\ j\ } H^{n-1}(\Omega_D B \times_D P_D g, P_D g;\ \hat{G})$$

$$\tau$$

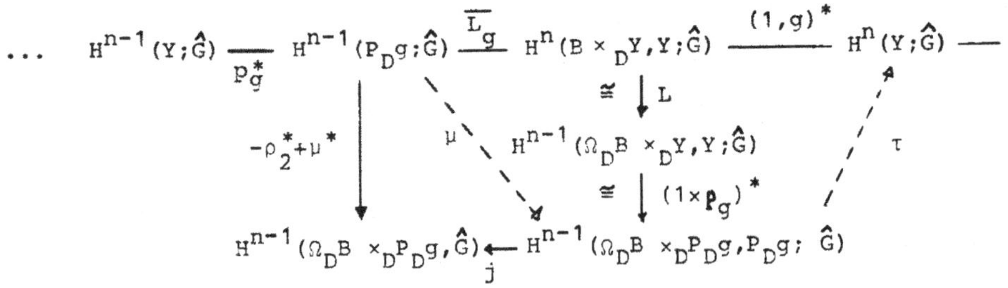

$\overline{L_g}$ is inverse to L_g, that is $\overline{L_g} = j(\lambda_g^*)^{-1}\delta$. (6.5.5) can be used

to show commutativity.

Remark: E. Thomas obtained in [125] such an exact sequence for the

case $D = *$. He defines operations μ and τ which are related to

our constructions L_g and L as indicated in the diagram. The op-

eration L_g is also to be found in a non-explicit form in McClendon's

work, see Definition 1.3 of [82].

In proving (6.4.6) we will use the

(6.4.8) *Lemma: Let* $A \subset X$ *and let* $p : \tilde{X} \to X$ *be a fibration with fiber* F. *Let* $\tilde{A} \to A$ *denote the restricted fibration, and* $\hat{G}$ *be a local group in* X. *If* (X, A) *is* n-*connected and* F *is* r-*connected, then*

$$p^* \;:\; H^k(X, A; \hat{G}) \;\to\; H^k(\tilde{X}, \tilde{A}; p^*\hat{G})$$

is an isomorphism for $k \leqslant n + r - 1$ *and a monomorphism for* $k \leqslant n + r + 2$.

Proof of (6.4.6): The fiber F of $B \to D$ is also b-connected. The diagram

$$\begin{array}{ccc} P_D B \times_D Y & \xrightarrow{\;p_o \times 1\;} & B \times_D Y \\ \uparrow{\scriptstyle \lambda g} & & \uparrow{\scriptstyle (g,\, 1)} \\ P_D g & \longrightarrow & Y \end{array}$$

is a pullback, and so by (2.2.4) $(p_o \times 1)$ is b-connected. Since $(g, 1)$ is a section of the fibration $B \times_D Y \to Y$, the fiber of $(g, 1)$ is b-connected. When we regard $(g, 1)$ as a fibration and again apply (6.4.8) we arrive at the statement of (6.4.6) $\Box$

Lemma (6.4.8) could be proved using a Serre spectral sequence in cohomology with local coefficients. We will however use a recently published theorem of M. Mather (theorem 47 of [73]) to turn out a more elegant proof.

(6.4.9) *Theorem : Let* $A \subset X$ *be a closed cofibration and let* $p : X \to X$ *be a fibration with fiber* F. *In the diagram*

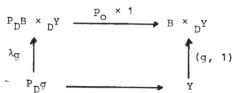

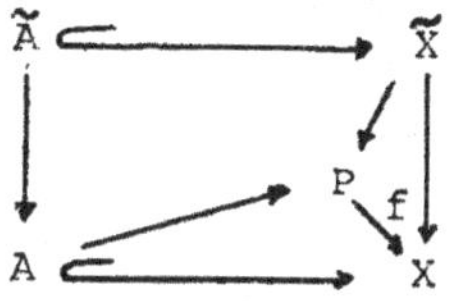

suppose that the outside square is cartesian, the inside square is co-

cartesian and that $A, X, \tilde{X}$ *are connected. Let* F *be the fiber of* $\tilde{X} \to X$ *and* G *the fiber of* $A \subset X$. *Then the fiber of* $f : P \to X$ *is homotopy equivalent to* $\Sigma F \wedge G$. *It follows that* $f : P \to X$ *is at least* $(n + r + 2)$*-connected when* (X, A) *is* n*-connected and* F *is* r*-connected.*

If $A = *$ this expresses a result of Ganea [34].

Proof of (6.4.8): Use the preceding theorem, (1.4.14) and the representation (5.2.4) of cohomology groups with local coefficients as section homotopy sets. □

(6.5) *The functional loop operation and the principal reduction of fibrations.*

It is a familiar fact that the mapping path space of a loop map Ωf is equivalent to a loop space, that is $P_{\Omega f} \simeq \Omega P_f$. We here show that the functional loop operation behaves similarly. As an application we obtain a classification theorem of James and Thomas. In an appendix we prove that a Postnikov decomposition has a principal reduction.

Consider a double relative principal fibration as in the commutative diagram

(6.5.1)
$$\begin{array}{ccccc} P_K f & & & & \\ \downarrow & & & & \\ P_D g & \xrightarrow{f} & \Omega_K A & \rightleftarrows & K \\ \downarrow & & & & \uparrow \\ X & \xrightarrow{g} & B & \rightleftarrows & D \end{array}$$

D is a space over K, and B and A are ex-fiber spaces over D and K respectively. This is the general situation which confronts us in the inductive procedure for the principal reduction of a Postnikov decomposition, as explained in the appendix to this section.

(6.5.2) Theorem (principal reduction of a double principal fibration): *If f in (6.5.1) is functional with respect to g, that is if there exists a $\xi_o \in [B \times_D X, A]_K$ trivial on X with $f \in E_g(\xi_o)$ then the double fibration $P_K f \to P_D g \to X$ is a principal fibration relative to X.*

We will need a somewhat more detailed version of this.

(6.5.3) Lemma: *Suppose the map $d = \rho g$ in (6.5.1) factors over a space L as $d : X \xrightarrow{e} L \to D$. Further let $\xi : B \times_D L \to A$ be a map over K trivial on L and such that $f \in E_g(\xi_o)$ where $\xi_o = (1 \times e)^*\xi$. Then the double fibration $P_K f \to P_D g \to X$ is a principal fibration relative L with the property that there is a section $\overline{\xi} : L \to P_K \xi$ (making $P_K \xi$ an ex-fiber space over L) and a map $\xi_g : X \to P_K \xi$ lifting e, for which $P_K f \simeq P_L \xi_g$ over X.*

Diagram (6.5.1) for $K = *$ is dual to diagram (3.5.1). The preceding theorem and the lemma are in this case the dual statements to (3.5.2) and (3.5.3). This justifies our leaving the proofs to the reader as exercises. They are somewhat complicated to write out, but formally they are strict duals to the proof of (3.5.3). The reader may care to formulate the duals of the other statements of (3.5) for himself. We will dualize only (3.5.7).

If $K = D = *$ the element Δf is defined as in (6.2.2). Similarly, for the maps f and g of (6.5.1) we obtain the element

(6.5.4) $$\Delta f = -f_*\rho_2 + f_*(\rho_2 + \rho_1) \in [\Omega_D B \times_D P_D g, \Omega_K A]_{K2}$$

trivial on $P_D g$.

(6.5.5) Theorem: If $f \in E_g(\xi)$ then $\Delta f = (1 \times p_g)^*(L\xi)$ where $(1 \times p_g) : \Omega_D B \times_D P_D g \to \Omega_D B \times_D Y$.

Again, the proof of this theorem is complicated to formulate but is dual to the proof of (3.5.7).

We will state the next corollaries only for $D = K = *$. The reader may generalize the operations of (6.2.1) and consider how the statements for non-trivial D and K must be formulated.

(6.5.6) Corollary: Let $P_f \to P_g \to X$ be such that $f \in E_g(\xi)$. If $u_o \in [U, P_g]$ then

$$-f_*(u_o) + f_*(u_o + \beta) = [\beta, u]^1_\xi$$

where $u = p_{g*} u_o \in [U, X]$. If $P_g \subset \overline{P}_g$ is the cofibration induced by a closed cofibration $P_f \subset \overline{P}_f$ and $u_o \in \langle \overline{P}_g, P_g \rangle$, then

$$-f_\#(u_o) + f_\#(u_o + \beta) = \langle \beta, u \rangle^1_\xi$$

where again $u \in \langle \overline{X}, X \rangle$ is induced by u_o.

This corollary shows that the secondary obstruction $\mathcal{H}_2(u)$ to lifting u to P_f is a coset of $\mathrm{Im}\,[\ ,\ u]^1_\xi$ or $\mathrm{Im}\,\langle\ ,\ u\rangle^1_\xi$ respectively. In (6.6) we will describe the Massey triple products as an instance of such secondary obstructions.

In light of (6.5.5), the classification formula (6.3.2) takes the form

(6.5.7) Corollary: Let $P_f \to P_g \to X$ be such that $f \in E_g(\xi)$. Let $u : U \to X$ be a map or $u : \overline{X} \to X$ be a retraction, and let $P_f \subset \overline{P}_f$ be a closed cofibration. Suppose that u can be lifted to P_f. Then there are bijections respectively

$$[U, P_f]_u \approx \text{Ker } [\ ,\ u]^1_\xi \times \text{Coker } [\ ,\ u]^2_\xi \ ,$$

$$P_f, P_f \ {}_u \approx \text{Ker} < ,\ u >^1_\xi \times \text{Coker} < ,\ u >^2_\xi \ .$$

Similarly to what was pointed out in the remark following (3.5.11), here $[U, P_f]_u$ and $< \overline{P}_f, P_f >_u$ are groups with u_o as zero element and can be imbedded in short exact sequences.

Remark: The formula for $[U, P_f]_u$ is essentially the classification theorem of James and E. Thomas in 2.2 of [57]. To see this, recall that the condition $f \in L_g(\xi)$ implies that $P_f \to P_g \to X$ is a 'stable decomposition' of $P_f \to X$ in the sense of [57]. This is so by definition exactly when $\Delta f \in \text{Im } (1 \times p_g)^*$ as in (6.5.5).

(6.5) Appendix: The principal reduction of a Postnikov decomposition

Theorem (2.3.4) on the principal reduction of a Postnikov decomposition can now be proved by induction. Essential for this are the isomorphism theorem (6.4.6) and the principal reduction procedure in (6.5.3). The proof proceeds dually to the proof of (3.5.13).

Let $\ldots \to Y_k \to Y_{k-1} \to \ldots$ be the Postnikov decomposition of a fibration and let for $k \geq 2$ the map $f_k : Y_{k-1} \to L(\tilde{\pi}_k, k+1)$ be the classifying map of $Y_k \to Y_{k-1}$ as in (5.3.2), (2.3.3). Let $n \geq r \geq 1$ and suppose $d : Y_r \to D$ is a $(r + 1)$-connected map.

We assume that for k with $n + 1 \leq k < n + r$ we have already found a map $g = g_k : Y_n \to B_k$ into an ex-fiber space B_k over D that is a classifying map for $Y_k \to Y_n$ and such that (B_k, D) is $(n+1)$-connected. For $k = n + 1$ let $g = f_{n+1}$. With the classifying map $f : f_{k+1} : Y_k \to L(\tilde{\pi}_{k+1}, k+2) = \Omega_K(A)$ for $Y_{k+1} \to Y_k$ we find ourselves in the situation considered in (6.5.1), where $K = K(\pi_1(Y_r), 1)$

and the map $D \to K$ induces the isomorphism between the fundamental groups.

It follows from (6.4.6) that $f \in E_g(\xi_o) \subset H^{k+2}(Y_k)$ for some $\xi_o \in H^{k+3}(B_k \times_D Y_n, Y_n)$. Since the map $1 \times e : B_k \times_D Y_n \to B_k \times_D D = B_k$ induces by (6.4.8) for $k < n + r$ an isomorphism

$$(1 \times e)^* : H^{k+3}(B_k, D) \xrightarrow{\cong} H^{k+3}(B_k \times_D Y_n, Y_n)$$

there exists a ξ such that $(1 \times e)^*\xi = \xi_o$. We now define $B_{k+1} = P_K\xi$ and $g_{k+1} = \xi_g$ as in (6.5.3). This completes the induction step and the proof of (2.3.4). $\square$

(6.6) Examples of secondary homotopy classification. Triple Massey products. The classification of vector bundles.

We now give some typical applications of the functional loop operation. First we dualize the development of triple Whitehead products in (3.6), in this way obtaining triple Massey products as secondary compositions. Then we treat an example of James-Thomas on the classification of vector bundles.

Keeping (6.3.6) in mind, we can see that similar considerations apply to the domain of definition of the functional loop operation L_g as to that of E_g in (3.6.1). We dualize definition (3.6.4) as follows.

(6.6.1) Definition: Let A be an H-group. Given maps

$$A \xleftarrow{\xi} B \times Y \xleftarrow{(g,1)} Y \xleftarrow{u} U$$

with $ug \simeq 0$, $\xi(g, 1) \simeq 0$ and ξ trivial on Y, we call $\langle \xi, g, u \rangle \subset [U, \Omega A]$ the secondary composition of (ξ, g, u). This is the set of all composites

$$U \xrightarrow{u'} P_g \xrightarrow{f} \Omega A$$

where u' lifts u and $f \in L_g(\xi)$. (6.5.6) implies that

(6.6.2) <u>Theorem:</u> <u>The secondary composition</u> $\langle \xi, g, u \rangle \subset [U, \Omega A]$ <u>is a coset of the subgroup</u> $u^* \pi_1^Y(A) + [\pi_1^U(B), u]_\xi^1$.

The secondary composition is a generalization of the Toda bracket of (3.6.3). This is because $\xi : B \times Y \xrightarrow{pr} B \xrightarrow{\xi'} A$ and $g\,\xi' = 0$ imply $\langle \xi, g, u \rangle = -\{\xi', g, u\}$.

We now use the method of Porter [98] to introduce triple Massey products as examples of such secondary compositions. Let X_1, X_2, X_3, X_{12}, X_{23}, X_{123} be spaces. We call a system $\{\mu\}$ of maps

$$\mu : X_1 \times X_2 \to X_{12}, \quad \mu : X_{12} \times X_3 \to X_{123},$$

$$\mu : X_2 \times X_3 \to X_{23}, \quad \mu : X_1 \times X_{23} \to X_{123}$$

<u>associative</u>, respectively <u>homotopy associative</u>, when the diagram

(6.6.3)
$$\begin{array}{ccc} X_1 \times X_2 \times X_3 & \xrightarrow{\mu \times 1} & X_{12} \times X_3 \\ \downarrow{\scriptstyle 1 \times \mu} & & \downarrow{\scriptstyle \mu} \\ X_1 \times X_{23} & \xrightarrow{\mu} & X_{123} \end{array}$$

is commutative, respectively homotopy commutative. We write $\mu(x, y) = x \cdot y$, and call μ a <u>pairing</u> in case $\mu(x, *) = \mu(*, y) = *$.

(6.6.4) <u>Definition:</u> Let $\{\mu\}$ be an associative system of pairings. We define maps m and M as in

$$\begin{array}{ccccc} & & P_M & & \\ & & \downarrow & & \\ & \overset{u'}{\dashrightarrow} & P_m & \xrightarrow{M} & \Omega X_{123} \\ & & \downarrow & & \\ U & \xrightarrow{u} & X_1 \times X_2 \times X_3 & \xrightarrow{m} & X_{12} \times X_{23} \end{array}$$

by

$m(a, b, c) = (a \cdot b, b \cdot c)$, $M(a, b, c, \sigma, \tau) = -\sigma \cdot c + a \cdot \tau$

for $\sigma \in P(X_{12})$ and $\tau \in P(X_{23})$ with $\sigma(0) = a \cdot b$ and $\tau(0) = b \cdot c$. $\sigma \cdot c$ is the path $t \mapsto \sigma(t) \cdot c$, similarly $a \cdot \tau$: $t \mapsto a \cdot \tau(t)$. $-\sigma \cdot c + a \cdot \tau$ is the addition of paths. The secondary obstruction

$$[u_1, u_2, u_3] = \{M_*(u') \mid u' \text{ is a lifting of } u = (u_1, u_2, u_3)\}$$

is called the <u>triple Massey product</u> for the system $\{\mu\}$.

The Massey-product map M can also be derived via the functional loop operation as follows. Let X_{123} be a loop space, and define

$$\xi : X_{12} \times X_{23} \times (X_1 \times X_2 \times X_3) \longrightarrow X_{123}$$

by $\xi(u, v, a, b, c) = a \cdot v - u \cdot c$. Then $\{\mu\}$ is trivial on $Y = X_1 \times X_2 \times X_3$ and the system $\{\mu\}$ of pairings is homotopy commutative if and only if there is a null-homotopy $\xi(m, 1) \simeq 0$. Thus for a homotopy associative system $\{\mu\}$ the functional looping $L_m(\xi) \subset [P_m, \Omega X_{123}]$ is well-defined, therefore also the secondary composition $\langle \xi, m, u \rangle$ with $um \simeq 0$. If $\{\mu\}$ is in fact associative, M and the triple Massey product $[u_1, u_2, u_3]$ are defined and we have the

(6.6.5) <u>Theorem: If $\{\mu\}$ is associative, then $M \in L_m(\xi)$ and for $u = (u_1, u_2, u_3) \in [U, X_1 \times X_2 \times X_3]$ and $m^*(u) = 0$ it is the case that $\langle \xi, m, u \rangle = [u_1, u_2, u_3] + u^*[X_1 \times X_2 \times X_3, \Omega X_{123}]$.</u>

The secondary composition thus has a larger indeterminancy than the triple Massey product, since the homotopies for diagram (6.6.3) enlarge the indeterminancy. In view of (6.5.6) we can characterize the indeterminancy of the triple Massey product by using the partial looping $L(\xi)$ and the fact that $M \in L_m(\xi)$. See in this connection the result 3.9 of [98] by Porter .

After this very general description of triple Massey products, we present the special case for cohomology groups. Let G_1, G_2, G_3 be finitely generated abelian groups. There exist Eilenberg-Mac Lane complexes and cup product maps making the diagram

(6.6.6)
$$\begin{array}{ccc} K_{n_1} \times K_{n_2} \times K_{n_3} & \xrightarrow{\cup \times 1} & K_{n_1+n_2} \times K_{n_3} \\ \Big\downarrow {\scriptstyle 1 \times \cup} & & \Big\downarrow {\scriptstyle \cup} \\ K_{n_1} \times K_{n_2+n_3} & \xrightarrow{\cup} & K_{n_1+n_2+n_3} \end{array}$$

commute. We have set $K_{n_i} = K(G_i, n_i)$ and $K_{n_i+n_j} = K(G_i \otimes G_j, n_i+n_j)$. We thus have an associative system of pairings as defined in (6.6.3). As before, we identify $H^n(X, G)$ with $[X, K(G, n)]$, and $\Omega K(G,n)$ with $K(G, n-1)$. Given $u_i \in H^{n_i}(U, G_i)$ satisfying $u_1 \cup u_2 = 0$ and $u_2 \cup u_3 = 0$, we have the triple Massey product

(6.6.7) $\quad [u_1, u_2, u_3] \subset H^{n_1+n_2+n_3-1}(U, G_1 \otimes G_2 \otimes G_3).$

By (6.5.6) and (6.1.12), $[u_1, u_2, u_3]$ is a coset of

$$u_1 \cup H^{n_2+n_3-1}(U, G_2 \otimes G_3) + H^{n_1+n_2-1}(U, G_1 \otimes G_2) \cup u_3$$

Remark: The original definition of the triple Massey product was for cohomology groups, see [75], [115] p.221. Let $C^*(U, G_i)$ be the singular cochain complex of U with coefficients in G_i. The Alexander-Whitney map is an associative pairing

$$\cup : C^*(U, G_1) \otimes C^*(U, G_2) \to C^*(U, G_1 \otimes G_2)$$

that induces the cup product. Let $\bar{u}_i \in C^{n_i}(U, G_i)$ be cocycles with $\bar{u}_i \in u_i$. Since $u_1 \cup u_2 = 0$ and $u_2 \cup u_3 = 0$ there are cochains $\bar{u}_{12}$ and $\bar{u}_{23}$ for which $\delta\bar{u}_{12} = \bar{u}_1 \cup \bar{u}_2$ and $\delta\bar{u}_{23} = \bar{u}_2 \cup \bar{u}_3$. Define $[u_1, u_2, u_3]$ to be the set of cohomology classes $\{\bar{u}_{12} \cup \bar{u}_3 + (-1)^{n_1+1} \bar{u}_1 \cup \bar{u}_{23}\}$.

Porter's definition, which we used here, has the advantage that it can be applied without any difficulty to generalized cohomology theories with products.

We now give some examples of how the classification theorem (6.5.7) and the Postnikov decomposition can be used to classify liftings. The method is due to Hermann [44] and James-Thomas [55], [56], [57].

Let $E \to Y$ be a fibration with fiber F. Let $1 < r < s \leq t$ and let $\pi_{r-1}(F) = \pi$ and $\pi_{s-1}(F) = G$ be the only non-trivial homotopy groups π_j of F for $j < t$. The Postnikov decomposition of $E \to Y$ leads to the commutative diagram

(6.6.8)

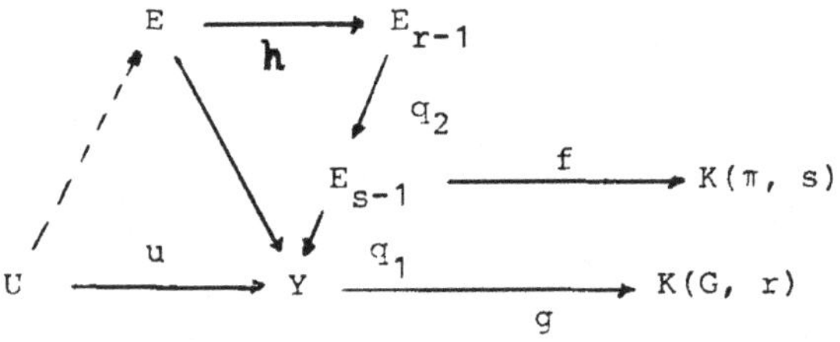

in which we assume that q_1, q_2 are principal fibrations for given f, g. The map h is thus t-connected, and if U is a CW-complex then h induces a bijection (see (1.4.13))

(6.6.9) $\quad [U, E]_u \approx [U, E_{r-1}]_u$ if $\dim U \leq t - 1$.

When f is functional with respect to g, we can use the classification formula in (6.5.7) to compute $[U, E]_u$. Thus suppose

(*) $\quad f \in L_g(\xi) \quad$ where $\quad \xi \in H^{s+1}(K(G, r) \times Y, Y; \pi)$.

If π is a ring, then the cup product induces the inclusion

$$\tilde{H}^*(K(G, r); \pi) \otimes H^*(Y; \pi) \longrightarrow H^*(K(G, r) \times Y, Y; \pi)$$

We now assume that ξ in (*) lies in the image of this inclusion, that is

$$(**) \qquad \xi = \sum_k x_k \otimes y_k, \quad x_k \in \tilde{H}^*(K(G,r);\pi),\ y_k \in H^*(Y,\pi).$$

We also regard an element $x \in H^k(K(G,r),\pi)$ as a cohomology operation $x : H^r(U,G) \to H^k(U,\pi)$, where $x(\alpha) = \gamma(\gamma^{-1}(x)_*(\gamma^{-1}(\alpha)))$
We define a cohomology suspension

$$\sigma = (\lambda^* S^*)^{-1} R^* : \tilde{H}^k(K(G,r)) \xrightarrow{R^*} \tilde{H}^k(S\Omega K(G,r)) \cong H^{k-1}(K(G,r-1))$$

where R is the evaluation map. Let $\sigma^2 = \sigma\,\sigma$, $\sigma^1 = \sigma$. With this notation, we have the

(6.6.10) Classification theorem: *Let* U *be a CW-complex with* $\dim U \leqslant t-1$. *If* $p : E \to Y$ *is a fibration,* $u : U \to Y$ *and conditions* $(*)$ *and* $(**)$ *are satisfied, then there is a bijection*

$$[U,E]_u \approx \operatorname{Ker}\,[\ ,u]^1_\xi \times \operatorname{Coker}\,[\ ,u]^2_\xi$$

where for $i = 1, 2$ *the homomorphism* $[\ ,u]^i_\xi : H^{r-i}(U,G) \to H^{s+1-i}(U,\pi)$ *is defined by*

$$[\alpha, u]^i_\xi = \sum_k (\sigma^i x_k)(\alpha) \cup u^*(y_k).$$

The reader may compare formula 6.5 in [57]

Proof: Set $\Omega A = K(\pi, s)$ and $B = K(G,r)$ in (6.5.7), then use the homotopy equivalence λ and the bijection γ of (0.5) and apply (6.1.12). □

The next example was computed with somewhat different methods by James-Thomas in [57].

Example: Let $Y = BO$ be the classifying space of the infinite orthogonal group O. The mod 2 cohomology $H^*(BO, Z_2)$ is a polynomial algebra generated by the Stiefel-Whitney classes $w_1, w_2, \ldots$

Now suppose n is odd and $n > 2$, and let $E = BO(n)$ denote the classifying space for $(n-1)$-sphere bundles. Let $p : BO(n) \to BO$ be the standard fibration with fiber $F = O/O(n)$. This fiber is $(n-1)$-connected, and we have $\pi_n(O/O(n)) = Z_2$ since n is odd. Moreover

$$\pi_{n+1}(O/O(n)) = \begin{cases} Z_2 & n \equiv 1 \mod 4 \\ 0 & n \equiv 3 \mod 4 \end{cases}$$

$$\pi_{n+2}(O/O(n)) = \begin{cases} Z_8 & n \equiv 1 \mod 4 \\ Z_2 & n \equiv 3 \mod 4 \end{cases}$$

If U is a CW-complex and $u : U \to BO$, we can apply the preceding classification theorem to compute $[U, BO(n)]_u$. We set $G = \pi_n(O/O(n)) = Z_2$ and $\pi = \pi_{s-1}(O/O(n)) = Z_2$ with

$$s = \begin{cases} n+2 & n \equiv 1 \mod 4 \\ n+3 & n \equiv 3 \mod 4 \end{cases}$$

The fibration p can now be approximated as in (6.6.8) by

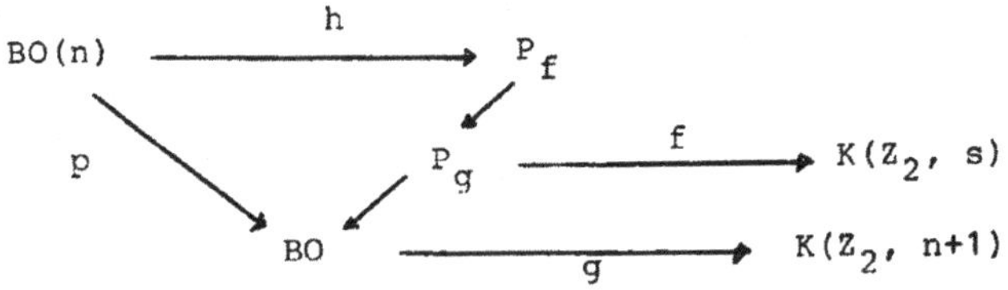

The operation of $\pi_1(BO) \neq 0$ is trivial on π and G since there is only one operation on Z_2. The map g represents the $(n+1)$-th Stiefel-Whitney class $g = w_{n+1} \in H^{n+1}(BO, Z_2)$. It follows immediately from theorem (6.4.6) that f is functional with respect to $g = w_{n+1}$, therefore an element ξ exists such that

$$(*)' \qquad f \in L_g(\xi) \quad \text{and} \quad \xi \in H^{s+1}(K(Z_2, n+1) \times BO, BO; Z_2).$$

We now want to compute ξ. This element is of the form (**) because of the isomorphism (coefficients Z_2 as always)

$$H^*(K(Z_2, n+1) \times BO, BO) = \tilde{H}^*(K(Z_2, n+1) \otimes H^*(BO).$$

Thus the preceding classification theorem is applicable. We need the following well-known facts, see [57].

I) For $K_n = K(Z_2, n)$, $H^n(K_n) = Z_2$ is generated by i_n, $H^{n+1}(K_n) = Z_2$ by Sq^1, $H^{n+2}(K_n) = Z_2$ by Sq^2 and $H^{n+3}(K_n) = Z_2 \oplus Z_2$ by Sq^2Sq^1 and Sq^1Sq^2. The Sq^i denote Steenrod operations. We have

II)
$$Sq^2\ H^n(O/O(n)) = 0, \qquad n \equiv 1 \mod 4,$$
$$Sq^2Sq^1\ H^n(O/O(n)) = 0, \quad n \equiv 3 \mod 4.$$

The Stiefel-Whitney classes satisfy in $H^*(BO)$ the Wu formulas

$$Sq^1 w_r = w_1 w_r + (r+1) w_{r+1}$$

$$Sq^2 w_r = w_2 w_r + r w_1 w_{r+1} + \tfrac{1}{2}(r-1)(r-2) w_{r+2}.$$

Now let $n \equiv 1 \mod 4$, so that $s = n + 2$. Then ξ is an element in

$$\xi \in H^{n+1}(K_{n+1}) \otimes H^2(BO) \oplus H^{n+2}(K_{n+1}) \otimes H^1(BO) \oplus H^{n+3}(K_{n-1})$$

therefore there are elements $\varepsilon_i \in Z_2$ for which

$$\xi = i_{n+1} \otimes (\varepsilon_1 w_2 + \varepsilon_1' w_1^2) + \varepsilon_2\ Sq^1 \otimes w_1 + \varepsilon_3\ Sq^2$$

II implies $\varepsilon_3 = 1$. $(*)'$ implies $(g,1)^*(\xi) = 0$, which is equivalent to the equation

$$0 = (g,1)^*(\xi) = w_{n+1}(\varepsilon_1 w_2 + \varepsilon_1' w_1^2) + \varepsilon_2\ Sq^1(w_{n+1}) w_1 + Sq^2(w_{n+1})$$

in $H^*(BO)$, since $g = w_{n+1}$. By III) we then have

$$0 = \varepsilon_1 w_2 w_{n+1} + \varepsilon_1' w_1^2 w_{n+1} + \varepsilon_2 (w_1 w_{n+1} + (n+2) w_{n+2}) w_1$$
$$+ w_2 w_{n+1} + (n+1) w_1 w_{n+2} + \tfrac{1}{2}\, n(n-1) w_{n+3}$$

$$= (\varepsilon_1 + 1)w_2 w_{n+1} + (\varepsilon_1' + \varepsilon_2)w_1^2 w_{n+1} + \varepsilon_2 w_{n+2} w_1$$

Since the last line is a representation in terms of basis elements, it follows that $\varepsilon_2 = 0$ and $\varepsilon_1' = 0$ and $\varepsilon_1 = 1$. This shows that

IV) $\xi = i_{n+1} \otimes w_2 + Sq^2$ for $n \equiv 1 \mod 4$.

Now let $n \equiv 3 \mod 4$ instead, so that $s = n + 3$. Then ξ is an element in

$$\xi \in \bigoplus_{i=0}^{3} H^{n+1+i}(K_{n+1}) \otimes H^{3-i}(BO)$$

so $\xi = \xi_0 + \xi_1 + \xi_2 + \xi_3$ with $\xi_i \in H^{n+1+i}(K_{n+1}) \otimes H^{3-i}(BO)$. It follows from II) as before that $\xi_3 = Sq^2 Sq^1 + \varepsilon Sq^1 Sq^2$ with $\varepsilon \in \{0, 1\}$. We can again represent the ξ_i in terms of basis elements and apply the equation $(g, 1)^*(\xi) = 0$ where $g = w_{n+1}$. III) and some calculation then imply

V) $\xi = Sq^2 Sq^1 + Sq^1 \otimes (w_1^2 + w_2)$ <u>for</u> $n \equiv 3 \mod 4$.

In view of the preceding classification theorem and the fact that $\sigma Sq^i = Sq^i$, we have proved with IV and V the following result

(6.6.11) <u>Corollary</u>: <u>Suppose</u> $u : U \to BO$ <u>can be lifted by a map</u> $U \to BO(n)$. <u>Then</u>

1) <u>When</u> $n \equiv 1 \pmod 4$ <u>and</u> dim $U < n + 1$ <u>there is a bijection</u>

$$[U, BO(n)]_u \approx \operatorname{Ker} \Gamma_1 \times \operatorname{Coker} \Gamma_2$$

<u>where</u> $\Gamma_i : H^{n+1-i}(U, Z_2) \to H^{n+s-i}(U, Z_2)$ <u>is defined by</u> $\Gamma_i(\alpha) = \alpha \cup u^*(w^2) + Sq^2(\alpha)$ $(i = 1, 2)$.

2) <u>When</u> $n \equiv 3 \mod 4$ <u>and</u> dim $U < n + 2$ <u>there is a bijection</u>

$$[U, BO(n)]_u \approx \operatorname{Ker}\Gamma_1 \times \operatorname{Coker} \Gamma_2$$

<u>where</u> $\Gamma_i : H^{n+1-i}(U, Z_2) \to H^{n+4-i}(U, Z_2)$ <u>is defined by</u>

$$\Gamma_i(\alpha) = Sq^2Sq^1(\alpha) + Sq^1(\alpha) \cup u^*(w_1^2 + w_2).$$

The necessary calculations for oriented sphere bundles and n even were carried out in [57]. If $U = M$ is a m-dimensional differential manifold and $u : M \to BO$ represents the stable normal bundle of $M \subset R^{m+n}$ then $[M, BO(n)]_u$ can be identified with the set of homotopy classes $[M \subset R^{m+n}]$ of immersions $M \subset R^{n+m}$, see [57].

Remark: Calculations like those in the example come up in the classification of immersions, see for instance, the work of D. Randall [102], [103] and J.C. Becker [21], C.A. Robinson [105]. Various authors have also used similar calculations in classifying vector bundles, see for instance the work of L.L. Larmore [63],[66].

List of Symbols

References

1. Adams, J. F.: On the structure and applications of the Steenrod algebra, Comment. Math. Helv. 32 (1958) 180 - 214.

2. " On the Non-existence of elements of Hopf invariant one. Annals of Math. 72 (1960), 20 - 104.

3. " Four applications of the self-obstruction invariants, J. London Math. Soc. 31 (1956) 148 - 159.

4. Barcus, W.: Note on cross-sections over CW-complexes, Quart. J. Math. Oxford Ser. (2) 5 (1954) 150 - 160.

5. Barcus, W.D.:Barratt, M.G.: On the homotopy classification of the extensions of a fixed map. Trans. Am. math. Soc. 88 (1958) 57 - 74.

6. Barratt, M.G.: Track groups (I), Proc. London Math. Soc. (3) 5 (1955)

7. Barratt, M.G.: Track groups (II), Proc. London Math. Soc. (3) 5 (1955).

8. Barratt, M.G.: Spaces of finite characteristic, Quart. J. Math. Oxford (2) 11 (1960), 124 - 36.

9. Baues, H.J.: Hindernisse in dem Produkt von Suspensionen, Math. Ann. 200, 11 - 23 (1973).

10. " Iterierte Join Konstruktionen, Math. Z. 131, 77 - 84 (1973).

11. " Whitehead Produkte und Hindernisse in dem Produkt von Abbildungskegeln, Arch. Math. 25, 184 - 197 (1974).

12. " Höhere Whitehead Produkte der zwei dimensionalen Sphäre, Comment. Math. Helv. 48, 116 - 118 (1973).

13. " Der Pontryagin Ring von Quotienten eines Torus, Math. Z. 134, 221 - 228 (1973).

14. " Relative Homotopiegruppen bei Abbildungskegeln, Compositio Math. 32 2 (1976) 169 - 183.

15. " Relationen für primäre Homotopieoperationen und eine verallgemeinerte EHP-Sequenz, Annales Scientifiques de l'Ecole Normale Supèrieure, fasc 4. 8 (1975) 509 - 533 .

16. " Identitäten für Whitehead-Produkte höherer Ordnung, Math. Z. 146, 239 - 265 (1976).

17. Baues, H.J.: Eine Vermutung über Schleifenräume von Sphären, Koninklijke Nederl. Akad. Wetensch. Proc. 79 A2 (93 - 99) 1976, Indagationes Math. 38, 2.

18. " Hopf-invariants for reduced products of spheres, Proceed. American Math. Soc. 59 1 (1976) 169 - 174.

19. " Rationale Homotopietypen, manuscripta mathematica 20, 119 - 131 (1977).

20. Baues, H.J., Lemaire, J. M.: Minimal models in homotopy theory, Math. Ann. 225 (1977) 219 - 242.

21. Becker, J.C.: Cohomology and the classification of liftings. Trans. Amer. Math. Soc. 133 (1968) 447 - 475.

22. Boardman, J.M. Steer, B.: On Hopf invariants, Comment. Math. Helv. 42 (1967), 180 - 221.

23. Bousfield, A.K. and Kan D.M.: Homotopy limits, completions and localizations, Lecture Notes in Mathematics 304, Springer New York (1975).

24. Dieck, T. tom; Kamps, K.H., Puppe, D.: Homotopietheorie, Lecture Notes in Mathematics No. 157, Springer, Berlin (1970)

25. Dold, A.: Halbexakte Homotopiefunktoren, Lecture Notes in Mathematics, No. 12, Springer, Berlin (1966).

26. Dror, E.: A generalization of the Whitehead theorem, Lecture Notes in Mathematics. 249, Springer (1971), 13 - 22.

27, Eckmann, B. and Hilton, P.S. : Homotopical obstruction theory. An. Acad. Brasil Ci. 40 (1968) 407 - 425.

28. Eggar, M.H.: The piecing comparison theorem. Proc. Koninkl. Nederl. Akad. Wetensch. 76 (4) (1973) 320 - 330.

29. " Ex-Homotopy theory. Compositio Math. 27 (2) (1973) 185 - 195.

30. " On structure-preserving maps between fibre spaces with cross-sections. J. London Math. Soc. (2) 7 (1973) 303 - 311.

31. Eilenberg, S.: Cohomology and continuous mappings, Ann. Math. 41 (1940) 231 - 251.

32. Eilenberg, S. and Mac Lane, S.: On the groups $H(\pi, n)$ III. Ann. of Math. 60 (1954) 513 - 557.

33. Ganea, T.: Fibrations and cocategory, Comment. Math. Helv. 35 (1961) 15 - 24.

34. " A generalization of the homology and homotopy suspension. Comment. Math. Helv. 39 (1965) 295 - 322.

35. Gitler, S.: Cohomology operations with local coefficients. American J. Math. 85 (1963) 156 - 188.

36. Gray, B.: Spaces of the same n-type, for all n, Topology 5 (1966) 241 - 243.

37. " On the homotopy groups of mapping cones. Proc. London Math. Soc. (3) 26 (1973) 497 - 520.

38. " Homotopy theory, an introduction to algebraic topology. Academic press, New York (1975).

39. Hardie, K.A.: On a construction of E.C. Zeeman, J. London Math. Soc. 35 (1960) 452 - 464.

40. " Derived homotopy constructions, J. London Math. Soc. 35 (1960) 465 - 480.

41. " Quasifibration and adjunction. Pac. J. Math. 35 (1970) 389 - 397.

42. Hardie, K.A. and Porter G.J.: The slash product homotopy operation to appear in Proc. London Math. Soc.

43. Hausmann, W.: HEF Faserungen und der klassifizierende Raum eines freien Produktes. Diplomarbeit Bonn (1975).

44. Hermann, R.: Obstruction theory for fibre spaces. Illinois J. Math. 4 (1960) 9 - 27.

45. Hill, R. O. Jr.: Moore-Postnikov towers for fibrations in which π_1(fiber) is non-abelian. Pac. J. Math. 62 (1976) 141 - 148.

46. Hilton, P.J.: On the homotopy groups of a union of spheres. J. London Math. Soc. 30 (1955) 154 - 172.

47. " On excision and principal fibrations, Comment. Math. Helv. 35 (1961) 77- 84.

48. " General cohomology theory and K-theory. London Math. Soc. Lecture Note Ser. 1 Cambridge (1971).

49. Hilton, P.J. and Mislin, G. and Roitberg, J.: Localizations of nilpotent groups and spaces. North Holland Math. Studies 15 Amsterdam (1975)

50. Hirzebruch, F., Hopf, H.: Felder von Flächenelementen in 4-dimensionalen Mannigfaltigkeiten. Math. Ann. 136 (1958) 156 - 172.

51. Hopf, H.: Abbildungsklassen n-dimensionaler Mannigfaltigkeiten, Math. Annalen 92 (1926) 209 - 224.

52. Hu , S.: Extensions and classification of Maps, Osaka Math. J. 2 (1950) 165 - 209.

53. " Homotopy theory, Academic press,New York, London (1959).

54. James, I. M.: On the homotopy groups of certain pairs and triads, Quart. J. Math. Oxford (2) 5 (1954) 260 - 270

55. James, I.M. and E. Thomas,: An approach to the enumeration problem for non-stable vector bundles. J. Math. Mech. 14 (1965) 485 - 506

56. James, I.M. and E. Thomas,: Note on the classification of cross-sections. Topology 4 (1966) 351 - 359.

57. James, I.M. and E. Thomas: On the enumeration of cross sections. Topology 5 (1966) 95 - 114.

58. James, I.M.: A relation between Postnikov classes. Quart. J. Math. 17 (1966) 269 - 280.

59. " Ex-homotopy theory I, Illinois J. of Math. 15 (1971), 324 - 337.

60 " Products between homotopy groups. Comp. math. 23 (1971) 329 - 345

61. James, I.M. and Whitehead J.H.C.: The homotopy theory of sphere bundles over spheres I. Proc. London Math. Soc. (3) 4 (1954) 196 - 218.

62. Kervaire, A. M. and Milnor J. W.: Groups of homotopy spheres I. Ann. Math. 77 (1963) 504 - 531.

63. Larmore, L.L.: Map cohomology operations and enumeration of vector bundles. J. Math. and Mech. 17 (1967) 199 - 208.

64. Larmore, L.L., Thomas, E.: Group extensions and principal fibrations, Math. Scand. 30 (1972) 227 - 248.

65. " Twisted cohomology theories and the single obstruction to lifting. Pacific J. Math. 41 (1972) 755 - 769.

66. " Real n-plane bundles over an (n+1)-complex. Osaka J. Math. 12 (1975) 325 - 342.

67. Lamotke, K.: Semisimpliziale algebraische Topologie. Grundlehren der math. Wiss. 147, Springer (1968)

68. Lemmens, P.W. H.: On ex-homotopy groups. Proc. Koninkl. Nederl. Akad. Wetensch. 74 (1971)

69. Lillig, J.: A union theorem for cofibrations. Arch. Math. 24 (1973) 410 - 415.

70. Lundell, T.A.; Weingram, S.: The topology of CW-complexes. Van Nostrand New York (1969).

71. Mac Lane, S.: Homology, Springer (1963).

72. Mahowald, M.: On obstruction theory in orientable fiber bundles. Trans. Amer. Math. Soc. 110 (1964) 315 - 349.

73. Mather, M.: Pull backs in homotopy theory. Can. J. Math. 28 (1976) 225 - 263.

74. Massey, W.S.: Some problems in algebraic topology and the theory of fibre bundles, Ann. of Math. 62 (1955) 327 - 359.

75. " Some higher order cohomology operations, Sympos. Inten. Topologica Alg. 145 - 154, Mexico City 1958.

76. Massey, W. S. and Peterson F.P.: The cohomology structure of certain fiber spaces. Topology 4 (1965) 47 - 66.

77. Maunder, C.R.F.: Cohomology operations of the n-th kind. Proc. London Math. Soc. 13, (1963) 125 - 154.

78. " Algebraic topology. Van Nostrand (1970)

79. Mc Clendon, J.F.: Higher order twisted cohomology operations Inventiones math. 7, 183 - 214 (1969).

80. " Obstruction theory in fiber spaces. Math. Z. 120, (1971) 1-17.

81. " On stable fiber space obstructions. Pac. J. Math. 36 (2) (1971) 439 - 446.

82. Mc Clendon, J.F.: Reducing towers of principal fibrations. Nagoya Math. J. 54 (1974) 149 - 164.

83. " A spectral sequence for classifying liftings in fiber spaces.

84. Meyer, J.P.: Relative stable homotopy. Proc. Conf. on algebraic topology, University of Illinois at Chicago Circle (1968) 206 - 212.

85. " Whitehead products and Postnikov systems. Amer. J. Math. 82 (1960) 271 - 280.

86. Milnor, J.: On spaces having the homotopy type of a CW-complex. Trans. Amer. Math. Soc. 90 (1959) 272 - 280.

87. Milnor, J. and Stashoff, J.: Characteristic classes, Ann. of Math. Studies, No. 76 (1973).

88. Moore J.C.: Semisimplicial complexes and Postnikov systems. Symp. Internat. de Topologica Alg. Mexico City (1958) 232 - 246.

89. Nakaoka, M., Toda, H.: On the Jacobi identity for Whitehead products, J. Inst. Polytechn. Osaka City Univ. 5 (1954) 1 - 13.

90. Nomura, Y.: On mapping sequences. Nagoya Math. J. (1960) 111 - 145.

91. " A non-stable secondary operation and homotopy classification of maps. Osaka J. Math. 6 (1969) 117 - 134.

92. Olum, P.: Obstructions to extensions and homotopies, Ann. of Math. 52 (1950), 1 - 50.

93. " On mappings into spaces in which certain homotopy groups vanish. Ann. Math. 5 (1953) 561 - 573.

94. " Mappings of manifolds and the notion of degree. Ann. Math. 58 (1953) 458 - 480.

95. " Factorisation and induced homomorphisms Adv. Math. 3 (1969) 72 - 100.

96. Pontryagin, L.: A classification of mappings of the 3-dimensional complex into the 2-dimensional sphere. Rec. Math. (Mat Sbornik) N. S. 9 (51) (1941) 331 - 363.

97. Porter, G.J.: Higher order Whitehead products, Top. Vol. 3 (1965) 123 - 135.

98. Porter, G. J.: Higher products, Trans. Amer. Math. Soc. 148 (1970) 315 - 345.

99. Postnikov, M.M.: Classification of continuous mappings of an n-dimensional polyhedron into a connected topological space, aspherical in dimensions greater than unity and smaller than n. Akad. Nauk. Gruzin SSR Trudy Tbiliss Mat. Inst. Rasmadoze 22 (1956), 165 - 202.

100. Puppe, D.: Homotopiemengen und ihre induzierten Abbildungen I. Math. Z. 69, (1958) 299 - 344.

101. Puppe, V.: A remark on 'homotopy fibrations'. Manuscripta Math. 12 (1974) 113 - 120.

102. Randall, A.D.: Some immersion theorems for projective spaces. Transact. Amer. Math. Soc. 147 (1970) 135 - 151.

103. " Some immersion theorems for manifolds. Trans. Amer. Math. Soc. 156 (1971) 45 - 58.

104. Robinson, C.A.: Moore-Postnikov systems for non-simple fibrations. Illinois J. Math. 16 (1972) 234 - 242.

105. " Stable homotopy theory over a fixed base space. Bull. Amer. Math. Soc. 80 (1974) 248 - 252.

106. Rutter, J.W.: A homotopy classification of maps into an induced fibre space. Topology 6 (1967) 379 - 403.

107. " Maps and equivalences into equalizing fibrations and from coequalizing cofibrations. Math. Z. 122 (1971) 125 - 141.

108. Samelson, H.: Groups and spaces of loops, Comment. Math. Helv. 28 (1954) 278 - 287.

109. Shimada, N., and Uehara, H.: Classification of mappings of an (n + 2)-complex into an (n - 1)-connected space with vanishing (n + 1) st homotopy group. Nagoya Math. J. 4 (1952) 43 - 50.

110. Shiraiwa, K.: Relation between higher obstructions and Postnikov invariants, Nagoya Math. J. 16 (1960) 21 - 33.

111. Siegel, J.: Cohomology and homology theories for categories of principal G-bundles. Trans. Amer. Math. Soc. 120 (1965) 428 - 437.

112. Siegel, J.: Higher order cohomology operations in local coefficient theory. Amer. J. Math. 89 (1967) 909 - 931.

113. " k-invariants in local coefficient theory. Proc. Amer. Math. Soc. 29 (1971) 169 - 174.

114. " Cohomology operations in local coefficient theory, Illinois J. Math. 15 (1971) 52 - 63.

115. Spanier, H.: Higher order operations. Trans. Amer. Math. Soc. 109 (1963) 509 - 539.

116. " Algebraic topology. Mc Graw Hill (1966)

117. Stasheff, J.: A classification theorem for fibre spaces. Topology, 2 (1963) 239 - 246.

118. Steenrod, N.E.: Homology with local coefficients, Ann. Math. 44 (1943) 611 - 627.

119. " Products of cocycles and extensions of mappings. Ann. Math. 48 (1947) 290 - 320.

120. " The topology of fibre bundles, Princeton University Press (1951).

121. Steenrod, N.E., Epstein D.B.A.: Cohomology operations, Ann. of Math. Studies Nr. 50.

122. Steenrod, N.E.: Cohomology operations and obstructions to extending continuous functions. Adv. Math. 8 (1972) 371 - 416.

123. Strøm, A.: Note on cofibrations II. Math. Scand. 22 (1968) 130 - 142.

124. Sullivan, D.: Geometric topology, part I: Localization, periodicity and Galois symmetry, MIT, 1970 (mimeographed notes).

125. Thomas, E.: Postnikov invariants and higher order cohomology operations. Ann. Math. 85 (1967) 184 - 217.

126. Tsuchida, K.: Generalized Puppe sequence and Spanier Whitehead duality. Tohoko math. J. 23 (1971) 37 - 48.

127. Toda, H.: On the double suspensions E^2, J. of Inst. of Polytechnics Osaka (1956) 7 No. 1 - 1, Ser. 1, 103 - 145.

128. Toda, H.: Compositions methods in homotopy groups of spheres Annals of Math. Studies Nr. 49.

129. " A survey of homotopy theory, Advances Math. 10 (1973) 417 - 455.

130. Whitehead, G.W.: Homotopy theory. MIT Press, Cambridge Massachusetts (1966)

131. " On the Freudenthal theorems. Ann. of Math. 57 (1953) 209 - 228.

132. Whitehead, J.H.C.: A certain exact sequence. Ann. Math. 52 (1950), 51 - 110.

133. " On the theory of obstructions, Ann. Math. 54 (1951) 68 - 84.

134. " The G-dual of a semi-exact couple, Proc. London Math. Soc. (3) 3 (1953) 385 .

INDEX

A

B

C

F

G

H

I

J

K

L

M

N

O

P

W

Z